Arianna Borrelli
Aspects of the Astrolabe

SUDHOFFS ARCHIV

Zeitschrift für Wissenschaftsgeschichte

- -

Beihefte

Herausgegeben von
Klaus Bergdolt
Peter Dilg
Menso Folkerts
Gundolf Keil
Fritz Krafft

Heft 57

Arianna Borrelli

Aspects of the Astrolabe

'architectonica ratio' in
tenth- and eleventh-century Europe

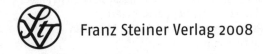 Franz Steiner Verlag 2008

Bibliografische Information der Deutschen Nationalbibliothek
Die Deutsche Nationalbibliothek verzeichnet diese
Publikation in der Deutschen Nationalbibliografie;
detaillierte bibliografische Daten sind im Internet über
<http://dnb.d-nb.de> abrufbar.

ISBN 978-3-515-09129-9

CONTENTS

LIST OF FIGURES AND TABLES

Figures:

Tables:

ACKNOWLEDGEMENTS

The present work combines my interests in the medieval world, in the epistemological aspects of mathematics, science and the manual crafts and in the emergence, interaction and disappearance of different modes of human thought and action. These interests would never have come to be without the present and former members of the Technische Universität Braunschweig who have accompanied me during my study of philosophy and history.

Among those who taught me, providing me with knowledge and professional example, I wish to offer particular thanks to Dr. Reinhard Loock, Prof. Dr. Claudia Märtl, Prof. Dr. Herbert Mehrtens and Prof. Dr. Dr. Claus-Artur Scheier. My gratitude also goes to Prof. Dr. Helmut Castritius, Prof. Dr. Ute Daniel, Dr. Goswin Spreckelmeyer and Prof. Dr. Reinhard Wolters, as well as to the fellow students with whom I shared my first academic discussions in the humanities. For her ever-friendly and helpful presence, I wish to thank Bärbel Girwert.

I am indebted to the University of Braunschweig for the material support which made it possible for me to pursue my newly found interests: this project would never have begun without the grant I received during the period 2002-2004.

I am especially grateful to Prof. Mehrtens for endorsing this project as my mentor and for his advice and help in bringing it to completion. Among those who assisted me in my research with expert advice, conceptual inspiration, material help and moral support, I wish to thank in particular Prof. Dr. Ludolf Kuchenbuch for his comments and suggestions. The conversation I had with him at the beginning of my project greatly helped me shape conceptual tools to deal with different sources on the Latin medieval astrolabe and in particular made me aware of the potential mixture of abstract and material elements characterising medieval tools.

In the difficult task of approaching artefact sources through literature, the help of Prof. Dr. David King was invaluable. I thank him for his expert comments, suggestions and advice on the complexity, dangers and rewards of the study of medieval astronomical instruments. A large part of the source material analyzed in my work was accessible to me at the Institute of the History of Science of the Ludwig-Maximillian-Universität in Munich, to whose members I wish to express my gratitude. My deepest thanks are due to Prof. Dr. Folkerts, who offered me his advice, allowed me to use the material stored in the Munich Archive and, once my work was completed, deeemed it worthy of publication in the present form. I am grateful to Prof. Dr. Paul Kunitzsch who, together with Dr. Richard Lorch, helped me find my way among medieval astrolabe texts and later contributed with precious corrections and advice. I thank Prof. Dr. Brigitte Hoppe, who kindly helped me with the first contacts, and Rosemarie Jakimenczuk for being always there ready to help me, and all members and guests of the Institute who interrupted their own work to offer me assistance. I regret being unable to list their names.

During my brief stays in Munich, I also enjoyed the hospitality of the Department of Medieval History of the Ludwig-Maximillian-Universität, where Konrad Frenzel, M.A., Prof. Eva Schlotheuber and Prof. Märtl were as kind and generous to me as they had been in Braunschweig.

I am deeply indebted to my advisor and to the two further referees of my thesis for their remarks and corrections on the final version of the dissertation. In particular I wish to thank Prof. Dr. Charles Burnett for taking it upon himself to act as a referee without having had any prior knowledge of me or my research project, thus offering a rare example of academic generosity. His engagement and his open-mindedness in evaluating the work of a young researcher venturing in a field new to her, but so well known to him, deserve great admiration.

Finally, I wish to thank the British Library in London, the President and Fellows of the Corpus Christi College in Oxford and the Library of the Stift St. Peter in Salzburg for generously waiving their fees when granting me permission to reproduce images of manuscripts in their collections.

1 INTRODUCTION

1.1 THE ASTROLABE AND MEDIEVAL MATHEMATICAL THOUGHT

In the ninth century of our era, in the Arabic-Islamic culture, the astrolabe was a well-known astronomical and computational instrument.[1] It was usually made out of metal and embodied a two-dimensional projection of the Ptolemaic cosmos. During the tenth century, knowledge of the astrolabe entered Latin Europe where, at the time, very little mathematical and astronomical knowledge was present. What was the astrolabe for Latin medieval Europeans? Why did they take an interest in it? More than one century of research has brought to light the complexity of this question.[2]

Historical studies often describe the medieval Latin astrolabe as an astronomical instrument, but, when it comes down to explaining what it was actually used for, they offer a varied list of possible roles: timekeeping device, abstract model of the cosmos, analogical computer, status symbol or didactic tool. The present study builds upon previous works in the hope of adding something to the richness of the picture.

Although my research subject is the astrolabe in tenth- and eleventh-century Latin Europe, this is at the same time an investigation of medieval mathematical thought, of its modes of communication and of the image (or images) of knowledge in which medieval mathematics was embedded.[3] This double focus is both necessary and fruitful: on the one hand, the assimilation in Latin Europe of astrolabe knowledge of Arabic origin can be fully appreciated only by taking into account the features of the mathematical culture in which that knowledge was being assimilated. On the other hand, the astrolabe offers a valuable key to explore and better understand the medieval mathematical arts and constitutes an ideal testing ground for hypotheses.

When speaking of the context of Latin medieval astrolabe studies, I will refer to 'high medieval mathematics', meaning the four arts of the quadrivium (arithmetic, geometry, astronomy and music) as they were being taught and practised in the tenth and eleventh centuries. Although the primary focus of this research shall be on the mathematics involved in the study of celestial phenomena, this subject cannot be confined to the discipline of astronomy: in the tenth and eleventh centuries, the borders between the four mathematical arts were

---

1. An overview of the early history of the astrolabe is given in 1.4. For an astronomical and mathematical introduction to the astrolabe, its structure and its possible uses see, for example: Hartner (1939), Michel (1947), North (1974), Hartner (1979), Poulle (1981) p. 29–34, Pingree (1987), Stautz (1999) p. 1–7 and p. 99–122, Kunitzsch (1996), Wintroub (2000), Maddison/Savage-Smith (1997) p. 186–199, King (2003b), Proctor (2005).

2. Historical research on the Latin medieval astrolabe started at the latest in 1899, with Nicolaus Bubnov's publication of the mathematical works of Gerbert of Aurillac Bubnov (1899).

3. On Yehuda Elkana's concept of 'images of knowledge', see 1.7.

ries, the borders between the four mathematical arts were constantly being re-drawn, bringing them closer to each other.[4]

In Latin medieval Europe, the astrolabe occupied a very special - if not unique - position as the focus of an early interest for Arabic knowledge and for its real or imagined ancient roots. Still, it is not easy to provide a satisfactory definition of it, and possibly even to say whether the Lain medieval astrolabe was a material object or an abstract geometrical pattern, as discussed in the following section.

1.2 THE ASTROLABE AND ITS ASPECTS

Behind the Latin terms which are translated as 'astrolabe' (among them 'astrolabium', 'astrolapsus', 'plana sphaera', 'wazalcora') lies one of the most complex conglomerates of evidence the Latin Middle Ages has left us, involving artefacts, drawings, manuscript descriptions, illuminations and philosophical reflections. Historians face the difficult task of weaving this material into a coherent picture, while still evaluating every single kind of evidence taking into account its own unique character. Evidence on the astrolabe has to be approached with much care, because some of it can be readily understood in modern terms, and may be too easily interpreted according to patterns which are not fitting to medieval thought and practice.

The problem is the following: at the immediate level of the sources, it is quite easy to distinguish between, for example: (a) astrolabe artefacts made out metal, (b) astrolabe lines drawn on parchment, (c) texts describing in words the drawing procedures, (d) texts commenting upon the importance of learning to perform the construction, (e) texts describing how to use an 'astrolabe'.

At the level of interpretation, though, problems immediately arise. For example: should texts describing procedures for drawing astrolabe lines be regarded as instruction booklets for making astrolabe artefacts out of wood or metal? They often are, but this might be incorrect, as I shall argue later on. Should preserved Latin astrolabe artefacts be regarded as tools which were used in the way described in contemporary texts? This, too, is a very problematic issue, as we shall see. Were the astrolabe lines drawn in Latin manuscripts actively constructed by Latin geometers, or had they been passively copied from Arabic originals?

More generally, it is tempting to distinguish Latin medieval knowledge about the astrolabe into a theoretical understanding of its principles on the one side, and the practical recipes for building and using it on the other. Making this kind of distinction has often forced the question, whether Latin scholars from the eleventh century did or did not 'really understand' the astrolabe, or whether they were 'only' interested in it for utilitarian reasons, for example, timekeeping.[5]

---

4   The state of the four mathematical arts in tenth- and eleventh-century Europe is discussed in
    4.1.5–7.
5   On this question, see 2.1.6, 4.1.7 and 4.4.2.

This question addresses a very important point, but it may be misleading in its clear-cut formulation. As I will argue, in the historical context studied here no distinction between theoretical and practical knowledge in the mathematical arts is possible.[6] In the eleventh century, a recipe-like list of instructions could be the most appropriate written form in which to store and transfer highly abstract knowledge. On the other hand, attention to abstract mathematical structures could be motivated by a desire to understand the rational structure ('ratio') behind carefully observed, and often also quantitatively estimated, natural phenomena. Because of this, it is not always hermeneutically productive to assume that in a medieval Latin text the term 'astrolabe' has to mean either an abstract pattern or a material object: it might mean both at the same time.

Another sharp distinction that may be misleading is the one between 'astronomical' and 'astrological' uses of the astrolabe. Medieval astrology was a form of rational natural philosophy and, therefore, if the astrolabe madeit possible to recognize an ordered, rational pattern ('ratio') in natural phenomena, its significance was at the same time astronomical and astrological.[7]

In conclusion, the Latin medieval astrolabe appears to carry different meanings which, by modern standards, are seemingly unrelated to or even incompatible with each other: an abstract geometrical construction and a practically useful timekeeper, a forbidden astrological tool and a device to compute the exact time of prayers, an exotic prestige object and a philosophical model for cosmic order. Since it is highly problematic to choose one among all these possible meanings, I will instead introduce, as hermeneutic tool, the idea of 'aspects' of the astrolabe. This means that I shall regard the research subject 'Latin medieval astrolabe' as a loose collection of co-existing 'aspects' corresponding to the various meanings listed above (and to a few more). For example, the mathematical aspects of the astrolabe will be abstract geometrical structures and methods, while its practical aspects shall be linked to sensory experiences such as holding an object in the hands and using it to obtain perceivable effects.

The various aspects of the Latin medieval astrolabe might or might not have been perceived and connected with each other by the historical actors: whether, when and how this was the case is the question which I will set out to answer in the following pages.[8]

---

6   See 4.1.7.
7   See 5.1.1–2.
8   The idea of considering the astrolabe not as an instrument, but rather as a loose collection of aspects, each of which might or not play a role according to the specific cultural and historical context, was suggested to me by the paper presented by John Michael Gorman at the 2004 Summer Academy of the Max-Planck-Institute for the History of Science (Berlin).

1.3 THE EARLY HISTORY OF THE ASTROLABE

The development of the geometrical structure of the astrolabe and its embodiment in an artefact were the work of Greek-speaking scholars and took place within the same context in which the main body of Greek-Roman astronomical and mathematical knowledge was produced.[9]

The earliest extant written and material evidence of a device with a geometrical structure similar to that of the astrolabe dates to the first centuries of our era.[10] The earliest extant treatise devoted to the geometrical construction of astrolabe lines was written by Ptolemy (Claudius Ptolemaeus, ca. 100–170) and is usually referred to as the 'Planisphaerium'.[11] The Greek original of this work is lost, and only its Arabic and Latin translations are preserved. The earliest extant description of an astrolabe artefact is a treatise by the Christian neoplatonic philosopher Johannes Philoponos of Alexandria (ca. 520–550). In the seventh or eight century of our era, the astrolabe entered the Arabic-Islamic world. The transmission of astrolabe knowledge from the Greeks to the Arabs was part of a wide process of knowledge transfer and assimilation involving, among other things, Ptolemaic astronomy, Euclidian geometry and neoplatonic philosophy.[12] The earliest preserved astrolabe artefacts were made in the tenth century in the Eastern part of the Arabic-Islamic world.[13]

In Latin literature, no reference to the astrolabe can be found before the late tenth or early eleventh century. It is important to note that not even late ancient Latin authors such as Martianus Capella (fl. ca. 410–439), Ancius Manlius Severus Boethius (d. 524) or Cassiodor (d. after 580) mention astrolabes in the works they have left us although, as already noted, devices based on the same geometrical projection as the astrolabe are attested in late ancient Europe. The presence of such devices might have contributed to keeping alive knowledge on how to project the surface of a sphere onto a plane, even though written Latin texts made no mention of it.[14]

Around the year 1000, evidence of the diffusion of astrolabe knowledge in Latin Europe appeared in the form of manuscripts discussing the astrolabe's geometric design ('mensura') and its possible uses ('utilitates').[15] These manuscripts also contained drawings of the steps needed for the geometrical construction of astrolabe-lines, as well as drawings of finished astrolabes. The authors make large use of Arabic terms. Only one astrolabe artefact exists which can be assumed to

---

9   The following overview on the origins of the astrolabe is based on: Neugebauer (1949), Stautz (1994), Anagnostakis (1984) p. 9–43. For further details and bibliography on the subjects mentioned below, see the relevant chapters of Part Two of this study.

10   It is the device known as 'anaphoric clock', discussed in 3.1.1.

11   On the 'Planisphaerium' see 4.5.1–2.

12   For an overview of Arabic astronomy and its Greek and Indian sources see: King (1999b) p. 3–46 (with numerous further bibliographical indications), Morelon (1996).

13   On preserved astrolabe artefacts, see 3.2.

14   North (1975) argues in this direction.

15   On early Latin astrolabe manuscripts, see 3.3.

have been produced by Latin scholars and craftsmen in the tenth or eleventh century.[16]

1.4 LATIN SCHOLARS AND THE ASTROLABE

The transfer of astrolabe knowledge from the Arab to the Latin world in the eleventh century has a very peculiar character. Astrolabe knowledge was transmitted
and assimilated into the Latin culture practically isolated from the body of astronomical, mathematical and astrological knowledge with which it had been regularly associated in the Greek as well as in the Arabic world. Apart from the astrolabe, there are only two further traces of the diffusion of new mathematical, astronomical and astrological knowledge in eleventh-century Latin Europe: Arabic-
Hindu numerals which, like the astrolabe, made their appearance in Latin manuscripts around the year 1000 and the Latin text known as 'Liber Alchandrei', that
deals with astrological subjects.[17] However, this should not be taken to imply that
the Latin interest in the astrolabe had nothing to do with mathematics, astronomy
or astrology: quite the contrary, as I shall argue. The point is that astrolabe knowledge seems to have appealed to tenth- and eleventh-century Latin scholars more
than other elements of Greek and Arabic natural philosophy which might also
have been within their reach.

The astrolabe was a device whose geometric structure, use and construction
went well beyond the limits of contemporary Latin mathematics and astronomy.
Moreover, it was a device that, although its roots could be traced back to classical
Antiquity, still carried unmistakable signs of its immediate Arabic-Islamic origin.
Even though the astrolabe can in principle be used as a timekeeping device, it is
very doubtful whether tenth- or eleventh-century astrolabe artefacts could perform
this function better than a simple sundial. In fact, one may wonder whether they
could perform it at all.[18] And yet, the astrolabe awakened the interest of Latin
scholars so much, that they devoted precious time and no less precious parchment
to the effort of acquiring and spreading knowledge about it.

More than one hundred years later, around the middle of the twelfth century, a
steady flow of mathematical and natural philosophical translations from Arabic
into Latin began. The centres of this translating activity were Northern Spain and
Southern Italy. In this period, along with other mathematical, astronomical and
astrological works, recently composed Arabic treatises on the astrolabe were
translated, too. At the same time, some Latin authors composed original works on
the subject. In historiography, the assimilation of astrolabe knowledge in Latin
Europe is often seen as a consequence of material imported or produced in the

---

16 On this unique artefact, see 3.2.5.
17 On Arabic-Hindu numerals, see 4.1.6. The 'Liber Alchandrei' and other high medieval Latin
     texts of astrological content have been studied and edited by David Juste: Juste (2000), Juste
     (2007), on this subject see 5.1.1.
18 On this question, see 2.3.4.

twelfth century, and not of earlier contacts.[19] One of the main arguments brought forward in favour of this view is that eleventh-century astrolabe texts, if judged by the standards of textbooks aimed at spreading astrolabe knowledge, appear to have quite a low pedagogical value, casting serious doubts on how far they might have actually contributed to the diffusion of knowledge.

In general, the diffusion of astrolabe knowledge in tenth- and eleventh-century Europe is often regarded as a sort of prelude to the twelfth-century flowering of Arabic-Latin cultural contacts. In my opinion, instead, it was a key stage in the development of Latin natural philosophy and mathematics, taking place under the influence of the Arabic neighbour and having far-reaching consequences in the following centuries. In this study, I will attempt to reconstruct some elements of the cultural background that in the first centuries of the high Middle Ages contributed to making the astrolabe the ideal strategy for transferring and assimilating knowledge from the Arabic into the Latin culture.

1.5 THE TWO MAIN THESES OF THIS STUDY

Although the written word contributed to making the astrolabe known in tenth- and eleventh-century Europe, I shall argue that other methods of knowledge transfer also played an essential role in this process of transmission, diffusion and assimilation. These methods included oral teaching, discussions, drawings, repeated exercise in imagining, drawing and practically employing geometrical structures, watching and performing demonstrations with models, constructing and taking apart those same models, memorizing phrases, structures and procedures. Each of these non-written and/or non-verbal strategies of knowledge transfer has characteristics which make it unique and must be investigated in its own right as to how it could be employed as an alternative or complement to the other ones. The first thesis which I shall discuss in this work is that the assimilation of astrolabe knowledge in Latin Europe was the result of a combination of written and non-written, verbal and non-verbal strategies of knowledge transfer. Using these methods much more knowledge was stored and diffused in Latin Europe than would appear from the rather poor contents of the earliest astrolabe texts. In fact, the apparent evolution of early Latin astrolabe texts cannot be simply interpreted as a result of a growing level of astrolabe knowledge, but was also a result of the changing attitude to the written word among tenth-, eleventh- and twelfth-century scholars.

To reconstruct the motives behind the Latin scholars' early interest in the astrolabe, it is necessary to take into account the specific nature and epistemological implications of those strategies. In the particular context of medieval mathematics, 'the astrolabe' could be at the same time: (1) a material device to be taken in the hands and used, (2) a geometrical structure to be imagined and manipulated in the

---

19 For an overview on research results concerning 12th-century translations, see Brentjes (2000), where the problems inherent to the subject are also discussed.

mind to grasp the necessity behind celestial phenomena and (3) an abstract pattern to guide natural philosophical reflections. Not only were these aspects not mutually exclusive, as would appear today, but it was exactly because of their co-existence that the astrolabe was particularly interesting for Latin scholars.

The modes of communication used in medieval mathematics were closely linked to practice (e.g. geometrical drawing, building of sundials, surveying, architecture) and thus could foster interest in those mathematical methods, constructions and devices which could bring about material results, i.e. results that could be perceived by sensory experience. These were, for example, the correct prediction of celestial movements or the numerical estimate of inaccessible heights and distances, as well as of experienced duration. Mathematical structures of this kind could be perceived as a natural and divine 'reason' ('ratio') governing phenomena and accessible to human reason not only through abstract reflection, but also thanks to the construction and use of devices like the astrolabe. In short, my second thesis is that high medieval astrolabe studies could be linked to an image of knowledge in which the material effects of what we today regard as 'applied mathematics' were epistemologically relevant.

This epistemological stance, which is all but obvious, may appear similar to the modern scientific one. However, as I shall argue in 5.5, the differences largely outweigh any similarities that might be present, and a comparison in that direction will be avoided, because it might lead to a misinterpretation of the sources.

In this work, non-written strategies of knowledge transfer and the images of knowledge associated to them play a central role, and I will briefly introduce them in the two following sections.

1.6 STRATEGIES OF KNOWLEDGE TRANSFER: POSITIVE AND NEGATIVE DEFINITIONS

The characterization of a strategy of knowledge transfer as 'non-written' is, at first, purely negative. A positive definition becomes possible only when taking into account the specific context in which any such strategy is employed. Even then, it usually remains difficult to fully grasp how it functions.

Research on literacy and non-literacy has taken its start from the study of oral poetry from ancient, medieval and modern times.[20] Because of this, much attention has been devoted to the opposition between the spoken and written word. One of the main results of these studies has been to underscore that the opposition between 'writing' and 'orality' is itself founded on premises given only in a culture with a high level of literacy. For members of a literate culture like ours it is difficult - if not impossible - to fully grasp how oral modes of communication work, since the relevant experience is limited or lacking. Thus, orality is inevitably de-

20   As a reference on the subject of orality, literacy and their implications I used: Ong (1982), Illich (1993). For a brief overview of the first stepping stones in the study of orality see: Ong (1982) p. 16–30.

fined in a negative way. In his seminal work on orality and literacy, Walter J. Ong remarked that:

> Thinking of oral tradition or a heritage of oral performance, genres and styles as 'oral literature' is rather like thinking of horses as automobiles without wheels. You can, of course, undertake to do this [...] explaining to highly automobilized readers who have never seen a horse all the points of difference in an effort to excise all idea of 'automobile' out of the concept of 'wheelless automobile' so as to invest the term with a purely equine meaning.[21]

Thus, even when dealing with non-written modes of communication, writing may implicitly remain the paradigm of knowledge transfer and storage. The idea itself of a clear-cut distinction between 'knowledge' and its 'storage form', between 'medium' and 'message' is strongly influenced by the paradigm of writing: for example, the function of memory in literate cultures tends to be assimilated to that of a page on which words are registered, whereas memory in oral or semiliterate cultures may have quite a different role, as we shall see later on.[22]

In the case of the Latin medieval astrolabe, the situation is even more complex, since not only the spoken and the written word come into play, but also drawings, geometrical and arithmetical patterns, artefacts and also the procedures to manipulate them.

When exploring the role of these modes of communication in medieval mathematics it is particularly important to distance oneself as far as possible from modern mathematical and scientific ways of reasoning. Because of this, I will make a special effort to introduce the mathematical aspects of the astrolabe without employing modern mathematical formalism and of axiomatic-deductive proofs. Since this approach is crucial for my arguments, I will devote part of chapter 2 to a discussion of prototypes of mathematical thought different from the modern one.

1.7 IMAGES OF KNOWLEDGE AS PREMISE AND SUBJECT OF RESEARCH

My two theses on eleventh-century astrolabe knowledge are formulated separately and will be discussed in different parts of this work, yet they are inextricably related to each other. This is because the modes of communication and storage of knowledge employed in a specific context are never completely independent from views relative to the possible sources, aims and justifications of that knowledge, for example, views on whether knowledge should or should not conform to sensory experience or spatial intuition.

In stating these theses, I take Yehuda Elkana's standpoint that "for each culture, society, group or community" there are "socially determined views on knowledge", ('images of knowledge'), determining such issues as the sources of

---

21 Ong (1982) p. 12.
22 Ong (1982) p. 57–68. On the medieval 'craft of memory' see below 4.1.2–4.

knowledge, its legitimacy, audience, location on the sacred-secular continuum, or its translatability into statements about nature.[23] I also agree with Elkana in considering Western science as a cultural system and that the difference between it and other modes of thought, for example high medieval natural philosophy, is no 'great divide', but rather a continuum, in which differences can be traced back to different images of knowledge.[24]

The connection between strategies of knowledge transfer and storage, on the one side, and images of knowledge, on the other, is very relevant to my arguments for two reasons: first, because gaining an insight into high medieval methods of knowledge transfer provides a key to understanding the related images of knowledge. Second, because using modern scientific knowledge as a reference point to study medieval mathematics, astronomy and philosophy inevitably brings with it epistemological implications that should be explicitly stated.

Both written formalism and the axiomatic-deductive style of exposition are part and parcel of modern science.[25] Consequently, whatever is 'non-written' also has to be considered as 'non-scientific' in the modern sense and so the gap between medieval natural philosophy and manual crafts, on the one side, and modern science and technology, on the other, may sometimes appear like a 'great divide' only because of the form in which medieval knowledge was produced, expressed and transmitted.

The contribution of non-written and non-verbal strategies of knowledge transfer to medieval philosophy and mathematics is for the modern historian a blind spot in two senses: first, because it cannot be directly grasped through extant sources and, second, because the gap between the modern literary experience and the medieval 'preliteral', 'illiteral', 'semiliteral' or 'quasiliteral' one is at risk of being too easily filled with oversimplifications.[26]

1.8 PLAN OF THE WORK

This study is organized in six chapters, of which the first one has an introductory character. Chapter 2 begins with a discussion of mathematical thought in general, explaining how it can exhibit great variety and be linked to very different modes of communication. After this, one specific kind of mathematical thought is described which is quite different from that linked to modern pure mathematics, and is instead nearer to the medieval mathematical arts (2.1). In section 2.2, the mathematical structures that can be embodied in an astrolabe artefact are introduced, using as far as possible a terminology fitting to that specific kind of

---

23 Elkana (1981). The issues determined by an image of knowledge, of which I only quoted some, are discussed on p. 15–21.
24 Elkana (1981) p. 6–10 and 29–42.
25 On the 'universality of science' and the problem of cross-cultural comparisons of knowledge production see: Turnbull (2000b).
26 On the interplay between different modes of communication in Latin medieval Europe see 3.1.6–7.

mathematical thought. After that, it will be shown how these structures were embodied in a planispheric astrolabe and in other material tools, and how they could in principle be employed to perform some specific functions (2.3). At the end of the chapter, I point out that it is important to distinguish between functions which an astrolabe artefact could in principle perform, and those for which the device could be successfully used in practice.

Chapter 3 offers a brief introduction to the cultural and historical context of early Latin astrolabe studies (3.1), followed by an overview of the different kinds of sources relevant to the subject: artefacts and drawings (3.2), and written texts (3.3). Since my analysis is for the most part based on manuscript sources (both texts and drawings), I offer a detailed analysis of some results of previous studies of astrolabe manuscripts. These results are the starting point to formulate my thesis that, in the assimilation of astrolabe knowledge in tenth- and eleventh-century Europe, the written word was complemented by other strategies of knowledge transfer (3.4).

In chapter 4 and in chapter 5, I argue in favour of the two main theses of this study, which have been summarily sketched in 1.5. At the beginning of each of these two chapters, the thesis to be discussed is stated in detail, to be followed by arguments supporting it. In chapter 4, I introduce the strategies of knowledge transfer that complemented the written word in medieval mathematics (the 'craft of memory', exercise, notes and drawings) (4.1) and then I offer evidence of the employment of these strategies in the diffusion of astrolabe knowledge. In particular, I will show that memory and drawings played a role in astrolabe studies (4.2 and 4.3), that the occurrence of the same labelled drawing in more than one early astrolabe text can be interpreted as a trace of a purely 'diagrammatic' form of knowledge transmission (4.4), and that in this way knowledge taken from Ptolemy's 'Planisphaerium' spread in Latin Europe (4.5). I will also argue that, when taking into account of the use of such strategies, eleventh-century astrolabe texts and drawings reveal that early Latin astrolabe knowledge was less 'defective' than usually assumed. As an example of this fact, I will show how early astrolabe texts and drawings dealt with the problem of the division of the zodiac circle on the astrolabe (4.6).

In Chapter 5, I will analyse some texts and drawings occurring in the eleventh-century manuscript BnF lat. 7412 (A.27).[27] In studying early Latin astrolabe literature, it is particularly important to discuss the texts not in isolation, but taking into account the structure of the manuscript in which they occur. The reason is that both texts and manuscripts often have a composite character, and it is only by considering the themes common to all the texts and drawings grouped in a single manuscript that the motivations of the authors and readers come to light. My analysis of BnF lat. 7412 (A.27) should offer evidence that the astrolabe was regarded as both a material and an immaterial instrument of rational natural philosophy. Thanks to the astrolabe, as a structure to be understood and as a tool to

---

27   For bibliographical references on this manuscript see app. A, item A.27.

be used, the human mind could grasp the 'architectonica ratio' according to which the Divine Artifex had created the world.

I chose to analyse the manuscript BnF lat. 7412 (A.27) because it is particularly rich in drawings, and also because it contains clear traces of Arabic-Latin contacts. However, the results of this analysis have a more general validity, since most texts and drawings present in BnF lat. 7412 (A.27) also occur in other eleventh-century manuscripts. I shall also attempt to connect the results of my analysis to artefact and pictorial sources, and more specifically to: (a) material evidence on medieval equatorial sundials (5.5.2), (b) some early European astrolabe artefacts (5.6.2) and (c) an eleventh-century illumination representing Abraham holding a compass and an astrolabe in his hands (5.6.3).

Finally, in chapter 6, after having summarized the previous results, I will suggest that, from the twelfth century onward, within the medieval mathematical arts a tension developed between two concurring images of knowledge in which respectively the written word and the experience of construction methods were regarded as the best - or even the only – way to gain knowledge about the 'ratio' of the natural world. According to the position taken with respect to this question, the astrolabe could appear as a forbidden tool or as a model to be imitated.

2 MATHEMATICAL AND PRACTICAL ASPECTS OF THE ASTROLABE

2.1 SOME CONSIDERATIONS ON MATHEMATICAL THINKING

2.1.1 Mathematical thinking as a set of prototypes

How can we define 'mathematical thinking' and 'mathematical understanding'? Research done in the last decades has shown that more than one answer to this question is possible. In 1996, a collection of essays was devoted to an interdisciplinary exploration of 'The nature of mathematical thinking' in modern Western mathematics.[28] Even within such a limited field of enquiry, the editor Robert J. Sternberg had to conclude:

> To the extent that one's goal is to understand *mathematical thinking* in terms of a set of clearly defining features that are individually necessary and jointly sufficient for understanding the construct, one is going to be disappointed. Indeed, it is difficult to find any common features that pervade all of the various kinds of mathematical thinking discussed in this volume.[29]

Sternberg suggested that a viable model for understanding mathematical thinking might be to consider it as a set of prototypes, characterized by typical, but not necessary features.

> The existence of multiple prototypes seems likely, if only because the people who seem to be the best statisticians are not necessarily the same as those who are best at calculus and vice versa.[30]

2.1.2 Mathematics across cultures and time

Considering mathematical thinking as a set of prototypes can be helpful not only in reconciling today's mathematical disciplines with each other, but also in bridging the gap between mathematics of different cultures and times.

Both in present and past cultures, practices and concepts can be found that lend themselves to being recognized as mathematical.[31] However, it is usually very difficult to go beyond a vague sense of recognition and specify what exactly in those concepts or practices can be regarded as mathematical. Mathematical knowledge conveys a strong impression of being context-independent and cumulative, so that the temptation of considering premodern or non-Western mathemat-

---

28 Sternberg/Ben-Zeev (1996).
29 Sternberg (1996), quote from p. 303.
30 Sternberg (1996) p. 304.
31 Wood (2000).

ics as primitive forms of the modern Western one is always present. In the last decades, though, this approach has come under attack from many fronts: philosophy, anthropology, pedagogy, cognitive psychology and biology.[32] In particular, historians of mathematics have recognized the need for taking into account the broader cultural context when analysing mathematical evidence from the past, although opinions diverge as to whether a reconstruction of past mathematical thinking is at all possible.[33]

In the present work, I take the viewpoint that such a reconstruction is possible at least in some cases and, in the following paragraphs, I will sketch the features of a prototype of mathematical thinking helpful for interpreting the mathematical aspects of the Latin medieval astrolabe.

2.1.3 Mathematical thinking and mathematical communication

Recent research has led to the recognition that mathematical thinking is in large measure shaped by the representational systems used to produce, store and communicate mathematical knowledge, even though it remains open to discussion how far results obtained in specific areas can be generalized to a broader field.[34] Working from the assumption that mathematics – at least as far as it can be historically investigated – is always a collective enterprise presupposing interpersonal communication, the form in which it is learned in a particular culture can be expected to be constitutive for that culture's mathematical thought. For example, the structure of number-naming systems specific to a language has a significant effect on children's learning to master mathematical tools.[35]

The way mathematical knowledge is stored and transmitted inside a community also plays an important role from the epistemological point of view, implicitly or explicitly contributing to the definition of what can or cannot be mathematically known. The outer aspects of mathematical knowledge have thus to be taken into account when studying the views held in a specific community about such questions as the fields of knowledge open or closed to mathematical investigation or the possible utility of mathematics to manual crafts and everyday life in general. Because of this, it is in general not possible to separate the history of a mathematical concept from the history of the form in which it has been expressed. Translating ancient mathematics into modern form, previously a standard practice

---

32 Among the many publications on this subject, I can quote only a few: Kitcher (1984), Ascher (1991), Noss (1996), Sternberg/Ben-Zeev (1996), D'Ambrosio (1999), D'Ambrosio (2000), Rotman (2000), Selin (2000), Turnbull (2000b), D'Ambrosio (2006).

33 See for example: Høyrup (1994), Bottazzini/Dahan Dalmedico (2001), Grattan-Guinness (2003), Netz (2003) (on the possibility of reconstruction p. 279).

34 Authors arguing in this direction with different accents are, for example: Kitcher (1984), Fischbein (1987) (esp. p. 19–24), Bloor (1991), Rotman (1998), Damerow (2007), Lefèvre (2004).

35 Miller/ Paredes (1996).

in mathematical historiography, has been recognized as a very tricky operation, to be avoided as far as possible.[36]

Modern mathematics is inextricably linked to the written formalism with which it co-evolved from the early modern period up to the middle of the nineteenth century, and it is therefore legitimate to assume that any culture in which such formalism was lacking was in all probability guided in its mathematical thinking by processes different from those governing modern mathematical thought. These considerations apply to all kinds of premodern mathematics, including the ancient Greek one, but high medieval mathematics may be assumed to differ even more from the modern Western one, because it did not even employ the patterns of axiomatic-deductive reasoning characteristic both of ancient Greek mathematics and of the modern one.[37]

2.1.4 Geometrical imagination

A very important tool of high medieval mathematics was what I shall call 'geometrical imagination'. Broadly speaking, it is the ability of constructing and manipulating the figures of Euclidian geometry not only by drawing and model-building, but also in the mind. Geometrical imagination played an important role in premodern mathematics and today still plays a role in architecture, engineering and didactics.[38]

Between the Late Middle Ages and the Renaissance, written statements about the employment of geometrical imagination in architecture, painting, mechanics and mathematics started appearing.[39] Such a prototype of mathematical thinking is more similar to the medieval mathematical arts than other forms of geometry and mathematics, especially because late medieval and Renaissance texts often attempted to codify in words and drawings knowledge and methods that in the previous centuries had been transmitted with non-written and non-verbal means.[40] Therefore, studies on late medieval and Renaissance drawings and models can be used as a help in understanding the Latin medieval astrolabe and its possible epistemological implications. However, although there are undisputable similarities, one should be weary of immediately identifying the geometrical imagination of high medieval mathematicians with the methods of Renaissance artists or with the techniques of modern engineers. In late medieval and Renaissance mathematics, sight and seeing came to play an increasing role, and drawings and models were often thought of as representing objects as seen by some idealized observer.[41] In

---

36 Grattan-Guinness (2003).
37 Evans (1977a), Mahoney (1987) p. 210.
38 See for example Mauersberger (1994), P. H. Maier (1999).
39 Popplow (2002), Camerota (2004), Peiffer (2004).
40 For some examples see Beaujouan (1974).
41 The construction of the astrolabe then came to be regarded as an operation of perspective drawing: the celestial sphere was drawn as it would appear to an observer on the south pole (Camerota (2004) p. 186).

high medieval mathematics, instead, it was the whole bodily experience of geo-metrical construction that offered a guide for mathematical thought.

At this point, a few remarks on the concept of 'mental images' are due: the question, what mental images are or if they exist at all is a very complex one.[42] Here, I wish to point out two important differentiations which must be kept in mind when discussing high medieval geometrical imagination, especially in com-parison to the early modern one. The first point is that mental imagery can be as-sociated to different senses, i.e. there are not only visual (see-like) images, but for example also auditory, olfactory and kinaesthetic (i.e. motion- and action-like) ones.[43] The latter played an important role in high medieval geometrical imagina-tion, as explained in the next paragraph. The second point is that, if we regard an image as representing something (e.g. a circle or the ordered cosmos), even within the visual experience, there are different modes of representation: not all visual images are pictorial, in the sense of 'looking like' what they represent. This dis-tinction is very important in the case of the astrolabe, which, as we shall see, can be considered as visually representing the movement of celestial spheres, but does not look like a spherical cosmic model.

2.1.5 Mathematics, the mind's eye... and hands

In the high medieval period, sight did not yet have the primary role it would gain with the advent of perspective drawing. In the tenth, eleventh and twelfth centu-ries, it was not so much the act of looking at the finished drawing or model, but rather the whole bodily experience of constructing and handling it that provided a guide for mathematical thinking.[44]

As an example of this attitude, I offer a quote from Philip J. Davis and Reu-ben Hersh's book on 'The Mathematical Experience' (1981). It is a quote not spe-cifically referring to the medieval mathematical arts, but very illustrative of the kind of mathematical thinking I believe played a key role in eleventh-century as-trolabe texts:

> In raw nature, untouched by human activity, one sees straight lines in primitive form. The blades of grass or stalks of corn stand erect, the rock falls down straight, objects along a common line of sight are located rectilinearly. But nearly all straight lines we see around us are human artefacts put there by human labour. [...] For example, when one

---

42 This subject is discussed in the collection of essays: Block (1981b). For an overview of the main open questions, see: Block (1981a).

43 The following remarks are based on: (Schwartz 1981). As Schwartz notes, images are vis-ual, auditory etc. in the (vague) sense of being 'see-like' or 'sound-like': the question of whether someone is really seeing or hearing something 'in the mind' is a further complication (Schwartz (1981) p. 110).

44 This fact was still reflected in fifteenth- and sixteenth-century treatises on drawing, which emphasized the importance of knowing how to draw a geometrical figure for being able to ac-tually understand the information conveyed by it (Camerota (2004) p. 175–177).

goes to build a house of adobe blocks, one finds quickly enough that if they are to fit together nicely, their sides must be straight. Thus the idea of a straight line is intuitively
rooted in the kinesthetic and the visual imaginations. We feel in our muscles what it is to
go straight towards our goal, we can see without eyes whether someone else is going
straight. The interplay of these two sense intuitions gives the notion of straight line a so-
lidity that enables us to handle it mentally as if it were a real physical object that we
handle by hand.[45]

I believe it is important, when speaking of geometry in the mind, to think not only
of the mind's eye, but also of its hands. This distinction is relevant for my arguments from at least two points of view. The first one is that a common approach
used in medieval texts to communicate mathematical knowledge in writing was to
describe the acts performed by a geometer. These recipe-like texts were not only
aimed at conveying information about the constructed figure, but also at sharing
with the reader the bodily experience of construction, which was in itself mathematically relevant.[46] A second implication of a mathematics guided by kinesthetic
imagination was that geometrical figures did not appear so much as eternal, abstract entities to be passively contemplated, but as products of rational activity,
whose existence could in turn be recognized only through rational activity. The
existence of these products pointed to the existence of a rational 'artifex'.[47]

While the modern passage quoted above deals with straight lines, in early
Latin astrolabe literature it was the circle that played the most prominent role.
Circles were thought of and described as figures drawn by pointing the compass
in a centre, choosing a radius, and turning the compass around. As we shall see,
by repeating this act following a specific pre-determined pattern, astrolabe-lines
could be drawn. It must be noted that in the eleventh century the compass became
the attribute of God the Creator.[48]

Taking into account the contribution of kinesthetic experience to high medieval mathematical thinking also helps understand the proximity between music and
the other three mathematical arts. As discussed in 4.1.5, in the eleventh century
music was not only closely linked to the other arts of the 'quadrivium', but it was
even the leading mathematical discipline. From the twelfth century onward, instead, the gap between music and its three sister arts increased. This development
took place parallel to the rise of axiomatic-deductive methods in theology, philosophy and mathematics and of perspective drawing in the visual arts.

45   Quoted taken from Ascher (1991) p. 124.
46   See 3.4.4.
47   See 4.1.7 and 5.1.2.
48   See 5.1.2.

2.1.6 Images of knowledge and mathematical truth: proof vs. evidence

A key difference between high medieval Latin mathematical texts, on the one side, and Greek and modern mathematics, on the other, is the former's apparent disinterest in axiomatic-deductive demonstrations. In Euclid's 'Elements' (ca. 300 BC), as in modern mathematics, a statement is true when proved by deduction from a given set of principles, the axioms. Axioms are not proven: they are presumed to be true.[49] The axioms of Euclidian geometry, though, have not always been regarded as mere suppositions, but have often been described as necessary and self-evident. Discussions about the meaning and epistemological implications of Euclidian geometry's alleged self-evidence and necessity exemplify very well the role played by images of knowledge: can there be something like an immediate perception of mathematical truth resulting in a mathematical knowledge which is not in need of deductive proofs?

If one regards Euclidean axioms as true because of their (alleged) self-evidence, as premodern and early modern mathematicians usually did, then the same criterion of validity might also be applied to other geometrical constructions, even when they can also be deductively demonstrated from the axioms. In other words, by performing a Euclidian construction, either with one's body or in the mind, the validity of its properties may be perceived as evident and therefore regarded as true, without feeling the need of a deductive proof.[50] For example, in Euclid's 'Elements' it is demonstrated that one side of a triangle can never exceed the sum of the other two (Elements I.20). In his commentary to Euclid's 'Elements', Proclus (ca. 510–585) informs us that Epicurean philosophers claimed that this statement was evident even to an ass that, when hungry, would always make a straight line for a bale of hay: in such a self-evident case, argued the Epicureans, a deductive proof was not only unnecessary, but even foolish.[51]

For those who work within the context of images of knowledge placing value on axiomatic-deductive structures, though, letting one's mathematical thinking be guided by even the most self-evident feeling of certainty - by 'intuition', a rather infamous term in mathematics - is regarded as incorrect and potentially misleading.[52] At the same time, though, even among modern mathematicians, the heuristic role of intuition, imagination and visualisation is widely recognized, provided that in the end the results obtained in this way can be reformulated into a deductive proof.[53] In a prototype of mathematical thinking guided by geometrical imagination, experiencing a construction procedure may provide convincing evi-

---

49 'Axiom' in: Oxford Companion (1995) p. 72.
50 In fact, there are indications that constructions in early Greek mathematics not only illustrated, but actually complemented deductive proofs (Harari (2003)).
51 Proclus (1873) p. 322 (com. prop. XX, theor XIII), quoted by Heilbron (1998) p. 6.
52 On the questions surrounding mathematical intuition, "one of the most overworked terms in the philosophy of mathematics", see for example: Kitcher (1984) p. 49–64, quote form p. 49. See also Fischbein (1987), especially p. 43–56 and 57–71.
53 See for example: Dreyfus/ Eisenberg (1996), esp. p. 272–274; on the importance of space orientation and visualisation in the didactic of mathematics: P. H. Maier (1999).

dence of its properties, because the mathematicians do not feel the need of a deductive proof. As may be expected, different modes of mathematical communication and thought may lead to different ideas of how mathematical knowledge can be attained.[54]

In conclusion, although there is ample evidence that the axiomatic-deductive method did not play a central role in high medieval mathematical arts,[55] this attitude should not be interpreted as a lack of interest in 'pure' mathematics, but rather as a specific feature of a particular prototype of mathematical thinking.

2.1.7 Geometrical imagination and 'tacit knowing'

Geometrical imagination is linked to non-verbal modes of communication, to bodily experience, and to a conception of mathematical evidence different from the axiomatic-deductive proof. As such, it seems to offer an example of what Michael Polanyi termed 'tacit knowing' or 'personal knowledge': a kind of knowledge which is closely linked to bodily experience, is acquired prior to being articulated in formal statements in some code (e.g. a verbal statement, a theorem, mathematical formulas), and which often cannot be fully expressed in such forms.[56] Polanyi investigated the role of this kind of knowledge in the modern scientific enterprise – both in mathematics and in other fields, even though the term 'tacit knowledge' is used today mostly to refer to craftsmen's skills rather than to mathematical thought.[57]

Polanyi's work and especially his considerations on the mathematical experience have been extremely useful to me for developing the focus of the present study. His concept of tacit knowing, though, is too closely related to the psychological experience of the individual to be applicable in this study, which cannot possibly address any other level than the collective mathematical experience of a (small) group of (anonymous) scholars communicating with each other. Moreover, the kind of knowledge transmission I shall discuss in the present work is 'tacit' only in a negative sense, like 'non-written' and 'non-verbal': such adjectives can only serve to express the limits of the understanding of the researcher, and not to positively denote any single, specific epistemological phenomenon.

In fact, one of the main thesis of this work is that the Latin medieval astrolabe only appears 'tacit', i.e. incapable of transmitting knowledge, to a modern audience, whereas Latin medieval mathematicians were rather deaf to the language of Euclidian style mathematics.

---

54   Ascher (1991) p. 125–128, Dreyfus/ Eisenberg (1996) p. 279–280.
55   Interest in reasoning 'more geometrico' grew only from the twelfth-century onward, and even then more in theological and philosophical than in mathematical context (Evans (1977a)).
56   A brief introduction to tacit knowing can be found in Polanyi (1967), especially p. 3-25; the main reference is Polanyi (1958).
57   Polanyi discusses personal knowledge in mathematics in Polanyi (1958) p. 124–131 and 184–193.

2.2 THE MATHEMATICAL ASPECTS OF THE ASTROLABE

The astrolabe can be seen as embodying two astronomical-geometrical structures which I shall indicate as 'shadow triangle' and 'flat sphere'. They are quite different from each other, but have been joined together in the astrolabe from the very beginning, i.e. in the oldest extant Greek treatises devoted to that instrument. In the following pages, I will describe those two structures avoiding as far as possible the use of modern terminology. I describe first the shadow triangle (2.2.1.), which is rather simple. After that, I introduce the basic features of the homocentric-sphere-model (2.2.2), which was at the core of ancient and medieval astronomy and was the context in which the method of the flat sphere was developed. Finally, I discuss the method used in ancient and medieval times for constructing a flat sphere (2.2.3.)

2.2.1 The shadow triangle

The shadow triangle is a geometrical pattern common to a series of methods for estimating heights, depths and lengths in cases where a measure by direct comparison, e.g. with a standard rod, is not possible. The shadow triangle can be employed to estimate the height of mountains and towers, the depth of wells or the width of rivers. The way the method works can be easily grasped by looking at the picture in fig. 1.

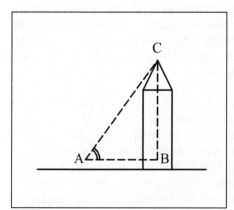

Figure 1: The figure shows how to use the geometrical construction of the shadow triangle to estimate the height of earthly objects. The eye of the observer (A), the base of the tower (B) and its top (C) can be schematically regarded as forming a triangle ABC. The difference in height between A and the base of the tower (i.e. the height of the observer) can at first be neglected: it will be taken into account at the end of the procedure. The angle BÂC comprised between the horizontal AB and the line of sight AC can be measured by the observer by sighting the tower's top through a straight tube from which a plumb line hangs. Thanks to the geometrical properties of triangles, knowing the value of the angle BÂC it is possible to deduce the ratio between the height of the tower BC (assumed as not directly measurable) and the distance AB between the observer and the tower's base (assumed as directly measurable). For example, in the simplest case, when the angle BÂC is 45°, we have AB = BC.

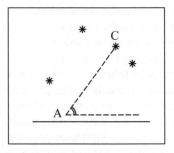 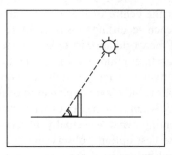

Figures 2a and 2b: The structure of the shadow triangle can be used to estimate the height of celestial bodies above the horizon in terms of an angle. For example, by sighting a star through a tube and then measuring the angle of the tube with respect to the horizon (2a). The height of the sun can instead be estimated by measuring the length of the shadow of a gnomon of known height (2b).

The geometrical principles on which the shadow triangle is based are the properties of similar triangles, i.e. triangles in which the angles occupying analogous positions are equal.[58] The shadow triangle can also be used to measure the height of celestial bodies above the horizon in terms of an angle(fig. 2a and 2b). Just as in the case of towers, the larger the angle, the higher the celestial body is in the sky. The angle can be used to indicate the position of sun or stars, and is usually called altitude. As figures 1 and 2a show, using the shadow triangle does not necessarily mean working with shadows, but the geometry of shadows does offer a very significant example of the unifying potential of this simple construction. The shadow thrown by a rod can be used to measure the height of a tower as well as the height of the sun.

In fact, it is only thanks to the shadow triangle that we come to speak of the 'height' of the sun at all, and this is an example of the unifying potential I referred to above: the height of celestial bodies above the horizon cannot even be thought of as something measurable through direct comparison with a standard rod or some other unit of length. It is only when sun, stars and horizon line are approached through the pattern of a shadow triangle that they appear to have heights, as towers do.

2.2.2 The homocentric-spheres-model

2.2.2.1 The celestial and terrestrial spheres

In Greek-Roman Antiquity as well as in the Latin, Greek and Arabic Middle Ages the sphere was the leading mathematical structure in astronomy and cosmology. The cosmos was thought of as a set of homocentric spheres, i.e. spheres having the same centre. The outermost sphere carried on its inner surface the fixed stars, while each of the inner spheres carried one of the cosmic 'wanderers': Saturn,

---

58   For a discussion of geometrical methods employed in finding heights, see Heilbron (1998) p. 97–108.

Jupiter, Mars, Venus, Mercury, sun and moon. The earth was a sphere, too, and stood unmoved at the centre of the whole structure, while the other spheres revolved around it, each according to is own (more or less) regular pattern.

This image of the world appears today hopelessly outdated, and might be seen as the product of abstract philosophical speculations, having little or nothing to do with reality. As a matter of fact, though, the homocentric-spheres-model offers a faithful representation and clever explanation of the way heavenly bodies are seen to move in the sky when time passes or when humans travel. Even today, when it comes to representing what is actually observed from the earth, homocentric spheres remain the best option: observational astronomical coordinates give the position of celestial bodies as if they were located on the surface of a sphere at whose centre the earth lies.

It is important to be aware of how close the geocentric model is to simple, naked-eye observations, such as those of seasonal changes in length of day, height of sun or constellations of the night sky. Thanks to the spherical model, the dynamics of moon phases and eclipses can be easily grasped and even the movements of some planets can be represented with good approximation. People living in today's industrialized areas are at most barely aware of these phenomena, which were instead familiar to ancient and medieval people. Land labourers had the greatest interest in paying attention to seasonal rhythms: moon phases were important for regulating the rhythm of life and work and also because the moon was the main source of light in the night hours. However, the homocentric-spheres-model could appeal not only to those interested in qualitative visualisation, but also, as we shall see, to the refined mathematical astronomer or astrologer who sought to understand the order of nature and perform exact astronomical computations.

2.2.2.2 Geometrical imagination and natural philosophy in the homocentric-spheres-model

The homocentric-spheres-model offers an example of the interaction between a form of mathematical reasoning based on geometrical imagination and natural philosophy. Geometrical imagination could help explain observations, while at the same time observing the patterns of natural phenomena and building devices such as sundials or armillary spheres could foster the development of new geometrical patterns to be employed in natural philosophy.

This two-way relationship is in my opinion very important for understanding the natural philosophical relevance of constructions such as the astrolabe. In a natural philosophy in which some geometrical structures help describe and predict phenomena, the distinction between a purely mathematical concept and its physical realisation is often blurred, as the use of words like 'sphaera' shows. In ancient and medieval literature, the term 'sphere' could be used to indicate the geometrical figure, a natural philosophical model and a structure assumed to actually exist in nature, such as the celestial or terrestrial sphere.[59] Probably the most famous example of the wide-ranging meaning of 'sphaera' is the title of John of Sacrobosco's (fl. ca. 1250) treatise on ptolemaic astronomy: 'Tractatus de sphaera' ('Treatise on the sphere'), a book which was often referred to simply as 'The Sphere of Sacrobosco'.[60]

2.2.2.3 From the solid spheres to the astrolabe

The homocentric-spheres-model allows in principle the reconstruction of celestial movements with a rather good quantitative approximation. However, it is not easy to use it in practice. It is difficult to visualize and manipulate in the mind the relative movements of concentric spheres, each eventually revolving around a different axis. On the other hand, building a three-dimensional model on which to perform computations can be quite complex and expensive. As Greek mathematicians had discovered, though, with the help of some more mathematics it is possible to draw on a surface a two-dimensional structure possessing the same astronomically relevant properties of the homocentric-spheres-model.

There are a number of ways to do this, but the method most used in ancient and medieval times was the one that today goes under the name of stereographic projection. I shall call it the method of the flat sphere (i.e. 'plana sphaera'). Before describing it, I will review those characteristics of the homocentric-spheres-model which are relevant in observational and mathematical astronomy.[61] I will discuss these characteristics only in a very elementary way, without trying to represent any specific natural philosophical theory of the past. My aim is simply to make the reader familiar with some of the mathematical and astronomical structures appearing in the oldest Latin astrolabe literature and, for the sake of clarity, I will discuss non-essential details only in the footnotes.

2.2.2.4 The poles and equator

Let us consider two homocentric spheres: the sphere of the fixed stars and the earth. The sun will be added later on, while the moon and planets can be left aside, since they played no role in early Latin astrolabe texts or artefacts. Both spheres, the celestial and the terrestrial one, have a north pole, a south pole and an equator: for simplicity, I will speak of 'poles' and 'equator' when the meaning is unequivocal. The celestial sphere revolves around its north-south axis in approxi-

---

59 Georges (1992) vol. 2, c. 2759; Oxford Latin Dictionary (1982) p. 1804.
60 Thorndike (1949).
61 The following astronomical explanations and numerical data are based on Moore (2001), especially p. 1–13 and Rigutti (2001).

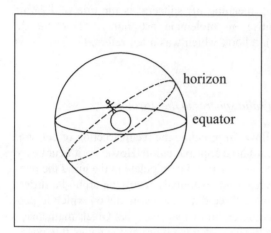

Figure 3: The celestial and the terrestrial sphere, with an observer standing on the latter one. On the celestial sphere, the horizon corresponding to that specific observer is marked (dashed line), together with the celestial equator (full line).

mately 24 hours or, to put it simply, in one day.[62] The earth stands still at the centre of the universe (fig. 3).

2.2.2.5 The horizon and the altitude circles

The earth is much smaller than the sphere of the fixed stars, but, to an observer standing on it, it still blocks the view of half of the sky (fig. 3), which therefore appears as a enormously large hemisphere reaching down to earth in all directions. The circle where the observer sees the (apparently) hemispherical sky meet the (apparently) flat earth is the horizon of that observer. The centre of the horizon circle is in principle the point on the surface of the earth where the observer is standing. However, since the radius of earth is negligible compared to that of the celestial sphere, the centre of the horizon can be assumed to coincide with the centre of the earth. Thus, within the geometrical model of the homocentric spheres, the horizon of an observer can be represented as a circle dividing the celestial sphere into two halves: a visible one and an invisible one. Since the earth is assumed to stand still, the horizon circle will always remain in the same position, while the celestial sphere - with the stars fixed on it - revolves on its north-south axis. The observer will thus see the stars rise and set with respect to his horizon.

---

62   Actually, the revolution time of the celestial sphere is less than 24 hours, which is the mean length of a solar day. Because of the movement of the sun with respect to the stars, a solar day is slightly longer than the period of revolution of the celestial sphere (i.e. of the earth's rotation around its axis), which is equal to ca. 23 hours and 56 minutes.

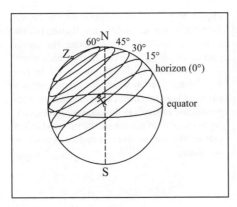

Figure 4: Altitude circles corresponding to a specific horizon for altitudes 15°, 30°, 45° and 60°. All celestial bodies on one of those circles have the same altitude above that particular horizon. The point Z is the local zenith.

According to their positions, some stars will remain more than half of the day above the horizon, while some others will rise and set in a matter of a few hours or even less. The nearer a star is to the celestial north pole, the longer it will remain above the horizon of any observer of the northern-hemisphere. Stars very near to the north pole will never set for him. On the other hand, observers on the northern hemisphere will never be able to see the constellations which are very near to the celestial south pole, as for example the Southern Cross. In general, the variety of the night sky depends on the position of the observer. More precisely, it depends on his angular distance from the north pole, i.e. his latitude.

As we have seen in section 2.2.1, it is possible to measure the altitude of a celestial body above the horizon in terms of the angle formed by the plane of the horizon (i.e. the horizontal plane) and the line connecting the observer to the celestial body. In this way, the position of stars and planets can be expressed in terms of altitude circles, i.e. circles lying on the celestial sphere parallel to the horizon. By constructions, each of these circles connects all the celestial points having a specific value of the altitude with respect to a particular horizon (fig. 4). As we shall see, altitude circles were a component of medieval Latin astrolabes.

2.2.2.6 The sun and the zodiac

Just like the fixed stars, the sun, too, rises and sets every day on the local horizon, moving as if it were one of the many fixed stars. Other than the fixed stars, though, the position of the sun with respect to the celestial sphere is not constant: the sun 'wanders' among the constellation. More precisely, it moves slowly but steadily along a circular path on the surface of the fixed-stars sphere. The time it takes for the sun to go once around the celestial sphere is the (solar) year, i.e. approximately 365 days and a quarter.

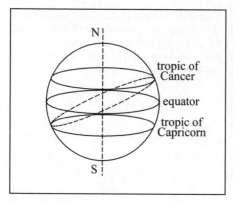

Figure 5: The celestial sphere with its North-South axis. On the sphere are drawn: (1) the zodiac (dashed line), representing the yearly path of the sun; (2) the two tropics, each representing the path of the sun on one of the two days of the year in which it occupies respectively its northernmost and southernmost position on the zodiac; (3) the equator, corresponding to the path of the sun on the two days of the year in which it occupies its middle position on the zodiac.

The yearly path of the sun is a maximum circle on the celestial sphere and has an inclination of about 23° 51' 20" with respect to the equator (fig. 5).[63] Twelve very famous constellations lie along this path: the twelve zodiac signs. I will refer to this circular path as the zodiac circle or simply the zodiac.[64]

2.2.2.7 Zodiac signs and constellations

Although the twelve zodiac signs owe their names and approximate positions to celestial constellations, the connection between those constellation and the astronomical/astrological entities known as zodiac signs had been lost already in Antiquity. In medieval astronomy and astrology the zodiac signs were defined as twelve equal portions of the zodiac circle.[65] The beginning of Aries was (and still is) defined as one of the two points where the zodiac meets the equator. As we shall see in the next paragraph, it is the point corresponding to the spring equinox. The other signs follow Aries in the well known order, each with an amplitude of 30°. The position of the sun on a specific day of the year was often indicated by stating at how many degrees of which zodiac sign it lay. Today, the position of the sun on the zodiac is usually given in terms of the angle between the sun and the beginning of Aries, measured eastwards: this is the sun's celestial longitude.

Thus, the zodiac signs were important in ancient and medieval astronomy not only because of astrological considerations, but also because they offered a reference point for determining the positions of the sun with respect to the fixed stars.

---

63 This is one of the values given by Ptolemy in the 'Planisphaerium' (Neugebauer (1949) p. 248).

64 The zodiac circle can also be referred to as the ecliptic, but I shall refrain from doing so, as this term is today understood to mean the path of the earth around the sun. The two circles coincide in the sense that, if seen from the sun, the earth appears to move along the zodiac. The name 'ecliptic' is due to the fact that a necessary (but not sufficient) condition for a sun or moon eclipse to occur is that in that moment the moon, too, should lie on the zodiac/ecliptic.

65 On the definition of the zodiac signs already in ancient Mesopotamia: von Stuckrad (2003) p. 65–67.

Moreover, the positions of moon and planets, too, could be indicated in this way. The reason is that the orbits of the earth and of the various planets around the sun almost all lie on the same plane. Therefore, when seen from the earth, the planets never seem to wander far from the path of the sun, i.e. the zodiac circle.

2.2.2.8 Seasonal variations in the length of daylight and in the night sky

On a particular day of the year, the sun will remain above the local horizon for a longer or shorter period of time depending on its position on the zodiac. The longest period of daylight for the northern hemisphere will be on the day in which the sun occupies its northernmost position on the zodiac (i.e. beginning of Cancer, summer solstice). In our calendar, this happens on June 22nd.

Analogously, the shortest period of daylight in the northern hemisphere corresponds to the day on which the sun reaches the southernmost point of the zodiac (beginning of Capricorn, winter solstice). In our calendar, that day falls on December 22nd. The circular paths described by the sun on the summer and winter solstice are called tropics, respectively of Cancer and of Capricorn. They are small circles on the celestial sphere and are parallel to the equator(fig. 5). The celestial equator corresponds to the path of the sun on those two days of the year on which daylight and night-time are equal everywhere on earth: the equinoxes. This happens when the sun is at the beginning of Aries (spring equinox, March 21st) and of Libra (autumn equinox, September, 23rd).

The longer the sun remains above a particular horizon, the higher it will appear to climb in the sky at midday. The higher the sun, the shorter the shadows thrown by objects. Thus, the length of the shadow thrown by a rod of standard height when the sun reaches its highest position can be used as a measure of the length of daylight. Thus, sundials are a combination of clock and calendar and, as we shall see, they are also useful for observing the movements of the sun.

The constellations visible in the night sky depend on the time of the year, too. On any day, the constellations nearest to the sun will be covered by its light, while those on the opposite side of the heavenly sphere will rise after it has set and be visible in the night sky.

2.2.2.9 The seven climates

While most Latin medieval people were familiar with the seasonal changes characteristic of the area they lived in, members of a relatively small group also had occasion to observe and appreciate how length of daylight, height of sun and constellations of the night sky varied when men travelled. Such changes had been of great interest to Greek-Roman and Arabic-Islamic astronomers and astrologers. The movements of sun and stars vary according to the angular distance of the

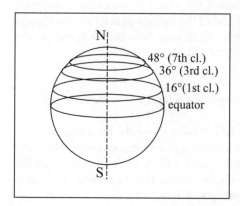

Figure 6: The (seven) climates: circles parallel to the equator dividing the inhabited part of the northern hemisphere into a number of circular belts, usually seven in number.

observer from the north pole (latitude) and one way to approximately keep track of such variations is a relatively simple scheme devised in Antiquity: the seven climates.[66] Different versions of the climate system can be found in Greek, Roman and Arabic literature. In the form in which it reached the Latin Middle Ages, the climate system was a mixture of different, sometimes incompatible, traditions, whose common feature was a division of the inhabited portion of the northern hemisphere into a number of horizontal belts, usually seven: the 'climates'[67] (fig. 6). Each climate was identified by the maximum length of daylight at its centre. The most widely used values were: 13 hours, 13½ hours, 14 hours, 14 ½ hours, 15 hours, 15 ½ hours and 16 hours. To each of these values corresponded a latitude, which could be computed with greater or lesser precision. An approximate set of values often used in astrolabe artefacts was: 16°, 24°, 30°, 36°, 41°, 45° and 48°. These values could be used to understand and approximately compute the variations of astronomic phenomena with latitude, but they could also serve as reference points for computing horoscopes for a place whose location was only approximately known.

However, the climates were not only part of astronomy, but also of geography. In fact, the climates constituted a link between the geometry of heaven and that of earth as well as between the mathematical and the geographical variety of the human world. High medieval Latin scholars could find in the Latin authorities various lists of cities and lands belonging to each climate, although these lists did not always make geographical sense. Lists of climate latitudes, maximum daylights and geographical extension would become an essential component of early Latin astrolabe texts.[68]

---

66   The main reference on this subject remains: Honigmann (1929). For a brief summary, see King (1999a) p. 6–9.

67   The term 'climate' is derived from the Greek κλίμα, i.e. inclination, and referred to the different inclination of the sun in the climates. In Antiquity the term never had the meaning that 'climate' has today (Honigmann (1929) p. 4–7).

68   See 5.3.4.

2.2.3 The flat sphere

Using the homocentric-spheres-model, it is possible to compute the period of day-light for any day of the year and for any latitude of the observer. It is also possible to predict when a particular star or constellation will rise, set or reach its highest position in the sky on any day of the year and in any place on earth. As already remarked, though, it is easier to perform such computations on a two-dimensional projection of the sphere model, i.e. on a 'flat sphere'. A flat sphere can be constructed in a few steps using ruler and compass. It represents in a plane the surface of a round sphere in such a way, that some particular properties of the three-dimensional original are preserved. For example, the rotation of a solid sphere around its axis is transformed into the rotation around the centre of the flat sphere.

2.2.3.1 Theses to be demonstrated by means of my description of the flat sphere's structure

In the following sections, I will outline the construction and properties of the flat sphere avoiding as far as possible the use of modern mathematical formalism.[69] I will offer almost no deductive proofs of my statements, since it is not my aim to explain the principles of stereographic projection, but only to provide the information necessary for discussing the oldest Latin astrolabe texts. Beyond conveying information, my description also aims at showing that:

(a) When described as a ruler-and-compass construction, the method for flattening the sphere is compact and quite easy to learn and use.
(b) With a single exception, the properties of the flat sphere can be recognized as valid thanks to a small effort in geometric imagination.

These two theses are a necessary premise for my argument that astrolabe knowledge could be stored in memory and that the flat sphere could be an invaluable help in applying geometrical imagination not only to the computation, but also to the philosophical understanding of the connection between celestial and terrestrial phenomena. Since it is difficult – if not impossible – to read a text and at the same time attempt to imagine geometrical figures, in my discussion I have made large use of pictures, commenting them as little as possible.

---

69 On stereographic projection, its properties, their proof and the history thereof, see: Neuge-bauer (1975) p. 857–860, Sergeyeva/ Karpova (1978), Jordanus de Nemore (1978) p. 44–71, Anagnostakis (1984) p. 1–5, Lorch (1995).

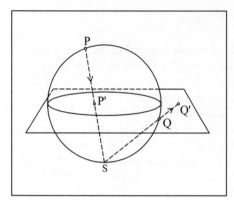

Figure 7: How to flatten the sphere: the method according to which points on the solid sphere are associated to points on a plane (modern term: stereographic projection). The points to be projected (P, Q) are joined with a line to the pole of projection (S). The projection (P', Q') of each point is the intersection of said line - or of its continuation - with the plane of projection. The pole of projection itself (S) projects onto infinity.

2.2.3.2 The general method for flattening the sphere

The method used to flatten a sphere is illustrated in figure 7, where we see the sphere, the pole of projection (the south pole S), and the plane of projection, which coincides with the equatorial plane.[70] This image does not have to be drawn, but only visualized in one's mind. All points on the surface of the sphere are projected onto the equatorial plane, on which the flat sphere shall be drawn. The south pole serves as pole of projection. We can imagine taking any point on the sphere and finding its projection according to a very simple procedure:

- points on the northern hemisphere (e.g. P): join the south pole to P with a straight line. The point in which this line passes through the equatorial plane will be P', the projection of P;
- points on the southern hemisphere (e.g. Q): join the south pole to Q with a straight line and prolong it until it meets the equatorial plane in a point. That point will be Q', the projection of Q.

According to the same procedure, any curve lying on the sphere can be projected onto the equatorial plane.

2.2.3.3 How to draw the flat sphere: the equator and the north pole

The general procedure for projecting points of the sphere onto the plane is easy to visualize, but it is still far from obvious how to actually draw on the sheet of paper the result of the projection, i.e. a flattened sphere. Only in a few special cases, the results can be easily drawn: the equator, for example, is already lying on the flat sphere plane and therefore each of its points projects onto itself. As a very small effort in geometrical imagination shows, the north pole projects onto the centre of

---

70   Any plane parallel to this one could be chosen as plane of projection for the same pole without changing the results, but this was the choice made in high medieval Lain texts and we shall conform to it.

the equator. It should also be evident to the mind's eye that there is one point on the sphere that cannot be projected at all: the pole of projection itself. There will be no south pole on the flat sphere. In an actual drawing, the whole portion of spherical surface near the south pole will be missing, because there is not enough space for all of it on any finite sheet of paper. This was no serious problem in the Middle Ages, since the celestial south pole and its surroundings were hardly ever visible to European astronomers.

The first elements of the flat sphere can now be drawn: circle (the projection of the equator) and its centre (the projection of the north pole). The choice of radius for the equator of the flat sphere is at this moment free, but it will determine the overall dimensions of the construction.

2.2.3.4 The tropics on the flat sphere

The tropics, as we have seen, are two small circles lying on the solid sphere parallel to the equator, one above and one below it, each at an angular distance from the equator of 23° 51' 20". We shall now project them onto the flat sphere, using a procedure that can best be understood by looking at fig. 8a and 8b. The three-dimensional constructions shown there do not have to be exactly drawn on paper: it is sufficient to sketch them or visualize them in the mind to see that the tropics of the flat sphere will be two circles having the same centre as the flat sphere's equator. As a consequence of the north-south asymmetry of the projection, the two circles, which are equal in three dimensions, will be different on the flat sphere, where the tropic of Cancer is smaller than the equator and the tropic of Capricorn is larger. The only information still needed to actually draw the tropics on the flat sphere is how large their radii ought to be once the radius AB of the equator is given. We can obtain this information by performing a very simple geometrical construction on paper, i.e by simply drawing the two-dimensional cross-section of the projection procedure (fig. 9). Once this has been done, the tropics on the flat sphere can be drawn by using a compass to construct the circles corresponding to OC and OD (fig. 11).

2.2.3.5 Ninety-degrees-tilting as a key trick in drawing the flat sphere

There can be no doubt as to the simplicity of the construction described above: it is relatively easy to visualize and very easy to draw. It is useful, now, to explicitly underscore the symmetry which is at the core of such simplicity.

Let us go back to fig. 9, which represents the cross-section of the projection procedure for tropics and equator. In that figure, the circle with diameter AB is a meridian circle, i.e. a maximum circle on the solid sphere which passes through its two poles. It is not the equator, which in this cross-section is the segment AB.

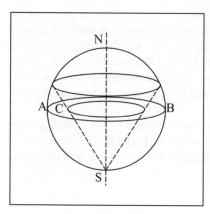

Figure 8a: Projection of the tropic of Cancer on the flat sphere. Since the tropic is a circle parallel to the equator plane, which is also the plane of the flat sphere, the projection construction is symmetrical with respect to a rotation around the north-south axis of the solid sphere. The rays of projection connecting the south pole to the tropic form a right cone. The projection of the tropic is an orthogonal section of this cone, and is therefore a circle. It is immediately apparent that its centre coincides with that of the equator.

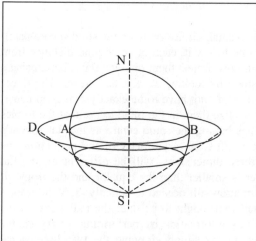

Figure 8b: Projection of the tropic of Capricorn on the flat sphere. Exactly the same considerations are valid as in fig. 8a.

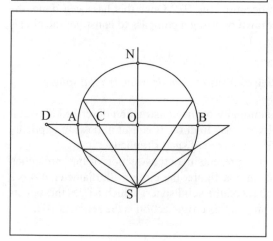

Figure 9: Two-dimensional cross-section of the projections shown in figure 8a and 8b: by connecting the south pole S with the extremes of the cross-sections of the tropics, the radii of their projections on the flat sphere are obtained: OC for Cancer, OD for Capricorn. AB is the diameter both of the equator and of its projection on the flat sphere.

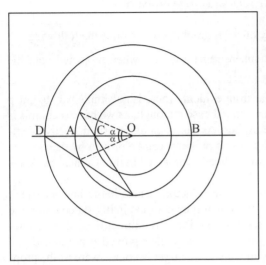

Figure 10: Using the construction already shown in fig. 9 to find OC and OD (i.e. the lengths of the radii of the projections of the tropics), we can draw the projections of the tropics on the flat sphere. At this point, the surface of the drawing is interpreted as the plane of the flat sphere, and the circle with diameter AB as the projection of the equator on it.

However, the meridian circle has the same radius as the equator, since both are maximum circles of the same sphere. Thanks to this fact, once we have drawn the cross-section and used it to find the segments OC and OD, we can draw the corresponding circles on the same sheet of paper by simply interpreting the maximum circle already drawn as if it were the equator of the flat sphere: this is what was done in fig. 10. In other words, we are virtually tilting the drawing by 90°, placing it on the plane of the flat sphere.[71]

The mental act of tilting the construction in one's mind while leaving it unaltered on paper (parchment, wax, wood or sand) is characteristic of a form of mathematical thinking based on dynamical geometrical imagination. As we have seen, the trick makes it possible not only to easily perform the construction, but also to convince oneself of its validity. In fact, it is probably easier to grasp the properties of the construction by reflecting on the drawing or by performing the construction than through verbal explanations. The trick of virtually tilting the construction is an essential ingredient of the standard procedure used to represent the various celestial circles on the flat sphere. The second ingredient essential to perform this task is a key property of stereographic projection: circles on the solid sphere are projected into circles lying on the flat sphere. I shall refer to this feature as the 'circle-into-circle property'.

71 Another way of looking at the same trick is to think of fig. 9 as if it were a projection made, not from the south pole, but from a point lying on the equator: thanks to spherical symmetry, the lengths of radii found is in the end the same as in the south-pole-projection. Again, a tilting of 90° is involved, but this time it is the whole projection procedure, and not only its results, which is thought of as tilted. The tilting is virtual in the sense that the drawing is not changed at all: only its interpretation.

2.2.3.6 The circle-into-circle property

By far the most important property of stereographic projection is the following:

- All circles lying on the solid sphere remain circles when projected onto the flat sphere.

The circle-into-circle property is far from evident. There is no simple way to grasp it by geometrical imagination in the general case although, as we have seen, this is possible in the special case of the tropics. As far as an axiomatic-deductive demonstration is concerned, the validity of this property can be proven by using some theorems on conic sections which were already contained in the 'Conics' by Apollonius of Perga (ca. 250–175 B.C.).[72]

Thus, the proof of this property was in principle accessible to ancient Greek mathematicians, but there is no evidence that they ever explicitly stated either the property or its proof. As we shall see later on, Ptolemy implicitly used the circle-into-circle property in his 'Planisphaerium', but neither proved it nor stated it.[73] As I have argued elsewhere, a way in which one might become aware of the property is by taking the model of a sphere made out of crossing circles, illuminate it with a point-like light source situated on the sphere's surface, and observe the shadow it casts on the wall (ceiling or floor) opposite to the light surface.[74] However, there is absolutely no hint that the ancient Greek had done this.

2.2.3.7 Consequences of the circle-into-circle property

The circle-into-circle property of stereographic projection has two very important consequences. The first one is:

(1) Circular motion around the north-south axis of the solid sphere is transformed into circular motion around the north pole of the flat sphere.[75]

In this way, as anticipated, the daily revolution of the celestial sphere can be easily represented on the plane. Although (1) is a very important feature of the construction, it is the second consequence of the circle-into-circle property that makes it easy to draw a flat sphere with ruler and compass. This second consequence is:

(2) Since the projection of a circle is itself a circle, it is fully determined once the position and length of one of its diameters are known.

---

72   Neugebauer (1975) p. 858–859.
73   See 4.5.3.
74   Borrelli (2006) p. 160–162.
75   When seen from the northern hemisphere of the earth, the revolution of the celestial sphere around its north pole takes place anticlockwise. However, the flat celestial sphere is drawn as if seen from a point above the north pole, i.e. outside of the sphere, so that the person holding the astrolabe in the hand has to be considered as standing outside of the celestial sphere. From that point of view, the rotation appears to happen clockwise, and so the rete of the astrolabe has to be rotated clockwise with respect to the horizon line.

Property (2) makes it possible to easily represent all remaining celestial circles on the flat sphere: the zodiac, any horizon and even the altitude circles associated with it, in short all circles shown in figs. 3–5.[76]

2.2.3.8 The zodiac and the horizons

The zodiac as well as all horizons are maximum circles of the solid sphere which are inclined with respect to the equator. As already seen, the inclination of the zodiac is 23° 51' 20", while each horizon has its specific latitude. All these circles can be projected onto the flat sphere by following the same procedure, which is analogous to that already employed for projecting the tropics. First of all, thanks to the circle-into-circle property, we know that zodiac and horizons will be circles on the flat sphere, too. Knowing this, it will be sufficient to find the centre and radius of their projection to completely determine it.

Let us imagine a generic maximum circle on the solid sphere, inclined with respect to the equator by a given angle δ. Now let us draw the two-dimensional cross-section of the procedure necessary to project it onto the equatorial plane, just as we have done in the case of the tropics. Fig.11 shows the cross section and the further steps of the procedure I am about to describe. The cross-section drawn is not chosen at will, but must be the one passing through the northernmost and southernmost points of the inclined circle.[77] As before, the equator is represented by the segment AB.

The inclined circle is represented in the cross-section by its diameter EF. The lines connecting the south pole to the points of the segment EF determine its projection E'F'. On the solid sphere, EF is a diameter of the inclined circle. Since the projection of the inclined circle is a circle, too, it follows from the symmetry of the situation that E'F' must be the diameter of the inclined circle (zodiac or horizon) on the flat sphere .[78] Once the diameter is given, the circle can be drawn.

As in the case of the tropics, we perform the virtual tilting of the construction, i.e. we draw the circle having E'F' as diameter, and consider the maximum circle as the equator.

---

76 On altitude circles see 2.2.2.5 and fig. 4.
77 In the case of the tropics, all north-south cross-sections look alike, but for an inclined circle this is not the case.
78 This statement is no axiomatic-deductive proof: it is evidence resulting from a visualisation process that might, eventually, be turned into an axiomatic-deductive proof. The symmetry of the construction performed assures us that the projections of the two halves of the circle cut by the diameter EF have to be equal and that, therefore, E'F' shall be a diameter of the projection. If we had not chosen E and F to be the northernmost and southernmost points of the circle, but simply the extremes of one of its diameters, their projections E' and F' would not have been the extremes of a diameter of the projection.

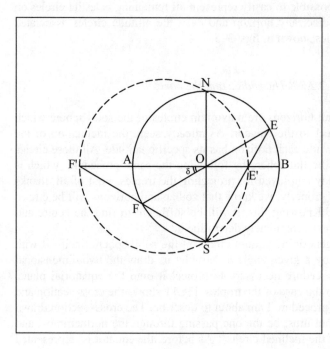

Figure 11: How to project onto a plane a maximum circle inclined at an angle δ with respect to the equator (here δ=30°):
(1) draw the cross-section of the circles (i.e. circle ANBS with diameters AB and EF);
(2) draw lines joining S to E and F and note the points E' and F' where they cut AB or its continuation;
(3) draw the circle having E'F' as diameter (dashed circle F'NE'S).

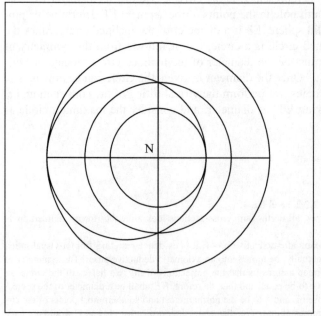

Figure 12: The tropics, the equator and the zodiac on the flat sphere. The zodiac, i.e. the circle off-centre, touches each tropic in one point (the solstices) and cuts the equator in two point (the equinoxes).

Taking the zodiac as an example and adding the two tropics to the picture, we see that the tropics touch the zodiac at its northern- and southernmost points, as it should be (fig. 12). Since the zodiac is not parallel to the equator, its projection on the flat sphere is not concentric to equator and tropics. On the solid sphere, equator and zodiac bisect each other, as all maximum circles do, but on the flat sphere this is not true. This is a fact to be remembered: although circles project into circles, equal arcs do not generally project into equal arcs.

For example, the twelve signs of the zodiac, which each occupy a 30° arc of the zodiac circle, do not extend over equal portions of the zodiac of the flat sphere. The problem of an appropriate division of the zodiac of the flat sphere is not trivial and in chapter 4.6. I shall discuss how Latin medieval scholars tackled it.

2.2.3.9 Conclusions: the basic three-step-procedure for projecting circles onto the flat sphere

Any circle on the solid sphere can be drawn on the flat one in three steps, which are a further generalization of those used to draw the zodiac and the horizons. Let us take a circle which is neither a maximum circle nor a parallel to the equator (fig. 13). It will be defined by the position of its northernmost and southernmost points E and F. The three-dimensional fig. 13. is only indicative of the structure and does not have to be drawn.

The first step of the actual construction is to draw a cross-section of the solid sphere, placing on it the points E and F (fig. 14a). The second step is to draw straight lines connecting E and F to the south pole, marking the points E' and F' where these lines cut the cross-section of the equator (i.e. the prolongation of A (fig. 14b). Thanks to the circle-into-circle property and to spherical symmetry, E'F' will be the diameter of the circle representing on the flat sphere the original one. The third step of the procedure is simply to draw the circle having E'F' as diameter, while considering the maximum circle already drawn as the equator: this is what I have called performing a virtual tilting of 90° (fig. 14c). That is all that is needed to construct circles on the flat sphere. By repeating this procedure, one can draw on it the tropics, the zodiac, any horizon and its altitude circles.

At this point, I hope to have offered enough arguments to show that the construction of a flat sphere is a relatively easy ruler-and-compass procedure, and that many of its astronomically relevant properties can be recognized as valid with the help of geometrical imagination. The only definitely non-evident feature of the flat sphere is the circle-into-circle property. All early Latin texts devoted to the construction of the flat sphere make use of this feature, but neither state nor prove it. This has been considered as evidence that the Latins only had interest in the practical use of the astrolabe, but I shall argue that it was not so.

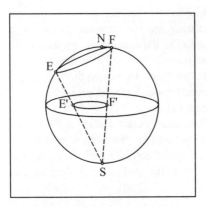

Figure 13: A generic non-maximum circle on the solid sphere, which will be projected onto the flat one.

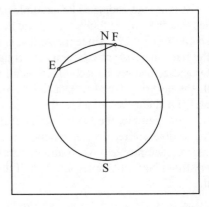

Figure 14a–c: The three steps to project a circle of the solid sphere onto the flat one.
Figure 14a: Step One: draw a cross-section of the solid sphere (here the circle with diameter NS) and of the circle to be projected (here the segment EF).

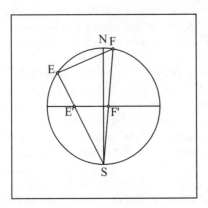

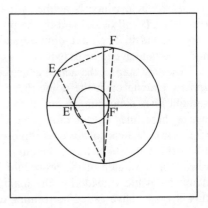

Figure 14b: Step Two: draw straight lines connecting the south pole S to the extremes E and F (i.e. the southernmost and north-ernmost points) of the circle to be projected. Mark the points E' and F' where the lines cut the cross-section of the equator (here the horizontal line).

Figure 14c: Step Three: draw the circle with diameter E'F' and consider now the maximum circle already drawn as the equator (virtual rotation of 90°). The dashed lines can be removed from the finished drawing, and the circle E'F' will represent the projection on the flat sphere of the circle EF in figure 13 above.

2.3 THE PRACTICAL ASPECTS OF THE ASTROLABE

The elements of mathematical and astronomical knowledge introduced in the previous sections can be embodied in a wide range of artefacts. These objects can be considered as devices built to apply a pre-existing theory to serve a specific purpose. However, as we shall see, the question of how and if an extant artefact could really have been used for a given theoretical purpose is all but simple. For example, the question of the functionality of astrolabes as timekeepers is extremely delicate.

Because of these problems, it is advisable to discuss the practical aspects of the astrolabe by clearly distinguishing theoretical functions from actually possible uses. In the following sections, I will discuss the planispheric astrolabe and its theoretically possible functions, first the astrolabe's back (2.3), then its front (2.3.3). Finally, I will address the question of which of these functions the planispheric astrolabe might really have performed (2.3.4).

2.3.1 The planispheric astrolabe

The device described below, which I shall indicate as a planispheric astrolabe, is a purely ideal artefact: a prototype representing some typical, but not necessary features of real astrolabe artefacts, such as those discussed in 3.2.[79] Only the main characteristics of early Latin astrolabes are discussed here: Arabic-Islamic artefacts were usually more complex. The attribute 'planispheric' indicates that the mathematical structure embodied in the artefact is the flat sphere described in 2.2.[80]

The planispheric astrolabe is a round, flat device made up of two or more concentric discs fitted one on top of the other and kept together by a pin passing through their common centre. Both sides of the astrolabe embody mathematical structures: on its front, lines are drawn (or engraved) which embody the flat sphere and can be used to perform astronomical computations; its back can be employed to measure the height both of celestial and terrestrial bodies with the method of the shadow triangle. I will start my description from the back of the astrolabe.

---

79 Detailed descriptions of the planispheric astrolabe and its various forms and uses can be found in: Hartner (1939), North (1974), A. J. Turner (1984) p. 1–9, Stautz (1999) p. 99–122, Proctor (2005).

80 Devices embodying other mathematical representations of the heavenly sphere are, for example, the spherical astrolabe and the universal astrolabe, with which this study is not concerned since they played no significant role in eleventh- and early twelfth-century Latin astronomy. Although some eleventh-century Latin texts are described as dealing with a 'spherical astrolabe', I shall argue that in fact they are only concerned with a simple celestial sphere, and not with the more complex device usually indicated as a 'spherical astrolabe' (see below, 5.3.5). For a general description of the spherical astrolabe and examples of extant instruments, see: Sezgin/ Neubauer (2003) p. 120–133.

2.3.2 The back of the planispheric astrolabe

2.3.2.1 The alidade as a sighting tool

On the back of the planispheric astrolabe - a round, flat surface - are drawn or engraved two almost concentric circular scales (fig. 15, (a) and (b)) and two short graduated lines forming a right angle (fig. 15, (c)). At the back's centre, a flat bar is pinned, but left free to rotate around it (fig. 15 (d)). The bar, called 'alidade', is provided with sighting devices at both ends: short, flat extensions with a hole in the middle.[81] The alidade of an Arabic-Islamic astrolabe is drawn in fig. 16, to the left. Letting the planispheric astrolabe hang down perpendicular to the ground from a line attached to its rim and sighting with the alidade a star, its altitude can be read on the outer scale, which is divided into sexagesimal degrees. To measure the altitude of the sun, one has to position the alidade so that a ray of light passes through both holes at the same time.

The alidade can in principle be used to measure the height of the sun and stars as well as that of earthly objects like towers or mountains - at least in principle. In practice, there is a catch: the precision of the measurement depends crucially on the dimensions of the disc and, especially if the altitude of a celestial body has to be estimated, these dimensions have to be quite large. Devices used for astronomical observation usually had a radius larger than the 10 or 20 centimeters typical of medieval astrolabe artefacts.[82] A further problem is that, during the measurement, the disc has to hang perfectly still and perpendicular to the ground: a condition that can not be easily fulfilled, even for metal devices, if they are too small.

2.3.2.2 The shadow square

An easy and quite precise way of estimating the altitude of the sun is to measure the length of a standard rod, the 'gnomon'.[83] The gnomon must be first marked, dividing it into a number of equal parts (usually seven or twelve), and is then set perpendicular to the ground. The length of its shadow is noted. One then estimates the ratio between the length of the gnomon and that of its shadow using as unit of measurement the parts in which the gnomon is divided (e.g.: the length of the shadow equals two parts out of seven).

---

81 On the term 'alidade' and its Arbic origin, see: Kunitzsch (1982) p. 527.
82 On medieval observational instruments: Poulle (1991), especially p. 253–262, for some examples of observational instruments: Sezgin/ Neubauer (2003) p.19–78.
83 The following description is based on: Hartner (1939) p. 302 and 304 (gnomon and shadow square) and Zinner (1967) p. 187–191 (geometrical square) and p. 226 (gnomon).

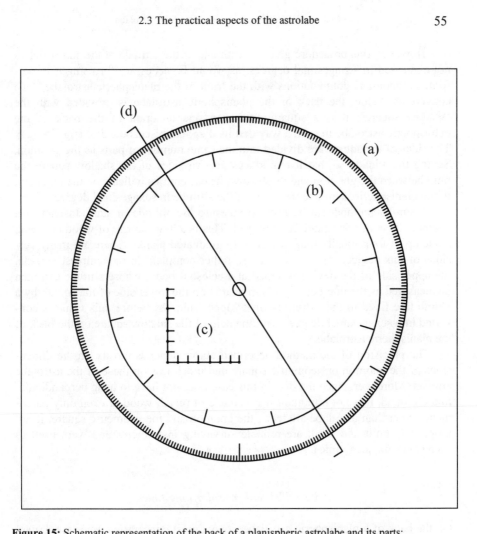

Figure 15: Schematic representation of the back of a planispheric astrolabe and its parts:
(a) outer circle, divided into twelve equal parts, each subdivided into 30 smaller ones, for a total of 360 equal parts (degrees). The twelve larger parts carry the names of the zodiac signs;
(b) inner circle, slightly off-centre, divided into twelve parts carrying the names of the months;
(c) the shadow square, i.e. the graduated lines forming a right angle in the lower left quadrant of the planispheric astrolabe's back.
(d) alidade, a moveable bar pinned to the centre of the planispheric astrolabe, represented here as a straight line with two jutting appendixes for sighting celestial bodies.
Combined, these features of the back of the planispheric astrolabe have the following possible uses:
–   360-degrees scale on outer circle + alidade: measure in degrees altitude of celestial bodies;
–   360-degrees scale on outer circle + shadow square (+ alidade): convert a measurement of celestial altitude done with a gnomon or quadrant into sexagesimal degrees (the alidade serves as a pointer in the comparison of the scales);
–   zodiacal scale on the outer circle + month scale on inner circle (+ alidade): determine in which degree of which sign the sun is on a particular day of the calendar year - and vice versa (the alidade serves as a pointer in the comparison of the scales).

However, this procedure gives an estimate of the altitude of the sun which is not expressed in sexagesimal degrees, as would be necessary for using it to perform astronomical computations with the front of the planispheric astrolabe.[84] To convert the value, the back of the planispheric astrolabe is provided with the 'shadow square': it is a square drawn on one quadrant of the back of the planispheric astrolabe in such a way that its diagonal bisects the disc (fig. 15, (c)). The sides of the square are divided into the same number of parts as the gnomon. Setting the alidade so that it reproduces on two sides of the shadow square the ratio between the gnomon and its shadow, its extremities indicate on the outer rim of the planispheric astrolabe the value of the altitude in sexagesimal degrees.[85]

A similar method can be used to measure the altitude of stars. Instead of a gnomon, a geometric square is employed. This is a large square of wood or metal, made up of four equally long and equally graduated planks. There are many versions of this device, some of which are rather complex. In the simplest version, the upper side of the device is aimed at a celestial body, whose altitude can then be measured as the number of parts indicated on the lower side of the square by a plumb line fixed to the extremity of its upper side. As before, this value is converted into sexagesimal degrees with the help of the shadow square on the back of the planispheric astrolabe.

The precision of the method described above depends mostly on the dimensions of the gnomon or geometric square and much less on those of the astrolabe artefact. Moreover, since the disc in this case does not have to hang perpendicular to the ground, it is not important if it is made of metal, wood or even only parchment. Other than astrolabe artefacts, the gnomon and the geometric square, if appropriately built and used, are reliable surveying instruments and were used as such since medieval times.

2.3.2.3 The scale of solar longitudes

On the back of a planispheric astrolabe, two almost concentric circles are drawn, both divided into twelve parts (fig. 15, (a) and (b)). The inner circle is inscribed with the names of the months, the outer one with those of the zodiac signs. The beginning of each zodiac sign corresponds approximately to the middle of a month, so that for each day of the year one can read the corresponding position of the sun in the zodiac (i.e. the solar longitude for that day), or vice versa. The sun does not pass at the same speed through all of the signs, and this effect is taken into account by making the two circles not exactly concentric.

---

84   The altitude circles on the front of the astrolabe are graduated in sexagesimal degrees (see 2.3.3).

85   When the shadow is longer than the gnomon, one can set the gnomon parallel to the ground and let it project its shadow on a vertical surface. The length of the shadow must in this case be read on the vertical side of the shadow square. The two sides of the shadow square are known as 'umbra recta' (horizontal shadow of a vertical rod) and 'umbra versa' (vertical shadow of a horizontal rod).

2.3.3. The front of the planispheric astrolabe

2.3.3.1. The disc of the celestial sphere and the disc of the local horizon

To represent in two dimensions the daily revolution of the heavenly sphere with respect to a chosen horizon, one takes two discs of equal dimensions and draws on the first one the projections of the celestial circles (i.e. equator, tropics and zodiac) and on the second one that of the chosen horizon (fig. 16). On the celestial disc (fig. 16, above to the right), a few stars can be positioned, too, and its outer rim usually coincides with the tropic of Capricorn. As can be seen in fig. 16, in a planispheric astrolabe the celestial disc was perforated, to be superimposed on the horizon disc.[86] On the horizon disc (fig. 16, below to the right), are also drawn the altitude circles corresponding to altitudes of 0° to 90°. Sometimes all ninety circles occur, but more often they are spaced at a distance of two to five degrees. On Islamic-Arabic astrolabe artefacts, the azimuth lines were also usually drawn, i.e. projections of maximum celestial circles passing through the zenith of the observer. However, these lines never appear in high medieval Latin astrolabe texts and drawings, so I shall not deal with them further.

By putting the celestial disc on top of the horizon disc and rotating it clockwise, the revolution of the celestial sphere with respect to the horizon can be simulated. In the planispheric astrolabe, the plate representing the celestial sphere, which is perforated, takes the name of 'rete' (net). Through the rete, the horizon on the plate under it can be seen and, when the celestial sphere rotates, the stars and zodiac signs rise and set on it. In the device called anaphoric clock, instead, it is the horizon plate that is perforated and put in front of a solid, revolving plate on which the constellations are drawn.[87]

2.3.3.2 The collection of horizon plates

Since all computations relevant for a specific latitude can only be performed with a horizon plate drawn for that particular latitude, astrolabe artefacts usually had several such plates, each engraved on both sides with horizon and altitude circles corresponding to a particular latitude. The choice of latitudes represented depended upon the preferences of astrolabe-makers and astrolabe-users: some items featured the latitudes of important cities, others those corresponding to the traditional seven climates.[88]

---

86 On the stars usually represented on astrolabes Kunitzsch (2005).
87 For references on the anaphoric clock see 3.1.1.
88 On the climates in general, see 2.2.2.9. On the climates in Latin medieval astrolabe texts and artefacts, see 5.3.4 and 5.6.1–2.

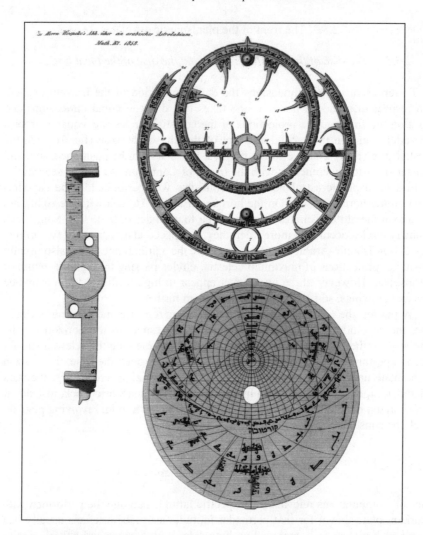

Figure 16: Modern drawings of parts of an Arabic-Islamic astrolabe artefact, made in Toledo in
   1029/1030 (420H) (**#116**, Berlin Staatsbibliothek Orientabteilung). The parts represented are:
- left: the alidade
- right top: the rete, i.e. a perforated disc representing the two dimensional projection of some
   circles on the celestial sphere (equator, tropics, zodiac), with pointers indicating the positions
   of various stars. The Arabic names of the stars and zodiac signs are engraved on the rete.
- right bottom: horizon plate for the latitude 38° 30' (Cordoba). The horizon line is clearly dis-
   tinguishable, since the geometrical patterns above and below it are different. Above the hori-
   zon-line, altitude circles form a sort of spider-web with azimuth lines, which are projections
   of maximum circles passing through the local zenith. Below the horizon, the lines corresponding to the twelve unequal (temporal) hours are drawn. When superposing the two plates
   and rotating the upper one clockwise with respect to the lower one, the stars rise and set at the
   horizon.

In most astrolabe artefacts, one of the horizon plates is larger than the other ones, has a high rim and can contain all other discs in itself. It is called the 'mater ('mother'). The various horizon plates can be taken out of the mater and rearranged: each time it is used the astrolabist can choose the horizon he needs.

2.3.3.3 Possible uses of the front of the planispheric astrolabe

By rotating the rete with respect to the local horizon, the revolution of the celestial sphere can be simulated, and, with it, the homogeneous flow of celestial time. As we shall see, Latin scholars were particularly interested in estimating the quantity (or the "parts") of time corresponding to some duration: for example, the quantity of daylight and night-time on a particular day of the year or the period between sunrise and the present moment.

To grasp with geometrical imagination the ordered motion of the spheres, the front of the planispheric astrolabe is sufficient. To become aware first hand of the way in which rational, cosmic order is mirrored in the regularity of natural phenomena, though, additional inputs are needed. These are provided by the back of the planispheric astrolabe.

The first additional information is the altitude of a star or of the sun above the local horizon at the present moment. As we have just seen, this quantity can in principle be measured thanks to the devices found on the back of the planispheric astrolabe. Alternatively, one might use other tools to perform the measurement and then convert the result into sexagesimal degrees by using the shadow square on the back of the planispheric astrolabe. The second necessary information is the actual position of the sun on the zodiac, which can also be read from the double circular scale present on the back of the planispheric astrolabe, as described in 2.3.2.3 and shown in fig. 15.

On the basis of these two pieces of information, the point of the rete representing the chosen celestial body can be superimposed on the altitude circle corresponding to the value of the altitude measured: in this way the present celestial configuration with respect to the local horizon is reproduced on the front of the planispheric astrolabe. At this point, the front of the astrolabe can be used for estimating the length of daylight for that particular day of the year, or to compute the present temporal or equinoctial hour, as explained in the next paragraph. Of course, the device can be effectively used only if the astronomical information it embodies is correct.

2.3.3.4 The hour lines and monastic timekeeping

The twenty-four hours we use today, usually referred to as equinoctial hours, were originally used only by astronomers.[89] For the purpose of regulating everyday life, equinoctial hours were never employed before the European Late Middle Ages, and, even then, it is doubtful how far they were used in practice.

In eleventh- and twelfth-century Europe, the 'hours' used to regulate the times of prayer and work were the temporal hours, which the Latin Middle Ages had inherited from the Roman. For every day of the year, the periods of daylight and night were each divided into twelve hours, whose length changed according to the season. Temporal hours cannot be easily measured with a mechanic clock, but it is not difficult to determine them using a sundial.

As we shall see, in astrolabe texts much attention was devoted to determining equinoctial and temporal hours and converting them into each other. On the front of the planispheric astrolabe, one full rotation of the rete represents a period of twenty-four equinoctial hours, so that each hour corresponds to a rotation of fifteen degrees.

Temporal hours are more difficult to represent, because their duration depends on the position of the sun on the zodiac and on the local latitude. On the planispheric astrolabe, they can be represented by taking into account the fact that, on the flat sphere, the circle corresponding to the projection of the path of the sun on a generic day of the year is a circle concentric to the equator and is divided by the horizon into two parts corresponding respectively to daylight (portion above the horizon) and night (portion below the horizon) (fig. 17). By subdividing each of these two parts into twelve equal portions, daylight and night can be divided into twelve temporal hours each. In this way, variable temporal hours are represented by arcs of circle having different lengths for different days, and can be related to the 15° arcs representing the fixed equinoctial hours.

In principle, this process should be repeated for each day of the year on the circle representing the corresponding solar path. However, a reliable approximation is obtained by performing the division only on the tropics and the equator, and then joining the division points corresponding to the same hour with lines or with an arc of circle: the resulting curves can then be used as a reference point to estimate the temporal hour for each position of the sun on the zodiac.

On astrolabe artefacts and drawings, lines representing temporal hours for a certain latitude are drawn on the corresponding horizon plate, usually in the space left free under the horizon. These hour lines, although in principle corresponding to night, can also be employed to estimate temporal hours during daylight, by using as a reference the point of the zodiac opposite to the actual position of the sun (i.e. its nadir).[90]

---

89   For a discussion of the questions related to the measure of time and its role in regulating society in the past and in the present see: Dohrn-van Rossum (1992), Gasparini (2001).
90   In the night, it is the position of the sun itself that indicates the hour.

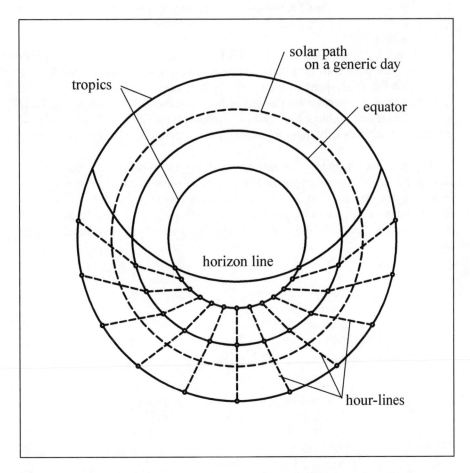

Figure 17: Representation of temporal hours on the horizon plate of a planispheric astrolabe. The portions of equator and tropics which lie under the horizon line have been divided into twelve equal parts, corresponding to the twelve temporal hours on the equinoxes and solstices. Joining these points with arcs of circle or (as done here) simply with straight lines gives an approximation for hour lines for each day of the year. The dashed circle corresponds to the path of the sun on a generic day: its portion lying below the horizon is approximately divided into twelve equal parts by the hour lines.

Superposing the rete to the horizon plate and knowing the actual position of the sun on the zodiac, as well as its altitude in the sky, it is possible to estimate which temporal hour of the day it is. The same can be done at night by measuring the altitude of some reference star.

2.3.4 Could astrolabe artefacts actually perform
the functions they were built for?

Some extant European astrolabe artefacts have such small dimensions, that they could not have been used for reliable measures of altitude.[91] This fact has led historians to believe that many of them were built to be prestige objects never to be used. However, in treatises on the use of the astrolabe, instructions are given to use the alidade as a sighting tool, and most medieval images of astronomers using astrolabes portray exactly that (unrealistic?) act. I shall come back to this subject in 5.2.5.

It is also not uncommon for astronomical data incorporated in astrolabe artefacts to be out of date or imprecise, especially in some European astrolabe artefacts.[92] Again, the question arises whether these objects were built and used by incompetents or were simply prestige objects or qualitative models. As I shall suggest later on, possibly these devices were built as a kind of material thought experiment.

---

91   Nallino (1930) p. 347, Poulle (1991) p. 260–262, Wintroub (2000) p. 45, King (2001) p. 363.
92   Maddison/Savage-Smith (1997) p. 186, Stautz (1997) p. 88–92 and p. 122.

3 SOURCES AND RESULTS OF PREVIOUS RESEARCH

3.1 INTRODUCTION: HISTORICAL AND CULTURAL CONTEXT

3.1.1 The earliest evidence on the method of the flat sphere and on the astrolabe

This study is primarily concerned with the astrolabe in Latin Europe in the tenth and eleventh century, and the main sources relevant to this focus shall be discussed in detail in chapters 3.2. and 3.3. In the course of the work, however, references will also be made to sources informing us on the astrolabe and on the method of flattening the sphere in other epochs and cultures. In this paragraph and in the following one, I shall therefore offer a brief overview and some bibliographical indications on earlier and non-Latin written sources as well as on the role of the astrolabe in non-Latin cultures.

The earliest evidence that the method of the flat sphere was known is linked to a water-driven device called anaphoric clock, in which the lines of a flattened horizon stood as a net in front of a revolving projection of the heavenly sphere. This device was described in Vitruvius' 'De architectura', written ca. 33–22 B.C.[93] Material evidence shows that anaphoric clocks were actually built in Europe at the latest in the first or second century AD: two fragments of such devices were discovered in Grande (Vosge) and near Salzburg.[94] Yet, as far as we know, no Greek or Roman mathematician described the method of the flat sphere before Ptolemy, who discussed it in the 'Planisphaerium'.[95] A further device apparently based on the same kind of projection was described in a letter by Synesius of Cyrene (ca. 370–414).[96] Synesius, who had studied in the neoplatonic school of Theon of Alexandria (ca. 350–400) and of his daughter Hypatia (ca. 370–415), claimed to have invented the device thanks to what he had learned from Hypatia. He had had a specimen of his invention made out of silver and was sending it as a present to the recipient of the letter, his friend Paionios of Constantinople. It is unclear what Synesius' device really looked like: in all probability it was no astrolabe artefact, but it was based on stereographic projection.

---

93 Vitruvius, 'De architectura', book 9, c. 8, 8–14. For a short description, see Neugebauer (1975) p. 869–870. For a longer discussion on this subject, see Drachmann (1954), A. J. Turner (2000).

94 Anagnostakis (1984) p. 10–12, A. J. Turner (2000) p. 539–547. Ptolemy used the term ἀστρολάβος to indicate not the planispheric astrolabe, but what we today refer to as an armillary sphere (Greek-English Lexikon (1996) p. 263).

95 On the 'Planisphaerium' see 4.5.1–2.

96 The text of the letter in German translation as well as a tentative reconstruction of the device can be found in: Vogt/Schramm (1970). On Synesius: Dannanfeldt (1976).

The first evidence of the astrolabe, i.e. of a device embodying both the flat sphere and the shadow triangle, can be found in two late ancient treatises, one written in Greek and the other in Syriac language:

- Treatise on the use of the astrolabe written in Greek by the neoplatonic and Christian philosopher John Philoponos (ca. 490–570).[97]
- Treatise on the use of the astrolabe written in Syriac by Severus Sebokht (d. 666–667), bishop of Qinnasrin (south of Aleppo).[98]

From this early evidence, historians have attempted to reconstruct the early history both of the astrolabe and of stereographic projection, but no conclusive agreement could be found on who was responsible for their invention. For the present study, this question is not particularly relevant, since in medieval Latin Europe the astrolabe was unanimously regarded as an invention of Ptolemy.[99] What may be of interest for our subject is the fact that, from the very beginning, the astrolabe was linked to philosophers belonging to the neoplatonic school. In late ancient neoplatonism, the study of the mathematical arts was regarded as a necessary premise to gain understanding of the divine 'ratio' governing the cosmos, which could not be perceived with the senses. Neoplatonism had a distinctive influence on Christian authorities such as Augustinus of Hippo (354–430) and Boethius and, through them, on Latin medieval philosophy.[100]

3.1.2 The astrolabe in the Arabic-Islamic, Byzantine and Jewish cultures

Astrolabes became known to Arabic-Islamic scholars at the latest in the eight century and, from then on, an Arabic-Islamic tradition both of composing astrolabe treatises and making astrolabe artefacts is attested.[101] Arabic-Islamic astrolabe treatises have not yet been systematically edited and analysed.[102] Even though

97 Neugebauer (1975) p. 253–254, Stautz (1994) p. 317–318. The treatise is edited in: Philoponos (1981). An English version is given in: Gunther (1931) p. 61–81. For further bibliographical references see Tihon (1995). On Philoponos and his natural philosophy: Sambursky (1973), Couloubaritsis (1998) p. 847–850.
98 Neugebauer (1975) p. 251–253, Stautz (1994) p. 318–319. The treatise is edited with French translation in: Sebokht (1899). An English translation of the French is given in Gunther (1931) p. 82–103. Further information and bibliography on Severus Sebokth can be found in: Sezgin (1978) p. 111–112.
99 For the opinion expressed in Arabic sources see: King (1981).
100 Fumagalli Beonio Brocchieri/Parodi (1998) p. 5–21, esp. p. 9–12, and Flasch (2000) p. 37–53. On the neoplatonic component of the Latin classics which were used as manuals of astronomy in the Middle Ages see McCluskey (1998) p. 114–127.
101 Sezgin/ Neubauer (2003) p. 79–85. For references on Arabic-Islamic astrolabe artefacts see 3.2.1–2.
102 For information on Arabic astrolabe treatises, see: Frank (1920), Sezgin (1978), King (1981), Kunitzsch (1981a), Abgrall (2000), Lorch (2005). A recent, most valuable contribution in this field is Richard Lorch's edition, commentary and translation of al-Farghānī's treatise 'On the astrolabe' (composed ca. 856–857 AD) al-Farghānī (2005).

there is no doubt that early Latin medieval astrolabe knowledge was of Arabic origin, up to now only two Arabic astrolabe-related texts could be identified as sources of the earliest Latin material: the Arabic translation of Ptolemy's 'Planisphaerium' and the treatise on the use of the astrolabe written by Muḥammad ibn Mūsā al-Khwārizmī (ca. 800–850).[103]

It is well beyond the scope of the present work to discuss Arabic-Islamic astrolabe treatises and artefacts or the role of the astrolabe in the Arabic-Islamic world, but a few remarks may be relevant to assess possible influences on the early Latin reception of Arabic astrolabe knowledge. There are indications that, in the Arabic-Islamic world, the astrolabe not only was an advanced astronomical-mathematical instrument, but also enjoyed a particular status, probably because of its complexity: in the signatures engraved on artefacts, astrolabe-makers showed special pride in their work, underscoring their skill not only in making the instrument, but also in performing the relevant calculations.[104] To express the fact that they had made an astrolabe, they eventually employed terms which were mostly used for authors of books, like 'writing' or 'composing'. Moreover, mathematicians, astronomers and astrologers often took pride in producing astrolabes, and in writing treatises describing old and new methods of flattening the celestial sphere.[105] The ability in planning and constructing an astrolabe artefact could be a measure of intellectual stature: according to his biographer, Muḥammad an-Nu'mān, who was Qāḍī of Egypt between the years 985 and 999 (375–389 H), had obtained his position by proving his capabilities in the construction of a silver astrolabe.[106] Thus, the philosophical aura which already surrounded the astrolabe in Late Antiquity also remained in the Arabic-Islamic culture. As I shall argue in chapter 5 of this work, this aspect of the astrolabe would play a central role in Latin Europe. At the same time, though, the astrolabe was also a tool for astrologers, and astrology played an important role in tenth-, eleventh- and twelfth-century Arabic-Islamic court life.[107]

It is difficult to say how much astrolabe knowledge remained alive in the medieval Byzantine empire. From the eleventh century onward, though, there is both written and material evidence that the astrolabe was known in Byzantium: an astrolabe artefact with Greek inscriptions made in the year 1062 (**#2**) is extant[108] and Anna Comnena (1083–ca. 1153/54), in the biography of her father, the emperor Alexis (r. 1081–1118), mentions a respected and feared astrologer, Seth from Alexandria, ὃς [...] ἀκριβέστατα προεμαντεύετο ἐν ἐνίοις οὐδὲ

103 See 4.5 ('Planisphaerium') and 3.3.8-9 (al-Khwārizmī's treatise).
104 My remarks on this subjects are based on: Mayer (1956) p. 14–15.
105 Sezgin (1978) p. 21–22.
106 Köhler (1994) p. 148 and p. 174.
107 Nallino (1944) esp. p. 1–41, Samsó (1979), Köhler (1994) p. 134–150. On the astrological use of astrolabes see for example Hartner (1939) p. 304–307.
108 Dalton (1926).

ἀστρολάβου δεόμενος'.[109] As was the case in the Arabic-Islamic world, in the Byzantine empire, too, interest in the astrolabe was linked to a revived interest in astrology.[110]

Finally, it was only in the early twelfth century that the first astrolabe treatises in the Hebrew language appeared, together with the first works of mathematics in the same language: before that time, Arabic-Jewish mathematicians had written their works in Arabic.[111]

3.1.3 The network of high medieval astronomical and mathematical studies

Historians have been able to reconstruct a broad and varied network of institutions and scholars which was the premise for the diffusion of interests and information in the mathematical arts in Latin Europe during the tenth and eleventh centuries.[112] The network extended from Christian and Muslim Spain to France, Southern Germany and Southern England. The individuals involved in the network were clerics, mostly monks. Not a few of them were also prominent political figures of their time, with leading roles in cloister reform as well as in secular events. The institutions in which we can positively say that exchange and storage of astrolabe knowledge took place were cloisters situated for the most part in Catalonia (Ripoll), Northern France (Fleury, Micy, Chartres), Lotharingia (Liège, Gorze), Southern Germany (Reichenau, Augsburg, Regensburg) and South England (Ramsey).

Since Latin astrolabe knowledge was without doubt of Arabic-Islamic origin, historians have tried to reconstruct the connections through which the initial exchange might have taken place. Two areas of Europe are possible candidates: Southern Italy, where Arabic, Latin and Byzantine cultures coexisted, and Muslim Spain (al-Andalus). As far as Italy is concerned, much work still has to be done in investigating possible sources on the exchange of astronomical knowledge in the tenth or eleventh centuries. We know that in the eleventh century Constantinus Africanus (d. 1087), who came from North Africa and had become a monk in Monte Cassino, translated from Arabic works on the medical art.[113] Up to now, though, no positive evidence could be found that Italian contacts contributed to the earliest stages of the diffusion of astrolabe knowledge in Latin Europe: all

---

109 Comnena (2001), vol. 1 p. 182 (Alexias VI, 7, 4), i.e.: "who [...] made the most accurate predictions, sometimes without using any astrolabe." On Byzantine texts on the astrolabe see Tihon (1995), especially p. 329–331.
110 Pingree/Kazhan (1991).
111 Gandz (1927), Levy (2002).
112 Works devoted to this subject are: Thorndike (1923–1958) vol.1 (1923) p. 697–718, Haskins (1924) p. 3–9, Thompson (1929), Millás Vallicrosa (1931), Van de Vyver (1931), Welborn (1931), Bergmann (1980), Bergmann (1985), Borst (1989), Burnett (1997), Burnett (1998). Some of them will be analysed later in more detail.
113 d'Alverny (1991) p. 422–426.

indications point to al-Andalus and Northern Spain, especially Catalonia, as the origin of the diffusion process.

3.1.4 al-Andalus, its school of astronomy of the tenth and eleventh centuries

During the tenth century and up to the beginning of the eleventh, al-Andalus was unified under the rule of the Caliphate of Cordoba. Cordoba was at the time also the centre of cultural life and of astronomical studies.[114] The development of an Andalusi school of astronomy was in large part due to Maslama al-Majrīṭī (d. ca. 1007/1008 = 398 H), mathematician, astronomer and astrologer who was active in Cordoba in the last part of the tenth century. Maslama taught a large number of students, some of whom later became astronomers and astrologers and wrote treatises on the astrolabe.[115] One student of Maslama was the mathematician and astronomer Abū l-Qāsim Aḥmad ibn aṣ-Ṣaffār (d. 1035). His treatise on the use of the astrolabe was translated into Latin twice, both times around 1150, and is still preserved in both languages. His brother Muḥammad "was famous for his construction of the astrolabe; in al-Andalus, none was better in building this instrument".[116] Astrolabe artefacts signed by Muḥammad ibn aṣ-Ṣaffār are preserved today, two of them were made respectively in 1026/1027 (417 H) (**#116**) and 1029/1030 (420 H) (**#3650**), while a third is only partly preserved and undated (**#4025**).[117]

Historians tend to agree that the growth of interest and competence in astronomy in al-Andalus was an important factor in attracting the interest of Latin scholars on this subject.[118] During the tenth and early eleventh centuries, the situation in Spain was favourable to intercultural exchange: at the court in Cordoba, Christians and Jews could take part in cultural life.[119] From the middle of the tenth century onward, the number of pilgrims travelling to the Christian north, especially to Santiago de Compostela, increased and, especially in the East Pyrenees,

---

114 The overview on the history of tenth- and eleventh-century al-Andalus and of its astronomical tradition is based on Samsó (1992) p. 80–110, Vernet/Samsó (1996) and Barton (2004).

115 A major source of information on science in tenth- and eleventh-century al-Andalus is the 'Book of the Categories of Nations' of Sāʿid al-Andalusī (1029–1070, i.e. 420H–462H). I have made use of the commented English translation of this work to be found in Sāʿid al-Andalusī (1991). This work also gives indications on the editions of the Arabic original. On Maslama and his students: Sāʿid al-Andalusī (1991) p. 64–66.

116 On the two brothers, see Sāʿid al-Andalusī (1991) p. 65, from where the quote is taken. On the Latin translations of ibn aṣ-Ṣaffār's astrolabe treatise, see app. C (**SJ and SP**). On the Arabic original and its Hebrew and Spanish translations see: Lorch/Brey/Kirschner/Schöner 1994) p. 125–131.

117 Frankfurt catalogue 1.3: The earliest Western Islamic astrolabes (tenth and eleventh century). See also Mayer (1956) p. 73.

118 Kunitzsch (1997).

119 Kassis (1999).

monastic movements inspired by the Cluniac reform led to the formation of new monasteries.[120]

We know that, around the middle of the tenth century, John, abbot of Gorze (ca. 900–974) spent three years in Cordoba on a diplomatic mission and on that occasion had contacts with Racemund, bishop of Elvira, who had an interest in astronomy, and also with the Jewish scholar and politician Abū Yusuf Ḥasdāy ben Isḥāq ben Shaprūṭ (fl. 940).[121] In the years between 967 and 970, Gerbert of Auril-lac (ca. 950–1003) was in Catalonia to study the mathematical arts. Later on, he went back to Latin Europe to become first bishop of Reims and then Pope Syl-vester II (r. 999–1003).[122] In one of his letters, Gerbert asked a certain 'Lupitus Barcinonensis' (i.e. Lupitus of Barcelona) to send him a book on astrology which he, Lupitus, had translated.[123] Lupitus has been identified with Seniofredus, also called Lobetus, who was archdeacon of the cathedral of Barcelona between 975 and 995, and has often been regarded as the translator or author of the earliest Latin astrolabe texts.

In February 1009, a coup d'etat took place in Cordoba, followed by a period of civil war (the 'fitna') that ended 1031 with the fragmentation of the Caliphate of Cordoba into a large number of small principalities (the 'taifas'). At that time, Maslama was already dead and his students had to flee Cordoba and build new lives for themselves as astronomers, astrologers, physicians or astrolabe-makers either in the taifas or in other parts of the Arabic world.[124] Although the taifas were continuously at war, often using Christian mercenaries from Northern Spain against each other, the region remained wealthy and the political crisis was not a cultural crisis. In fact, the level of Andalusi natural philosophy, mathematics and medicine increased after the fall of the Caliphate, with Saragossa, Toledo and Sevilla becoming the new centres of academic life.

The situation also remained favourable to intercultural contacts, in particular with the Christian principalities of Catalonia which, at least during the first half of the eleventh century, had close political and economic contacts with the Muslim

120 Mundó (1971).
121 Thompson (1929) p. 187–192 and McCluskey (1998) p. 166–170. On Ḥasdāy see Zuccato (2005) p. 751. Against the possibility that John of Gorze might have had anything to do with the diffusion of astrolabe knowledge Borst (1989) p. 56–58. Racemund was one of the au-thors of the Calendar of Cordoba (ca. 960), a calendar combining up-to-date astronomical data with Christian feasts, pre-Islamic Arab meteorological predictions and Greek medicine.
122 A recent sudy on Gerbert's study in Catalonia and al-Andalus is Zuccato (2005). A collection of commented sources on Gerbert of Aurillac and his cultural environment is: Guyot-jeannin/Poulle (1996). Further collections of essays devoted to this brilliant, multiform figure are: Charbonnel/Iung (1997) and Tosi (1985). On Gerbert of Aurillac, see also 3.1.5 (his stu-dents), 3.3.6 and 3.4.2 (his disputed authorship of J), 4.1.6 (his letter on triangles and triangu-lar numbers), 4.3.4 (his astronomical devices), 5.1.4 (his letter on the sundial), 6.5 (his leg-end).
123 "Itaque librum de astrologia translatum a te mihi petenti dirige" (Bubnov (1899) p. 101–102 (Ep. 24)). On Gerbert and Lupitus: Martinez-Gasquez (2000), especially p. 243–244.
124 On astrolabe-makers and authors of astrolabe treatises in eleventh-century al-Andalus, see: Sā'id al-Andalusī (1991) p. 64–65 and 69 .

south. In general, during the late tenth and the eleventh centuries, the close contacts between Latin and Andalusi culture were very important for the development not only of Latin astronomy and mathematics, but also of music and literature.[125]

The possibility of protracted intercultural, multilingual exchange is a very important premise for my argument, since I will argue that it was not primarily through exchange of a few written texts that astrolabe knowledge was transmitted to the Latin West, but rather thanks to discussions and oral teaching taking place in an orally multilingual context, in which Latin- and Arabic-speaking scholars were able to communicate directly thanks to the spoken word, if not to the written one.

During the eleventh century, the influence of Cluniac monasteries in the Christian north increased and, as the northern principalities slowly extended their power to the south, the influence of Cluny spread to the whole northern half of Spain. The fall of Toledo in 1085 marked a shift in the balance of power to the advantage of the Christian north. After that, tensions between the Arab and the Latin worlds became increasingly intense, especially when the taifa-rulers turned for help to the Almoravids, Muslim Berber rulers of North-Africa, who would eventually establish their rule in the southern part of Spain.

3.1.5 Prominent figures in the diffusion of astrolabe knowledge in Latin Europe

While the political constellation in Spain evolved, astrolabe knowledge had been spreading in Latin Europe along the network of cloisters described in 3.1.3. Apart from the persons already mentioned, three important figures who are often mentioned in connection with the diffusion of astrolabe knowledge are:

– Abbo, abbot of Fleury, (940/945–1004), mathematician, astronomer and political enemy of Gerbert of Aurillac. I shall discuss his work in 4.3.4;
– Abbo's student Byrhtferth of Ramsey (ca. 960–1012), also discussed below in 4.3.4;
– Fulbert, bishop of Chartres (ca. 960–1028), who was a student of Gerbert of Aurillac. I shall refer to him in 4.2.3.

Two further scholars who might have played a role in the transmission process were:[126]

– Constantine, dean of Fleury (active ca. 988–996), who worked under Abbo but also exchanged letters with Gerbert about mathematics and astronomy;
– Bern (ca. 978–1048), who was a monk in Fleury and became abbot of cloister Reichnau in the year 1008.

---

125 Chejne (1980), Boase (1992), Zuccato (2005).
126 On Constantine: Borst (1989) p. 66–69, Burnett (1998) p. 333–334. On Bern: Borst (1989) p. 70–77 and Schmale (1999).

Possibly apart from Fulbert, there is no definitive proof for any of these persons that they ever came into contact with astrolabes.

From the first decades of the eleventh century onward, written evidence permits ientification of the names of some Latin scholars who came into contact with and diffused astrolabe knowledge: Ascelin of Augsburg (early eleventh century?) and Herman, monk in Reichenau (1013–1054), both of whom wrote treatises explaining the construction of astrolabe lines.[127] In the second half of the eleventh century, we have evidence that astrolabe knowledge might have been used in the context of observational astronomy by William, abbot of Hirsau (1026–1091) and Walcher, prior of the monastery of Great Malvern (d. 1135).[128]

By the early twelfth century, Christian rulers had gained full control of the northern half of Spain and in this context literary translations from the Arabic started playing a prominent role for transmitting knowledge in the mathematical arts.[129] In this second stage of the process of assimilation of astronomical knowledge, Jewish scholars from Andalusi Spain played a very important role as mediators of the knowledge exchange. Not only did they largely contribute to the translations as bilingual interpreters, some of them also travelled in Latin Europe, spreading their knowledge among the Latin.[130]

Southern Italy was involved in this second stage of knowledge transfer, too: various translations of mathematical and astronomical works were made there. Yet, somehow, the influence of the Italian connection on twelfth-century astronomical and mathematical studies remained very limited, if compared to its potential. For example, Ptolemy's Almagest was translated from Greek into Latin in Sicily around the year 1160, but this translation found no diffusion in Latin Europe.[131] In general, twelfth-century translations are considered the main vehicle by which knowledge was transmitted from the Arabic-Islamic into the Christian-Latin culture, while the role of earlier contacts is often minimized. However, the situation is less clear-cut than it would at first appear.[132]

3.1.6 The interplay of modes of communication in the Latin Middle Ages

I will conclude my brief introduction to the cultural and historical background of high medieval astrolabe studies with a few remarks on orality, literacy and other modes of communication in that age. The Latin Middle Ages were no primary

---

127 On Ascelin of Augsburg: Bergmann (1985) p. 99, Burnett (1998) p. 333–334. On Herman of Reichenau and his treatise on astrolabe construction see 3.3.7. and 4.6.7.
128 See 5.1.3 and 5.5.2.
129 On twelfth-century translations from the Arabic and from the Greek, see: Carmody (1956), d'Alverny (1991), Brentjes (2000). Especially on astronomy and astrology: McCluskey (1998) p. 188–190.
130 On translation partners in general, and in particular on the Jewish ones, see Brentjes (2000) p. 295–303. For some examples, see 6.2.
131 Haskins (1924) p. 157–168.
132 Brentjes (2000) esp. p. 303–305.

oral culture, but in many ways a highly literate one. This applies in particular to the scholarly elites involved in the spread of astrolabe knowledge.[133] At the same time, though, the context of eleventh-century learning was essentially oral, even as far as reading and writing were concerned, since scholars usually read aloud - to others or to themselves - and often wrote in dialogical form.[134] In fact, what makes the Middle Ages particularly interesting is the variety in which written and non-written modes of communication were combined.

Studies on medieval communication have led to an increasing appreciation of its variety. At the end of his review of new approaches to medieval communication, Mario Mostert remarked that:

> Research is centered ever more on the question of the relative importance of writing, seen as part of the whole of medieval forms of communication. This change of perspective constitutes a clear challenge for future research, and suggests a first approach to the subject.[135]

Franz H. Bäuml devoted a paper to the 'Varieties and consequences of medieval literacy and illiteracy', proposing to consider them as "determinants of different types of communication rather than as personal attributes", thus making possible "an examination of the characteristics of medieval preliteracy, illiteracy, quasi-literacy, and literacy".[136] In his survey on work on medieval orality and reading, D. H. Green has stressed the need to further investigate "the symbiosis of orality and writing in the Middle Ages, including the questions when and how changes come about within this interrelationship".[137]

However, the spoken word was not the only alternative to writing: non-verbal modes of communication such as images, gestures, colours and even smells may have to be taken into account. For example, in investigating written and oral law, ritual, symbolic gestures are very important.[138] Again, it is difficult, if not impossible, to make general statements on such modes of communication without characterizing them in a negative way. As Michael Clanchy wrote:

> 'Non-verbal' communication is too negative a way to describe the power of liturgy and ritual because it suggests that words alone might have done a better job. 'Non-verbal communication' is used here for lack of any better term for describing the indescribable.[139]

---

133 As introductions to the subject of modes of communication in the Middle Ages I have used: Bäuml (1980), Ong (1982), Green (1990), Illich (1993), Mostert (1999a).
134 Ong (1982) p. 95, Illich (1993) p. 51–73.
135 Mostert (1999b) p. 35.
136 Bäuml (1980), quotes from p. 265.
137 Green (1990) p. 279.
138 Hageman (1999).
139 Clanchy (1999) p. 11.

3.1.7 The 'rise of literacy' in eleventh- and twelfth-century Europe

In the eleventh and twelfth centuries, the relationship between orality and literacy in Latin Europe underwent a change which is often described as the rise of literacy.[140] The volume of written documents increased in all fields and the production became more differentiated. New styles of writing were developed and also new techniques of page lay-out. New fields were opened to the written word, such as romance vernacular literature and the practice of the mechanical crafts. Most relevant for our subject, in the eleventh century the first Latin translations of Arabic natural philosophical and medical texts appeared and, in the course of the twelfth century, Europe was literally flooded with translations of Arabic works on philosophy, mathematics, astronomy and astrology. In the thirteenth century, the focus of translating activity shifted from the Arabic to the Greek language.[141]

The term 'rise of literacy' is somehow misleading, since in the course of this process orality did not actually lose ground: it just happened that the functional interplay of orality and literacy was transformed. It would be wrong to imagine this change as a linear process, in which the written work simply took over functions previously carried out by the spoken one. The rise of literacy was paralleled by changes in various spheres, such as religion, politics, economy and education. Textbooks would become a central characteristic of thirteenth-century universities. Many theses have been formulated on how the changes in orality and literacy were connected not only to socio-economical and political transformation, but also to a change in the modes of thought.

In my analysis of early Latin astrolabe texts, I will argue that such texts should be interpreted by assuming that they were complemented by other modes of communication, and that they were first attempts at expressing in writing elements of a body of knowledge that had up to then been stored and transmitted primarily through non-written and non-verbal strategies. Being first attempts, these texts were characterized by a variety of different styles and, when judged by modern standards of written communication, they often appear very unclear.

---

140 Stock (1983) p. 12–59, Illich (1993), Bourgain (1994).
141 d'Alverny (1991).

3.2 MATERIAL AND PICTORIAL SOURCES: CATALOGUES, ANALYSES AND OPEN QUESTIONS

3.2.1 Arabic-Islamic and European astrolabe artefacts

There are approximately 1400 extant astrolabe artefacts in museums and private collections.[142] Almost all of them are either Arabic-Islamic or European ones, a few are inscribed with Hebrew characters and one is a Byzantine artefact dated to the year 1062 (**#2**).[143] Early Arabic-Islamic astrolabes can be distinguished into Eastern and Western ones. For the present subjects, mostly European and early Western Islamic (especially Andalusi) artefacts are of interest.

Even though written sources suggest that astrolabe artefacts were already made in Late Antiquity, no specimens from that epoch are extant. The earliest artefacts preserved are Eastern Islamic ones from the tenth century. The earliest extant Western Islamic astrolabes are from eleventh-century al-Andalus. The earliest European astrolabe (**#3042**) was constructed in tenth-century or early eleventh-century Catalonia (**#3042**).[144] Arabic-Islamic astrolabes are often engraved with the name of their maker and with the date and place of production. This has greatly facilitated their study and classification as Eastern or Western Islamic, even making it possible to connect written evidence on schools of astrolabe-makers with their extant products.[145]

In the study of early European astrolabe artefacts, instead, no classification is as yet possible: specimens are usually unsigned and undated and too few of them have been systematically studied. Placing and dating early European astrolabes has proven extremely difficult. In principle, each of them has to be regarded as unique and studied in all its peculiar features: in this way, possible grounds for connections might come to light. The most singular piece is probably the earliest European astrolabe (**#3042**), whose features suggest a much earlier date (tenth century) than that of other European astrolabe artefacts, which are usually regarded as produced at the earliest around 1200.[146]

Apart from its date and provenance, many questions can be asked of an astrolabe artefact: the script, numerals, language and contents of its inscription, whether it can be associated to some group of artefacts sharing a specific set of features, whether on the contrary it has one or more unique characteristics, which

---

142 Overviews on extant astrolabe artefacts are: Stautz (1999) p. 9–98 (with many images), King (2003b). A choice of the most recent research results can be found in: King (1994), King (2001) p. 364–405 (with ample bibliography). The number of extant astrolabes artefacts is taken from Stautz (1999) p.125. Images and descriptions of European astrolabe artefacts can be found in the online catalogue Epact, which lists objects from the collections of museums in Oxford (Museum of the history of science), London (British Museum), Florence (Museum of the history of science) and Leiden (Museum Boerhaave).

143 Dalton (1926).

144 On this artefact see below 3.2.5.

145 Mayer (1956), Mayer (1959).

146 King (2003b) p. 54.

astronomical and geographical data have been used as a basis for its rete and horizon plates, the level of technical competence of its maker and whether it bears signs of intercultural exchange.[147] I will not attempt to describe here the difficulties and rewards of the study of astrolabe artefacts: I will only give a short overview of the research done on cataloguing extant astrolabe artefacts and then address those results and problems that are directly relevant for the present research.

3.2.2 The Frankfurt catalogue of medieval astronomical instruments and previous cataloguing projects

Extant astrolabe artefacts are being catalogued in the context of the project for a Catalogue of medieval Islamic and European (up to 1550) astronomical instruments at the Institute for the History of Science at Frankfurt University.[148]

Each instrument included in the catalogue is indicated with a number in the format: #[number], for example **#3042** for the oldest extant European astrolabe. Drawings of astrolabe artefacts are assigned an entry in the Frankfurt catalogue, too: for example, the set of drawings found in the manuscript BnF lat. 7412 (A.27) on f. 19v–23v and discussed in 3.2.5 has the number **#4042**. These unique denominations are extremely useful, particularly when dealing with early European astrolabes, whose provenance and date of production can very rarely be identified with precision.

The first survey of astrolabe artefacts was the work of Robert T. Gunther (1931): his catalogue listed about 330 astrolabes, was a great achievement for its time and is still the only reference for some pieces.[149] However, it contains not a few errors and only features about a quarter of the artefacts now known. In particular, Gunther's classification of European astrolabes according to their alleged provenance (e.g. as 'Italian', or 'Hispano-moresque') is often unfounded.

A very useful tool for studying Arabic-Islamic artefacts is Leo A. Mayer's catalogue of Islamic astrolabe-makers (1956).[150] Mayer collected both the maker's names often engraved on Islamic astrolabes and the evidence from written sources, and from there reconstructed a chronological and geographical panorama of Arabic-Islamic astrolabe-making. Mayer's task was planned to be continued by

---

147 Examples of how some of these questions can be formulated and definitely or tentatively answered are: Saliba (1991), King (1993), King/Turner (1994), K. Maier (1996), Stautz (1997), King (2001), King (2003a).
148 On the Frankfurt cataloguing project: King (1991), King (1997). The Catalogue can be consulted on-line: Frankfurt catalogue. The checklist of Islamic astronomical instruments is also printed in: King (2005) p. 993–1019.
149 Gunther (1931). My overview on previous cataloguing projects is based on: King (1994) p. 147–148, Stautz (1999) p. 123–126.
150 Mayer (1956).

the late Alain Brieux and Francis R. Maddison, but their work has not yet been published.[151]

In the year 1955, Derek de Solla Price began a 'Computerized checklist of astrolabes', which was updated in 1973.[152] The updated list contained about 1100 entries. The entries 1 to 4000 of the Frankfurt Catalogue correspond to those of de Solla Price's checklist.

3.2.3 Difficulties in dating European astrolabe artefacts

Dates and signatures start appearing on European astrolabes only from the fourteenth century onward, but there are at least a dozen European artefacts which are probably earlier than that.[153] There are various methods for trying to determine the date and place of production of European astrolabes, for example philological analyses of the words inscribed on them, in particular the names of the months and of the stars.[154]

It is also possible to date the astronomical data used for the construction of an astrolabe, such as star positions or equinoctial and solsticial dates. However, comparisons of dated instruments with the date of the astronomical data engraved on them has shown that this method cannot always be relied upon.[155] Performing stylistic comparisons helps, but only if a relatively large number of artefacts can be compared with each other, and the same is valid for analyses of the metal. This is not always possible for early European astrolabes.

A metal-analysis of the oldest European astrolabe **#3042** was able to show that it is not a modern fake, but could not help in the exact dating.[156]

3.2.4 European astrolabe artefacts and astrolabe drawings used as evidence in this study

Since this study is devoted to the Latin astrolabe, European astrolabe artefacts are in principle a very important source. Because of the difficulty of dating them, though, I have used only the oldest one (**#3042** of Frankfurt catalogue) as direct evidence for the state of early Latin medieval astrolabe knowledge. This artefact will be discussed more in detail in the next paragraph.

A main source for the following discussions are the geometrically accurate figures which, in some Latin manuscripts, illustrate both the construction proce-

---

151 Brieux/Maddison (unpublished). Doubts on the complete publication of the results of this project are expressed by Stautz (1999) p. 125–126.
152 de Solla Price (1955), de Solla Price (1973).
153 Some of the earliest dated European astrolabes are listed in K. Maier (1999) p. 696. A number of early European astrolabes are discussed in King (1995).
154 K. Maier (1994), K. Maier (1996).
155 G. L'E. Turner (2000).
156 See 3.2.5.

dures needed to draw astrolabe parts and the finished products, e.g. horizon plates drawn for specific latitudes. I do not regard these drawings as 'pictorial sources', but as something in between written and artefact evidence. The reason is, that it is not only their visual component which is relevant to my analysis, but above all the mathematical knowledge implicit in their drawing procedure. In other words, these drawings were not made to be simply looked at, but to evoke with their structure the whole procedure through which they had come to be. As I shall argue, they are written traces of kinaesthetic memory.

In 5.6, I will discuss at length a specific set of astrolabe drawings (**#4042**), which is found in an eleventh-century Latin manuscript. I shall point out some similarities between these drawings and a number of early European astrolabes. Since those artefacts have to be regarded as later than the eleventh century, my aim in doing so is simply to point out that the Latin tradition of astrolabe-making may have felt the influence of the first stages of assimilation of astrolabe knowledge.[157] The early European astrolabes I shall refer to are: **#161**, **#166**, **#167**, **#202**, **#303**, **#420**, **#558**, **#589**, **#4556**.[158]

3.2.5 The earliest European astrolabe (#3042)

The oldest extant Latin European astrolabe (**#3042**) came to the attention of historical research only in the year 1962, when Marcel Destombes published a study describing it and dating it to the tenth century.[159] Destombes characterized it as a 'Carolingian astrolabe' and this was the name used for it in the following years, although this attribution was hotly disputed.[160]

The artefact had been acquired by Destombes in the year 1961 from an antiquarian dealer, to whom it had come from an unnamed Spanish collector. It has proven impossible to determine where the astrolabe had been kept for the previous centuries and initially this fact even led some historians to doubt its authenticity. This point, at least, is now cleared: the astrolabe **#3042** was actually produced in the Middle Ages.

In 1995, a conference was devoted to discussing Destombes' astrolabe, combining expertise from different branches of historical research.[161] The results of paleographic studies presented in this meeting tended to confirm Destombes' dating, showing that the engravings on the mater and plates of the astrolabe **#3042** are related to the characters found on stone inscriptions and manuscripts produced in the late tenth century.[162] The provenance is more difficult to establish: evidence

---

157 See 5.6.2.
158 Bibliographical references on these artefacts are given in 5.6.2.
159 Destombes (1962).
160 For an overview of the history of and discussions on Destombes' astrolabe, see: Beaujouan (1995).
161 The studies presented to the meeting were published in the journal issue: *Physis* 32 (1995). A summary of the results is: Beaujouan (1995).
162 Mundó (1995) (dating ca. 975–1000), Stevens (1995) (dating ca 930–1069).

strongly suggests Catalonia, and a Catalonian origin is considered as certain by Anscari Mundó, an opinion shared also by David King.[163] Wesley Stevens, on the other hand, considers the evidence for Catalonia as not conclusive and does not rule out Northern Italy, or even Southern Italy and Southern France.[164] It is also possible, as Prof. Kunitzsch suggested, that the Latin received the instrument from al-Andalus as a blank, and therefore that only the inscriptions, and not the artefact itself, stem from Latin hands.[165]

The inscriptions on the rete, on the other hand, seem to be more recent (ca. 1200). Studies of the craftsmanship of **#3042** and comparisons with other early astrolabe artefacts confirmed the unique character of Destombes' find, showing that there is no reason to doubt its early attribution.[166] Metal analyses proved that the instrument is no modern fake.[167] The strongest argument against dating **#3042** in the tenth or early eleventh century remains the fact that extant astrolabe texts from that age do not appear to have been able to provide enough information to allow Latin astronomers or craftsmen to build such an artefact.[168]

I share the opinion that the astrolabe **#3042** was made in the tenth century, and therefore I believe that it is evidence of the earliest stage of reception of astrolabe knowledge in Latin Europe. In 4.6.10–11, I shall argue that the comparison of the division of the zodiac on **#3042** and in early Latin texts does not oppose an early dating, as has been claimed.

3.2.6 Early Andalusi astrolabe artefacts with additional non-Arabic engravings as evidence of cultural exchange

Andalusi astrolabes of the tenth and eleventh century are interesting for this study because they might have come into the hands of Latin scholars, thus contributing to the process of knowledge transfer. They might also have been used as gifts for Latin potentates, making the astrolabe known to European courts. Extant early Western Arabic-Islamic astrolabes which are engraved with their date and place of production were all made in al-Andalus after the period of the civil war (ca.1030).[169] No astrolabe carries an inscription stating that it was produced in tenth-century al-Andalus, i.e. in the period in which we assume that Latin scholars first came into contact with astrolabe knowledge.[170]

---

163 Mundó (1995) p. 316 and David King, both in the Frankfurt catalogue and in a private communication.
164 Stevens (1995) p. 276–277.
165 Kunitzsch (1991/1992) p. 3 note 3.
166 d'Hollander (1995), King (1995), G. L'E. Turner (1995). Also the stars on the rete of this astrolabe were analysed: Kunitzsch/Dekker (1996), Stautz (1997) p. 84–85.
167 Grautze/Barrandon (1995).
168 Poulle (1995).
169 Frankfurt Catalogue I.3: The earliest Western Islamic astrolabes, Stautz (1999) p. 24–30.
170 On this subject, see 5.6.1–2.

Some of the earliest extant Andalusi astrolabes are of particular interest because they are engraved not only with Arabic, but also with Latin or Hebrew characters. For example, European symbols for the zodiac signs and European names of the months appear on two astrolabes made respectively in Toledo in 1067/1068 (460 H) (**#118**, Oxford, Museum of the History of Science) and in Saragossa in 1079/1080 (472 H) (**#1099**, Nürnberg, Germanisches National-museum).[171] An astrolabe made in Cordoba in 1054/55 (446 H) (**#3622**, Cracow, Jagellonian University Museum) is also inscribed in the Catalan dialect of Valencia.[172] An astrolabe made in Toledo in 1029/1030 (420 H) (**#116**, Berlin, Staats-biblothek Preußischer Kulturbesitz) is also engraved with Hebrew characters both on the zodiac circle and on the latitude plates for Toledo and Cordoba.[173] This artefact is interesting because, as already mentioned, Jewish astronomers and mathematicians played a very important role in the early twelfth century as mediators between the Arabic and the Latin culture.

Astrolabe artefacts with multiple layers of inscriptions are historical sources of great value because they can inform us about exchanges of mathematical, astronomical or astrological knowledge that cannot be grasped through written sources. The fact itself that European owners added engravings to the objects suggests that they did not purchase them only as exotic products, but had an actual interest in using them. However, interpreting these sources is a difficult task. For example, it is not yet possible to effect philological comparisons of inscriptions on metal capable of providing a reliable dating of the additional engravings, because the script used for engravings on metal cannot be dated exclusively by comparisons with the characters used when writing on parchment or paper.[174]

An example of the richness of the picture that can be gained from the study of astrolabes with multiple inscriptions is the discussion of an astrolabe from fourteenth-century Christian Spain which is engraved in Latin, Hebrew and Arabic (**#4560**, private collection).[175] However, like most other astrolabe artefacts, this one, too, is much later than the period discussed here. Therefore, it can at most serve as a generic hint at the complexity of knowledge exchange surrounding the circulation of artefacts – possibly also artefacts made out of more perishable materials like copper, wood or parchment. In the Late Middle Ages and Renaissance there is ample evidence that astrolabe and similar devices were made out of wood, parchment and, later on, paper.[176]

---

171 K. Maier (1996) p. 260–261, Gunther (1931) p. 253–256, description of **#1099** by D. A. King in Bott (1992) p. 568–570.
172 K. Maier (1999).
173 K. Maier (1996) p. 260, Woepcke (1858) (**#116**).
174 K. Maier (1996) p. 260.
175 King (2003c).
176 Biagioli (2006) p. 160–166.

3.2.7 Additional pictorial sources

I indicate as additional pictorial sources all representations of astrolabes or of other mathematical and astronomical instruments which appear in a context different from manuscripts primarily devoted to astrolabe knowledge. I assume that, other than astrolabe manuscripts, such images were directed to a broader, non-specialist audience.

There is only one eleventh-century image of an astrolabe artefact appearing in a context not immediately related to mathematics and astronomy:

- image of Abraham with an astrolabe in his hands occurring in the manuscript BnF lat. 12117 (A.30). I shall discuss this source in 5.6.3.

Further eleventh-century images of particular relevance for my arguments are:

- images of 'deus geometra' with compass and scales occurring in eleventh-century manuscripts. They are discussed in 5.1.2.

While eleventh-century images are rare, there are a number of later astrolabe images which I shall quote in the following pages:

- angels holding astrolabes in the sculptures of Chartres Cathedral (5.1.2)
- representations of the liberal art Astronomy with an astrolabe (6.4)
- representations of astronomers and astrologers using an astrolabe to measure celestial altitudes (6.3, 6.4).

3.3 TEXTUAL SOURCES: RESEARCH RESULTS AND OPEN QUESTIONS

3.3.1 Extant astrolabe manuscripts

Manuscripts containing passages devoted to the 'astrolabium' or 'astrolapsus' were circulating in Latin Europe at the latest around the beginning of the eleventh century. Twelve to fifteen astrolabe manuscripts from the eleventh century are extant, and we possess at least twenty-two that were written in the twelfth century.[177] From the thirteenth century onward, the volume of such manuscripts increased and only rough estimates of it exist. The most popular thirteenth-century Latin text on the astrolabe, which is usually referred to as the 'treatise of Pseudo-Messahalla', is preserved in ca. 200 manuscripts.[178] In the fifteenth century, printed editions started appearing.[179]

Much effort has been devoted to the study of Latin medieval astrolabe texts, in particular to the analysis of the oldest manuscripts and of their contents. These studies have concentrated on investigating the channels through which knowledge of the astrolabe penetrated Latin Europe and on reconstructing the earliest Latin

177 Extant manuscripts are listed in app. A and B.
178 Kunitzsch (1982) p. 499.
179 Kunitzsch (1982) p. 507–508; A. J. Turner (2005).

European texts on this subject, eventually locating their geographic provenance and identifying their authors.

3.3.2 Astrolabe manuscripts as a source for this study

The present study is aimed at investigating, first, the strategies of knowledge transfer involved in the assimilation of astrolabe knowledge into the medieval Latin culture and, second, the motivations of the historical actors taking part in the process. With this aim in mind, I have taken the results of previous studies as a starting point: the collection and classification of manuscripts; the reconstruction and editions of some of their contents; the evidence of a network of astronomical and mathematical studies in Latin Europe, and, finally, the various theses on the diffusion of astrolabe knowledge which have been proposed and debated in the course of time. This material provided me with a standpoint from which to turn back and take a second look at some of the oldest Latin manuscripts containing astrolabe texts. The questions I have asked are: how much knowledge could these written pages actually transmit? What interests and needs may have moved the Latin scholars who compiled them?

Since these questions are not the same ones as those on which previous studies have focused, my approach to the sources is in part different from theirs. For one thing, I will devote much attention not only to the single astrolabe texts, but also to the manuscripts as a whole. Their structure and choice of themes can reveal much about the capabilities, interests and worldview of their writers and readers. Moreover, in the case of the earliest material, it has proven difficult to isolate single astrolabe texts from their manuscript context, as discussed at length in 3.3.6.

Since the present study focuses on the tenth, eleventh and early twelfth centuries, attention is primarily devoted to astrolabe manuscripts from that period, which are listed in appendix A. In discussing the influence of early astrolabe texts on late medieval astronomical and mathematical studies, however, the diffusion of earlier texts in later manuscripts has to be taken into account. Appendix B lists later manuscripts containing astrolabe texts from the eleventh and twelfth century. The overview in appendix B does not include late medieval manuscripts in which only astrolabe treatises written after 1200 occur.

All dates of the manuscripts are taken from secondary literature and, where dating is disputed, diverging indications and their sources are given.

3.3.3 Additional textual evidence

It would be desirable to complement the evidence from astrolabe manuscripts with accounts on the astrolabe coming from other sources than texts primarily devoted to the device itself. Unfortunately, the number of eleventh-century references to the astrolabe outside of astrolabe texts is very small and there are no contemporary accounts of its introduction into Latin Europe. Only from the twelfth century onward a series of legends surrounding the astrolabe and the figures of Gerbert of Aurillac and Herman of Reichenau developed: they tell us much about the role of the astrolabe in the twelfth and thirteenth centuries, but probably cannot be trusted as far as astrolabe knowledge of the tenth and eleventh centuries is concerned. I will discuss them in 6.5.

Therefore, the written sources on which my analysis is based are essentially astrolabe manuscripts. For completeness, I list here the few additional eleventh-century textual sources which I have used, along with the paragraphs in which they are discussed:

- mnemonic verses and lists of astrolabic terms by Fulbert of Chartres (4.2.3);
- mention of an astrolabe in the correspondence of Radulf of Liège and Regimbold of Cologne (4.2.3);
- Petrus Damiani's account of a simoniac priest's use of an astrolabe (5.2.3);
- Anna Comnena's mention of the use of the astrolabe by astrologers (3.1.2);
- accounts of the astronomic studies of William, abbot of Hirsau (5.5.2);
- account of astrolabe use by Walcher of Malvern (5.1.3).

3.3.4 Preliminary remarks on earlier and later Latin texts on the astrolabe

A comparison between manuscripts from the eleventh century and later ones allows the distinction between texts which were present right from the beginning and those which only appeared after 1100. As anticipated in the introduction, there are differences both in style and content between the earlier and the later texts. For example, earlier texts are for the most part anonymous, while later ones often bear the name of their author. Thanks to this fact, it has been possible to establish that most later texts were composed or translated from the Arabic in the period 1130–1160. Earlier, anonymous texts are instead very difficult to date with precision. This work will focus on the earlier texts.

These texts can be found in manuscripts of all periods, albeit in a large number of variants. They were produced in the eleventh century, a few perhaps even in the late tenth century. Some of them remained popular until the Late Middle Ages. Although this material is often referred to as a 'corpus of older astrolabe treatises', I will refrain from doing so. The meaning implied by the term 'treatise' is that of a literary composition with clearly defined wording and order, written by a single (possibly anonymous) author with a unity of purpose. Since not all eleventh-century texts fulfil these criteria, I believe it is somehow misleading to refer to them indiscriminately as 'treatises' and shall therefore call them simply 'eleventh-century astrolabe texts'. However, there are some of the earliest texts which can actually be regarded as treatises, for example the works on astrolabe construction ascribed to Herman of Reichenau (**h**) or the one attributed to Ascelin of Augsburg (**a**).[180]

The later, twelfth-century Latin texts on the astrolabe, on the other hand, can be safely described as treatises. In the last decades, a good number of them have been edited.[181] Most astrolabe texts from the eleventh century, instead, are printed only in editions not based on all extant manuscripts.[182] The reason is that reconstructing eleventh-century texts has proven to be a very difficult task.

In my discussion of early Latin astrolabe texts, I have used as a reference the versions printed in modern editions. As discussed in detail later on, a large number of these printed texts have a composite character, and I have accordingly discussed parts of them as separate compositions. The abbreviations used are summarized in appendices C and D. Appendix C offers an overview of the printed material and of those parts of it which I have discussed in this study. To indicate printed texts, I have made use of already established abbreviations (e.g. **J**, **J'**, **h**, **h'**, **h''** etc.), but I have introduced new ones to specify those portions of printed texts which I regard as independent compositions (e.g. **h1**, **h2**, **dz1** etc.). Appendix D lists the main astrolabe treatises which were written in the twelfth century: these are for the most part unitary compositions and many of them have been edited from all extant manuscripts. To indicate these texts I have used the symbols introduced by Paul Kunitzsch in his 'Glossar der arabischen Fachausdrücke in der mittelalterlichen europäischen Astrolabliteratur'.[183]

3.3.5 The reconstruction of eleventh-century astrolabe texts

Astrolabe texts from the eleventh century present a picture which is problematic at two levels. The first level is the philological reconstruction of the text itself: attempts at using preserved manuscripts from all periods to reconstruct eleventh-century astrolabe texts have often run into problems. In some cases, historians have been able to reconstruct a text that can be traced back to the eleventh century; but it often seems that the oldest manuscripts only contain mutilated versions of the modern reconstruction, for example, versions in which chapters are presented in variable order, or interspersed with passages devoted to other subjects. I shall discuss this kind of problem in 3.3.6.1–5.

Problems at a second level arise when trying to interpret the contents of the reconstructed texts. Even in those cases in which an eleventh-century core could be safely identified, the question has often arisen, how far – or if at all – such texts might have contributed to an early diffusion of astrolabe knowledge in Latin Europe. Historians tend to agree – and I share their opinion – that most early

---

180 On the texts **h** and **a**, see app. C. On Herman of Reichenau and **h**, see 3.3.7 and 4.6.7.
181 See app. D.
182 For printed texts, see app. C. On the necessity of studying further their manuscript tradition: Bergmann (1985) p. 96–97.
183 Kunitzsch (1982) p. 475–476.

Latin texts on the astrolabe are unclear, ambiguous and sometimes simply wrong.[184] In short, it is very doubtful how much knowledge on the astrolabe a reader could have extracted from them without further help. If judged by modern standards, as they have often been, such texts have to be considered as more or less incompetent attempts at explaining the design and use of the astrolabe. This fact has in turn been taken as evidence of the low level of astrolabe knowledge in eleventh-century Latin Europe.[185] Instead, I will argue that the mathematical competence of the authors should be judged separately from their capability - or willingness - to express their knowledge in written words.

In the next paragraphs, I will focus on what I have described as problems at a first level, i.e. questions relating to the philological reconstruction of the texts in what is assumed to be their original version. I shall argue that the problems encountered at the philological level suggest that, in some cases, the idea of reconstructing a single, original version of the text may have to be abandoned in favour of a more flexible approach admitting the possibility of equivalent variants. Moreover, as we shall see, it is only in a first approximation that philological problems can be distinguished from questions relative to the quality of the contents.

In the past, there have been many fruitful attempts at reconstructing eleventh-century astrolabe texts, and it would be impossible for me to summarize here all relevant results. Therefore, I shall discuss only three very important cases:

- the reconstruction of the text on astrolabe use usually referred to as **J**;
- the problem of establishing which sources were used by Herman of Reichenau in composing his treatise on astrolabe construction (**h**);
- the question of the relationship between the Latin text **J'** on the use of the astrolabe and its Arabic sources.

In the end, I will summarize the questions which in my opinion remain largely open.

3.3.6 Problems in the philological reconstruction of the earliest Latin astrolabe texts: the example of **J**

3.3.6.1 Nicolaus Bubnov's edition of 'de utilitatibus astrolabii' (J)

The most closely analysed and widely discussed eleventh-century text on the astrolabe is the one known under the title 'de utilitatibus astrolabii' and usually indicated as **J**. This text was published in 1899 by Nicolaus Bubnov among Gerbert of Aurillac's 'opera incerta'.[186] Ever since Bubnov's edition, the question of the

---

184 Bergmann (1985) p. 116, Borst (1989) p. 39–40, Poulle (1995) p. 229.
185 So for example: Borst (1989) p. 40–41, Bergmann (1985) p. 113–115, Poulle (1995) p. 227, p. 229 and p. 234–235.
186 Bubnov (1899) p. 114–147.

authorship of **J** has been discussed.[187] I will not attempt to summarize the whole debate, but I wish to highlight some moments of it, because the discussion surrounding **J** offers a good example of the problems that research on eleventh-century astrolabe texts must face.

The text published by Bubnov is composed of 21 short chapters describing the astrolabe and its possible uses.[188] Bubnov's detailed analysis of **J**'s manuscript tradition shows that eleventh-century manuscripts often contain **J**'s chapters (or part of them) in varying order, whereas twelfth-century sources tend to present **J** in the same form as in Bubnov's edition.[189]

In addition to this, a good number of twelfth-century manuscripts group **J** together with two other texts of astronomical and mathematical interest.[190] The first one, indicated by Bubnov with the letter **h** and also known as 'de mensura astrolabii', explains how to draw astrolabe lines. While **J** is anonymous, the author of **h** tells us his own name, 'Hermannus', and could be identified as the monk Herman of Reichenau.[191] The second text often associated with **J** in twelfth-century manuscripts describes the construction of a vertical sundial and also discusses other subjects, such as the measure of the earth's circumference. It is known as 'de horologio viatorum', **k** or **hv**. In twelfth- and thirteenth-century manuscripts, the sequence **h+J+k** appears so often that in their oldest edition the three texts were presented as a single work by Herman of Reichenau.[192]

Bubnov realized that the three texts had separate manuscript traditions and had been grouped together at a later stage.[193] He showed that the attribution of all of them to Herman of Reichenau was incorrect and discussed whether **J** should be attributed to Gerbert of Aurillac or not. The text itself contains no mention of its author, and in eleventh-century manuscripts it is anonymous. In a handful of twelfth-century manuscripts it is instead ascribed to Gerbert, but this late evidence was not enough to convince Bubnov, who edited **J** among Gerbert's 'opera incerta'.[194] In the following years, various opinions for or against Gerbert's authorship were expressed. Lynn Thorndike and Charles H. Haskins proposed to ascribe **J** not to Gerbert, but to his correspondent Lupitus of Barcelona.[195]

In the year 1931, two independent, detailed studies of the manuscript tradition of **J** and of the oldest Latin astrolabe texts appeared: a paper by André Van de Vyver on the first Latin translations of Arab treatises on the astrolabe and a

187 For a detailed overview, see Bergmann (1985) p. 66–72.
188 The contents of **J** are discussed in more detail in 5.2. and 5.3.
189 Bubnov (1899) p. 109–114.
190 Bubnov (1899) p. 111–112.
191 On Herman of Reichenau, see 3.3.7.
192 Herman of Reichenau (1853). Today, Herman is generally assumed to be the author of **h**, and Werner Bergmann expresses the opinion that he wrote **k** as well (Bergmann (1985) p. 168–172).
193 Bubnov (1899) p. 109–110.
194 Bubnov (1899) p. 112. An edition of the text attributing **J** to Gerbert can be found in Poulle (1996b), along with a French translation and commentary.
195 Haskins (1924) p. 8–9. On Gerbert and Lupitus see 3.1.4.

monograph by José Maria Millás Vallicrosa: 'Assaig   d'història de les idèes fisiques i matemàtiques a la Catalunya medieval'.[196]

3.3.6.2 André Van de Vyver's analysis of the earliest Latin astrolabe manuscripts

Van de Vyver's study was written as a preliminary report on a research project that the author could unfortunately never complete. The paper offered a sketchy but wide-ranging panoramic of the differences and similarities of Latin astrolabe manuscripts.[197] The manuscripts were grouped into two classes, according to whether they contain **J** in the 'proper' (i.e. Bubnov's) order or not. Van de Vyver interpreted the variants of **J** not as corrupt versions, but as testifying to successive stages of composition:

> On pourrait être enclin à accorder la priorité à ce second groupe de mss., parce que plusieurs de ses représentants formant une classe distincte [...] présentent les chapitres du De utilitatibus astrolapsus [i.e. **J**] dans l'ordre régulier, qui a été abandonné pour des extraits par tous les mss. du premier groupe [...]. Bien au contraire, ces modifications témoignent d'une activité et d'un intérêt, que nous ne retrouvons non plus dans les manuscrits récents (s. XII–XIII) réunissant les trois traités édités par Pez.[198]

Van de Vyver suggested that **J** was a reworking of another eleventh-century text with quite similar contents, known under the title 'Sententiae astrolabii' (i.e. **J'**).[199]

3.3.6.3 Millás Vallicrosa's studies on ms. Ripoll 225

Millás Vallicrosa's study was more ample and detailed than Van de Vyver's. He did not focus his attention on the text **J**, but on one particular manuscript, which had not been known to Bubnov and had been found in the library of the cloister Ripoll in Catalonia (ms. Ripoll 225, now in Barcelona).[200] At the time, the manuscript was thought to have been written in the tenth century, and Millás Vallicrosa assumed that it had been composed in Ripoll itself, where at the time Lupitus of Barcelona was active. Taking up Thorndike's and Haskins' thesis, Millás Vallicrosa regarded the manuscript Ripoll 225 (A.2) as a direct testimony of the transmission of astrolabe knowledge from the Arabic culture into the Latin one and devoted his attention to editing it and discussing its contents. Some time after Millás Vallicrosa's study was published, though, it was established that the manu-

196 Van de Vyver (1931), Millás Vallicrosa (1931).
197 See in particular the summary table on Van de Vyver (1931) p. 280–281.
198 Van de Vyver (1931) p. 282–283.
199 See app. C.
200 Millás Vallicrosa (1931) p. 150–211.

script Ripoll 225 (A.2) has to be dated to the eleventh century and that it was produced in France and not in Catalonia.[201]

In the Ripoll manuscript, some – but not all – chapters of **J** can be found in an order different from that of Bubnov's edition. In the same manuscript, the text 'Sentientiae astrolabii' (**J'**) also occurs, and Millás Vallicrosa expressed the same opinion as van de Vyver, considering it as the source on whose basis **J** had been composed. It was to underscore its relationship to **J** that Millás Vallicrosa indicated the 'Sententiae astrolabii' as **J'**.

3.3.6.4 Millás Vallicrosa's edition of early astrolabe texts

On the basis of Ripoll 225 (A.2) and of other eleventh- and twelfth-century manuscripts, Millás Vallicrosa reconstructed a series of astrolabe-related texts and printed them in his monograph. He considered them as either direct translations from the Arabic or reworked versions of translations into a more literary Latin form. Since he considered Ripoll 225 (A.2) as the ultimate source of Latin astrolabe material, Millás Vallicrosa took its contents as reference point for the edition. At times, though, he integrated them with passages taken from other manuscripts. Among the texts he reconstructed, some are quite short and deal with the geometrical construction of the astrolabe (**h'**, **h''**), while others describe the possible uses either of the astrolabe (**J'**) or of other models and geometrical constructions (e.g. **J'a** on a celestial globe, and some passages on the universal horary quadrant[202]). These latter texts have the same structure as **J**: they are composed of a series of very short chapters, not always appearing in the same sequence in the manuscripts.

Millás Vallicrosa saw the texts he published in a genealogical relationship to each other: according to him, **J'** was a translation from the Arabic later reworked into **J**. Similarly, the text **h''** on astrolabe construction was supposedly a translation from the Arabic which was later reworked into **h'**. Thus, a distinction was introduced between a first (**J'**, **h''**) and a second (**J, h'**) stage in the written transmission of astrolabe knowledge.

Millás Vallicrosa's edition has been an invaluable tool for later studies on eleventh-century Latin astrolabe literature, but a detailed analysis of the manuscript tradition of the published texts is still lacking. Because of this, it is in some cases difficult to estimate how the printed texts relate to the versions actually found in manuscripts. This is particularly true in the case of the texts grouped by Millás Vallicrosa as 'chapters relating to the use of the spherical astrolabe' or 'texts relating to the quadrans'.[203]

---

201 For the various references concerning the dating of the manuscript Barcelona, Archivo de la Corona de Aragón Ripoll 225 see app. A, item A.2.

202 See 5.3.5–6 and 5.4.5.

203 'Capítols relatius a ús de l'astrolabi esfèric (**J'a**), els quals en els mss. apareixen relacionats amb el text J'' Millás Vallicrosa (1931) p. 288–290; 'textos relatius al quadrant amb cursor' (Millás Vallicrosa (1931) p. 304–308).

3.3.6.5 Werner Bergmann's thesis of the two-stages composition of J

In 1985, Werner Bergmann again took up the analysis of the manuscript tradition of **J**. Based on a thorough analysis of earlier and later manuscripts, he argued that a further distinction was necessary. Namely, one should distinguish between, on the one hand, an older version of **J** containing only 19 chapters (**J(19)**) and, on the other, the 21-chapter-text edited by Bubnov (**J(21)**).[204] In Bergmann's opinion, the **J(19)** version was a work of Gerbert of Aurillac, whereas the 21-chapter-text **J(21)** must be seen as a reworking of **J(19)** due to Herman of Reichenau. Bergmann expressed the opinion that Gerbert had derived his knowledge on the astrolabe from a translation of an Arabic treatise on the astrolabe, which was later eventually lost.[205] Thus, in the case of **J**, a written transmission in three stages would have to be assumed: (1) a (lost) translation form the Arabic into Latin, (2) a first reworking of that translation by Gerbert (i.e. **J(19)**) and (3) a second reworking due to Herman of Reichenau (i.e. **J(21)**).

Bergmann's reconstruction of the process of composition of **J** has been criticized by Arno Borst and Paul Kunitzsch, who remarked that Bergmann's clear-cut classifications of the manuscripts as containing either **J(19)** or **J(21)** could not be maintained. Kunitzsch concludes that:

> In the present state of exploration of the history and transmission of the text **J**, I think, it would be premature to give definite explanations of how, when and by whom the various corruptions and deficiencies were brought into the text of this chapter [i.e. **J 17**].[206]

Once again, the material found in the oldest manuscripts does not allow a reconstruction of an 'original' text.

Moreover, it is important to note that Bergmann felt the need to assume a lost translation as the source of **J**, because he did not believe that the text **J'** contained the knowledge found in **J**. Here, as anticipated in 3.3.5, problems at the philological level combine with questions at a second level, i.e. relating to the information actually carried by the reconstructed text. In the case of **J**, which is quite a long text, the problems on the philological level dominate. Texts devoted to the construction of the astrolabe, instead, are shorter and easier to reconstruct from the manuscripts. In their case, it is the problems at the second level that dominate. A question of this kind which has often been discussed is which sources Herman of Reichenau used in composing his work on astrolabe construction (**h**). I shall discuss it in the following paragraph.

---

204 Bergmann (1985) p. 173–174.
205 Bergmann (1985) p. 174.
206 Kunitzsch (2000b) p. 249.

3.3.7 The sources of **h**: the problem of the lost original?

The earliest astrolabe text whose author could be identified with certainty is the one usually indicated as **h**. It was written in the first half of the eleventh century in the cloister Reichenau by the monk Herman, who ranged among the best scholars of his time.[207] Herman never visited Spain: in fact, he was afflicted since child-hood by an illness impairing his motion and in all probability spent most of his life on the island of Reichenau. How could he have learned enough about astro-labes to write the text **h**, which is the clearest early Latin description of how to construct astrolabe line? This question has been discussed as often as the author-ship of **J**. In general, it has been assumed that Herman became familiar with the astrolabe by oral teaching, but that his main source of information on its construc-tion was a written text, either one of those published by Millás Vallicrosa or an-other one, now lost.

In 1980 Werner Bergmann published a detailed analysis of the manuscript transmission of the star table included in **h**, using some errors of transcription as guidelines for ordering the manuscripts in classes. From his analysis, he concludes that Herman's source was a translation of which today only the first part remains, which is the text **h''**.[208] According to Bergmann, this translation was part of a cor-pus of astrolabe texts which had been put together in France or Lotharingia, using knowledge brought from Spain by Gerbert of Aurillac.

Almost ten years later (1989), Arno Borst published a monograph in which he analysed a fragment of parchment on which parts of the older astrolabe texts oc-curred and interpreted it as the remains of a very early stage of transmission of astrolabe knowledge.[209] Borst shared Bergmann's opinion that most parts of the material translated in the first stage are lost, and that Herman had at his disposi-tion 'better' astrolabe texts than **h''** and **h'**.[210] He disagrees with Bergman on the identity of the mediator who had provided Herman with the texts: according to him, the mediator was Bern, abbot of Reichenau.[211]

In 1998, Charles Burnett offered his analysis and solution of the problem. He accepted Millás Vallicrosa's theory according to which **h''** was a direct translation from the Arabic and **h'** its literary reworking. He did not comment on the quality of these texts and did not feel the need to postulate the existence of other, lost translations.[212] According to Burnett, the texts had reached Herman through a chain involving not Gerbert of Aurillac, but Abbo of Fleury, Constantin of Fleury and Ascelin of Augsburg.[213]

---

207 On Herman of Reichenau and the astrolabe, see Borst (1989) p. 77–84. For further bibliogra-phy: Struve (1999).
208 This is only a sketch of Bergmann's conclusions, which are summarized in Bergmann (1980) p. 98–102, Bergmann (1985) p.120–121 and p.173–174.
209 Borst (1989).
210 Borst (1989) p. 27–29, p. 43–44 and p. 50–52.
211 Borst (1989) p. 58–59 and p. 70–77.
212 Burnett (1998) p. 330 and 338–339.
213 Burnett (1998) p. 330–339.

I have discussed the various solutions to the question of Herman's sources in some detail because I believe they are a good example of the problems of implicitly assuming that the written word was the main vehicle through which astrolabe knowledge spread in Latin Europe. Bergmann and Borst felt compelled to assume the existence of lost treatises because they estimated that the quality of extant texts was too poor for them to have served as a basis for Herman's treatise on astrolabe construction. However, one has to note that, in the beginning of his treatise, Herman explicitly states that he will try to put clearly into writing the geometrical construction of the astrolabe, which was usually found in a form "muddled, obscure and sometimes mutilated".[214] In other words, Herman's main task was writing down the geometrical construction in a proper way, because the extant versions were considered unsatisfactory already in his time. One may assume that, if 'better' texts had been at hand, there would have been no need for Herman to write his treatise.

Both Bergmann and Borst somehow suggest that eleventh-century Latin scholars actually had to make an effort to gather some understanding of the structure and function of the astrolabe from the written material – and possibly artefacts – circulating in Europe.[215] If this were true, the evolution of the contents of astrolabe texts would be evidence of a process of knowledge production, and not only of knowledge transmission. I will instead argue that the problem Herman was facing was not that of attaining mathematical knowledge, but that of making this knowledge readable. This enterprise was by no means easier: in fact, it was possibly more difficult, but Herman was an expert in the field and his efforts were thoroughly successful.[216]

3.3.8 The problem of relating the earliest Latin astrolabe texts to Arabic sources

After it was established that Ripoll 225 (A.2) had not been written in tenth-century Catalonia but in eleventh-century France, no other manuscript was found to replace it as direct evidence of a connection between Andalusi and Latin culture. None of the earliest manuscripts appears to be significantly older than the others, except possibly for the fragment discovered by Arno Borst in the Town Archive in Constance.[217] Moreover, none can be unequivocally brought in connection with areas of close Arabic-Latin contact, such as Northern Spain or Southern Italy. This is particularly perplexing, because the contents of the manuscripts indicate very clearly their Arabic origin. A very large number of Arabic

---

214 "Cum a pluribus saepe amicis rogarer, ut mensuram astrolabii, quae apud nostrates confusa, obscura et passim mutilata vulgo invenitur, lucidius pleniusque scribere temptarem [...]" (Herman of Reichenau (1931) p. 203).
215 Bergmann (1980) p. 100–101, Borst (1989) p. 28–30 and p. 40–45.
216 On Herman's expertise in making mathematical knowledge readable Borst (1989) p. 77–78. For an example of Herman's capabilities see 4.6.7 and 4.6.9.
217 Borst (1989). References to this fragment are given in appendix A, item A.12

words written in Latin characters are used and sometimes even Arabic characters appear.

A key assumption shared by all theories discussed up to now is that the translation of a number of texts from Arabic into Latin was at the origin of the diffusion of astrolabe knowledge in eleventh-century Latin Europe. Yet researchers agree on the fact that eleventh-century manuscripts do not bear direct testimony to the introduction of astrolabe knowledge into Latin Europe, but only to its diffusion in a broad network connecting the centres of Latin scholarship.[218]

The employment of literary means of transmission already in the earliest period can hardly be doubted, yet no trace of it remains: this lack of evidence may be the result of chance, but it could also indicate that the role of the written word in the earliest stages of knowledge transfer was more limited than in the subsequent period of diffusion within Europe.

3.3.9 Paul Kunitzsch's analysis of the relationship of the text **J'** to Arabic sources

The merit of having shed some light on this key aspect of the process of knowledge transfer is due to Paul Kunitzsch, who has analysed the earliest Latin literature on the astrolabe, tracing some portions of it to its Arabic sources. The results of his analysis of the text **J'** on astrolabe usage are particularly illuminating and I will quote a longer excerpt from them, because I believe Kunitzsch's judgement to be very significant.

> The text of the 'Sententiae' [i.e. **J'**] comprises three sections: (i) an introduction, (ii) a description of the astrolabe and its parts, and (iii) the 'use' of the astrolabe. Of these, the introduction shows no distinct signs of being a translation. As it seems, it is styled in the own words of the (Latin) author, or compiler, of the whole treatise. Sections ii and iii together represent, though in a modified manner, what is usually contained in the Arabic treatises on the use of the astrolabe. [...] My impression is [...] that section ii of the 'Sententiae' is not a direct translation from the Arabic, but that it was written in Latin by someone who had an astrolabe in front of him (and perhaps also Arabic text material on the astrolabe), and who had at his side a person with sufficient knowledge of Arabic to help him in identifying all the Arabic names of the individual parts of the instrument and their descriptions, and in putting them down in Latin in a reasonable way.[219]

Here I may add that knowledge of the Arabic language would hardly have been enough to identify the parts of an astrolabe and that the 'person' who Kunitzsch assumes stood beside the Latin author must have also been an astronomer. As I shall argue later on, Kunitzsch's opinion on section (ii) of **J'** is well compatible with the possibility that the text was written from notes taken while a person competent both in Arabic and in astronomy was explaining the structure of the astrolabe.

Even more interesting is section (iii), in which Kunitzsch could identify a number of passages as a literal translation from an Arabic treatise by Muḥammad

---

218 Bergmann (1980) p. 101–102; Borst (1989) p. 43.
219 Kunitzsch (1987) p. 231–232.

ibn Mūsā al-Khwārizmī and some others as closely related to it. However, all this material of Arabic origin is interspersed with original Latin comments and explanations:

> Different again is the case of section iii, on the use of the instrument. Here [...] we have really a translation from the Arabic. [...] But of a total of 276 lines [...] only roughly a seventh could be proved to be a literal translation from al-Khwārizmī's text as we have it [...]. If we include the sections also that show a close relationship to al- Khwārizmī's text, we arrive at a maximum of roughly a quarter of the text lines of section iii of the 'Sententiae' that have an Arabic counterpart in the treatise of al-Khwārizmī. It is evident that the Latin author, or compiler, has added a great many explanatory phrases.[220]

Again, this structure could be very well explained by assuming that the Latin text was composed on the basis of notes taken either in writing or in memory from a lecture based on al-Khwārizmī's text. The student (or students) had paid attention and understood the subject and he or they were later able to discuss and eventually fix into writing what they had learned. The lecture might have taken place in Latin or in Arabic: the fact that astrolabe texts so often contain Arabic words (and sometimes even whole sentences) written in Latin characters might indicate that the author was capable of understanding spoken Arabic (so that it made sense to note down the Arabic words), but was unable to write Arabic characters.

3.3.10 Conclusions: the theory of written transmission of astrolabe knowledge and its problems

As shown in the previous chapters, it is often assumed that the transmission of astrolabe knowledge from the Arabic to the Latin culture took place by means of the written word: in a first stage, Arabic texts were translated into Latin, in a second stage they were reworked into a more literary form and circulated among Latin scholars.[221] The main problems this theory has to face are:

(1) No manuscripts have been found testifying to the first stage of transmission, i.e. originating around the year 1000 in areas of Arabic-Latin contacts, such as Northern Spain or Southern Italy.

(2) Only fragments of eleventh-century Latin astrolabe texts could be shown to be translations of Arabic works and they are always combined with material of probable Latin origin: at the level of the text, too, the first link in the chain of transmission is missing.

(3) In the oldest manuscripts, longer texts (e.g. **J, J', J'a**) appear in a significant number of variants with chapters ordered in different sequences or missing.

---

220 Kunitzsch (1987) p. 232.
221 Bergmann (1985) p.120–121 and p.173–174, Borst (1989) p. 27–29, p. 43–44 and p. 50–52, Burnett (1997) p. 4–5, Burnett (1998) p. 334–339.

(4) Shorter texts (e.g. **h''**, **h'**) or single chapters of the longer ones present a more stable form, but their contents are often unclear or apparently incomplete.

(5) Although texts devoted to the same subject often present similarities, there are no clear indications as to how and if they directly relate to each other as original and reworked version (e.g. **h''**, **h'**, **h** and **J'** , **J**).

These questions can be solved by assuming the existence of lost originals, of which only corrupt copies were preserved. However, although loss of material surely contributed to shaping extant eleventh-century astrolabe texts, other factors have to be taken into account as well.

The many variants of longer texts can be explained by assuming a process of progressive reworking, as suggested by van de Vyver and also, in a way, by Borst and Bergmann. If it is so, it must be asked who was taking part in this process and how far it is possible to attribute the reworking to a single author. The fact that different versions of the same text have survived suggests that the rewriting was a collective enterprise.

Another factor which has to be taken into account is the possible contribution to the diffusion process of oral and non-verbal modes of communication. A local pool of astrolabe knowledge could receive input not only in the form of manuscripts, but also through discussions based on exchanging and commenting short notes, drawings, solid models or memorized experiences in geometrical imagination. With the aim of discussing such questions in chapter 4, I will report here some of Catherine Jacquemard's results on the process of textual reworking of parts of a text known as 'Geometria incerti auctoris' (**GIA**) This case-study is particularly interesting, not only for methodological reasons, but also because the contents and tradition of the texts analysed are closely related to those of astrolabe texts. Therefore, this example offers an ideal basis to start discussing the subject of readability in high medieval mathematical texts.

3.4 MEDIEVAL MATHEMATICAL TEXTS AND READABILITY: THE EXAMPLE OF THE 'GEOMETRIA INCERTI AUCTORIS' (**GIA**)

3.4.1 Catherine Jacquemard's analysis of the **GIA** and its relevance for the present study

Catherine Jacquemard is preparing an edition of the 'Geometria incerti auctoris' (**GIA**), an eleventh-century mathematical text in which the astrolabe makes a brief appearance in the context of various surveying procedures.[222] The manuscript tradition of the **GIA** has close links to that of the earliest astrolabe texts, since most early astrolabe manuscripts also contain passages from the **GIA**. The history of modern editions of the **GIA** and of astrolabe texts is also similar: the chapters now

---

222 The following overview on the **GIA** relies on: Jacquemard (2000) p. 81–83.

seen as constituting the **GIA** were first published as part of a larger work on geometry attributed to Gerbert of Aurillac: the 'Geometria Gerberti'.

In his edition of Gerbert's 'Opera mathematica', Bubnov established that only some parts of the 'Geometria Gerberti' could be attributed to Gerbert. The chapters he regarded as spurious – among them, all those containing references to the astrolabe – were grouped together as books three and four of a 'geometry by an unknown author' (**GIA III** and **GIA IV**). Bubnov assumed that the other books of the same work had been lost, and printed what remained among Gerbert's 'opera incerta'.[223] In reconstructing the **GIA**, he left out a few chapters that in his opinion could be attributed neither to the 'Geometria Gerberti' nor to the **GIA**. This decision was later criticized by other historians.

Jacquemard's paper investigates the manuscript tradition of one of these, so to say, doubly spurious chapters, indicated with the title 'de profunditate maris vel fluminis probanda' (**dpp**), which discusses a method for estimating the depth of rivers and of the sea.[224] The text **dpp** is particularly interesting for the present study, because it describes how the astrolabe can be employed in surveying not only as a tool to measure heights, but also as a device to estimate time duration. The passage will be discussed from this point of view in 5.3.7. Here, I shall only summarize Jacquemard's results of the analysis of its manuscript tradition, which are interesting for two reasons. The first one is that they support the thesis of a contribution of non-written strategies of knowledge transfer to the diffusion of knowledge of Arabic origin. The second reason is that they offer an example of the problems arising when attempting to capture in written words mathematical knowledge expressed in other forms, and of the different solutions found by Latin authors.

3.4.2 The **GIA** as product of progressive rewriting

Jacquemard's thesis is that the so-called books three and four of the **GIA** were the result of successive rewriting of a number of separate kernels, which were collected together only at a later stage.[225] In particular, chapters 20–25 of **GIA III** and the text **dpp** had their origin in Arabic sources, from which they were imported along with elements of astrolabe knowledge.[226] According to Jacquemard, this material was later reworked in its Latin form and was collected together, first, with the other chapters of the **GIA III** and, later on, with those of the **GIA IV**.[227] Not all parts of the **GIA** had an Arabic origin.

---

223 Bubnov (1899) p. 317–365.
224 The text exists in two versions: a longer one (28 lines in Jacquemard's edition) and a shorter one (21 lines in Jacquemard's edition). Both versions are printed in Jacquemard (2000) p. 105–106. The results of the analysis of the manuscript tradition of **dpp** are also discussed in Haire/Jacquemard (2000b).
225 Jacquemard (2000) p. 88–92.
226 Jacquemard (2000) p. 83–94.
227 Jacquemard (2000) p. 95–96 and p. 98–99.

In Jacquemard's opinion, the progressive reworking should be regarded as a collective enterprise, leading to a 'Geometriae incertorum auctorum'.[228] The extant manuscript collections of geometric material can be seen as the final result of a process of knowledge transfer, assimilation and production taking place simultaneously on more than one front. The 'defects' in the transmission of the **GIA** should be interpreted not as a proof of incompetence or decay, but rather as evidence of the birth of a new discipline:

> Ce n'est pas un travail monolitique et maladroit qu'ont enregistré les manuscrits contemporains, mais les coups d'essai, sans cesse retouchés, inhérents à une discipline nouvelle en pleine gestation.[229]

Jacquemard formulates hypotheses on the historical actors involved in the process: although the contribution of Gerbert of Aurillac to the earliest stages cannot be directly grasped, the memory of his teaching dominates the tradition of the **GIA**.[230] As far as the later stages are concerned (i.e. those of literary reworking and archiving), evidence points to the cloisters of the Mosel area, in particular to the school of Fulbert of Chartres.[231]

Jacquemard does not address the question of how far non-written strategies of knowledge transfer may have contributed to the production and assimilation of mathematical knowledge. At the same time, though, her remarks on the memory of Gerbert's teaching and on the collective character of the enterprise suggest that such strategies may have played a relevant role.

3.4.3 Were the single stages of rewriting of the **GIA III** an individual or a collective enterprise?

At the level of the single stages of rewriting of smaller groups of chapters, anyway, Jacquemard reverts to the thesis of a single author. In her opinion, the chapters **GIA III 1–19** were in part a rewritten version of **GIA III 20–25** and were the work of a single author. However, she also admits that the work of that author does not coincide with the texts in the extant manuscripts:

> Aucun des manuscrits existants n'a conservé exactement son travail. Les représentants de la classe E [i.e. one of the groups of manuscripts defined by Bubnov] donnent le témoignage le plus fiable de cette seconde strate de rédaction , mais il est entaché d'anomalies. En effet, a côté de séquences très logiques [...] on constate des incohérences qui semblent dues à la combinaison maladroite de deux recensions légèrement différentes d'une même oeuvre. De plus, les collections fournies par les manuscrits de la classe E ne sont jamais exactement identiques. Les copistes successifs ont innové à chaque fois en retouchant, supprimant, déplaçant des chapitres ou en ajoutant des fragments qu'il faut rapporter à d'autres mains.[232]

228 Jacquemard (2000) p. 102–103.
229 Jacquemard (2000) p. 102.
230 Jacquemard (2000) p. 96–98.
231 Jacquemard (2000) p. 100–102.
232 Jacquemard (2000) p. 90–91.

Thus, even on the smaller scale of a few short chapters, similar problems of transmission appear as for the longer texts. Once again, the existence of a lost original has to be assumed. In my opinion, although the 'rewritten' texts (i.e. **GIA III 8–11–12–13–15**) were probably the work of a single author, there is little evidence that they were part of a work composed to replace the 'original' texts (i.e. **GIA III 20–21–23–24**). On the contrary, the **GIA III** was in the end put together out of material of both kinds, even when the passages dealt with the same subject. This is a feature typical of early Latin astrolabe texts and manuscripts, which often look like uncritical collections of passages offering variations on the same theme.

Once again, the question of modes of communications alternative or complementary to the written word surfaces. The disunity of transmission as well as the uncritical, repetitive structure of the written end product could be interpreted as evidence that the rewriting process was a collective enterprise in an even broader sense than that assumed by Jacquemard. Written texts and manuscripts may have been the result of lectures and group discussions involving a multiplicity of modes of communications, whose results were partly fixed into writing by more than one participant. Later, manuscripts containing different, eventually diverging solutions to the same problem might have been collected and could have provided a basis for further discussion. Especially in the case of knowledge transfer from the Arabic into the Latin culture, multilingual communication may have been much easier in speaking than in writing.

3.4.4 The question of readability and its possible solutions: the literary and the 'recipe' style

A key result of Jacquemard's analysis is the attention she has drawn to eleventh-century efforts to attain 'readability' (lisibilité) in texts devoted to the mathematical arts. Having stated my objections to the theory of straightforward rewriting, I shall now discuss the differences remarked by Jacquemard between the two sets of chapters of the **GIA III**, which she regards respectively as 'original' and 'rewritten' versions of the same material. I believe those differences are very significant, because they testify to different approaches to readability in mathematical texts. Instead of speaking of 'original' and 'rewritten' version, I will simply refer to texts written either in a 'recipe' style or in a 'literary' one. Jacquemard draws a parallel between the recipe style of **dpp, GIA III 20–25** and astrolabe texts of a (supposed) first stage of transmission (**h'', J', J'a**) and the literary style of **GIA III 1–19** and of astrolabe texts of a (supposed) second stage of transmission (**J, h'**).[233] The most important differences between the texts written in the recipe

---

233 Jacquemard (2000) p. 88–91 and p. 94–95. Here it has to be remarked that Jacquemard considers the manuscript BnF lat. 7412 (A.27) as a witness to the texts of the first stage, whereas BnF lat. 7412 (A.27) is actually a mixture of the two stages, as it contains **dpp, GIA III 20–**

style (**GIA III 20–25, dpp, J', J'a** and **h''**) and those written in the literary one (**GIA III 1–19**, in some measure **J** and **h'**) are:

(1) Texts in recipe style usually address the reader in the second person, whereas those in the literary style describe a third person, e.g. the 'geometra', while he carries out the procedures.[234]

(2) Most texts in the recipe style have the form of a sequence of instructions bound by a simple structure of the kind: "if you want to [...] do such-and-such [...] and you will see that [...]"("quando quaeris [...] construe [...] et videbis [...]").[235] Texts in the literary style, instead, combine the 'if-then' structure with a more complex syntax, expressing not only the sequence, but also the functional or causal relationship of the various steps to be performed. For example, ablative absolutes are used to underscore that some steps are preliminary to others (e.g. "cacumine invento, spatium [...] diligenter mensuretur").[236]

(3) In recipe texts, a significant part of information is conveyed through drawings, because geometrical elements (e.g. segments, points) are identified in the text using the letters indicating them in the drawings. In texts of the literary kind, instead, points are indicated by verbal descriptions, such as "the centre of the circle" or "the height to be measured".[237]

Points (1) and (2) seem to indicate that texts in literary style contain a narrative element, making explicit the role of the actor and the causal and functional structure of the procedure. What was fixed in writing were systematic, reproducible connections linking a series of steps into an abstract methodical procedure thanks to which trees, rivers, towers or fields – elements of God's Creation and of the man-made world alike – could be related to each other in geometrical or numerical proportions.

The literary narrative enhanced the readability in the sense that it drew the attention of the reader on the systematic elements of the action described. It may be asked, though, to what degree this was a pedagogical improvement as far as understanding the procedure was concerned. Texts in recipe style, on the other hand, focussed on the performance of the various steps, which was described in the second person, inviting the reader to share the experience. For a mathematical text, this is a very important pedagogical feature. Sharing the experience, either by actually performing the procedures or by just imagining doing so, is crucial for becoming aware of those functional relationships explicitly stated in the more literary texts. A certain amount of experience is necessary to understand the relation-

---

25, J'a and **h''**, but **J** instead of **J'** (see app. E). Jacquemard's indications on: Jacquemard (2000) p. 92 –93 are in this respect wrong: BnF lat. 7412 (A.27) contains **J 1–19** and **J 21**, but only one chapter of **J'**.

234 Jacquemard (2000) p. 110–111 and p. 114–115.

235 Jacquemard (2000) p. 110 and p. 112.

236 Jacquemard (2000) p. 115–116.

237 Jacquemard (2000) p. 112 and p. 116–117.

ships.[238] In a sense, texts in different styles were pursuing different aims: the texts in literary style attempted at describing abstract mathematical structures, while recipe-texts rather tried to make the reader aware of them by using them in practice. This is a difference similar to the one existing between stating the rules of Euclidian geometry and teaching how to use them to solve problems.

The different approaches are particularly evident in the role of drawings: in texts in recipe style, the construction and labelling of drawings usually takes place in the beginning. From then on, labelled drawings are assumed as constantly present to the reader, dispensing the writer from the necessity of describing the elements labelled each time they are mentioned. The text is thus less readable, in the sense that it cannot be understood without the drawings. In literary texts, on the other hand, drawings tend to be simple illustrations, and the text conveys verbally the information included in them (e.g. "a, that is the centre of the circle").

However, for understanding the procedure discussed, it is usually necessary for the reader to have the relevant construction present either in front of his eyes or in his mind. The letters used in labelled drawings provide an efficient, written but non-verbal formalism which is essential for storing and transferring elements of mathematical knowledge. In some cases, the procedure described can be immediately grasped by simply glancing at a drawing, while verbal explanations rapidly become cumbersome and unclear.

In conclusion, I believe that texts written in different styles should not always be regarded as rewritings of one another: they could also be parallel attempts at capturing in writing the same elements of knowledge, perhaps with a different accent. It must also be noted that a clear-cut distinction between the two styles is not always possible, particularly in the case of astrolabe texts.[239] In the third part of this study I shall offer examples of this phenomenon in texts devoted to astrolabe constructions.

3.4.5 Conclusions: different criteria for evaluating eleventh-century astrolabe texts

For Latin scholars of the eleventh century, making new mathematical knowledge readable was a novel and difficult task, possibly as hard as that of assimilating the new geometrical and mathematical-astronomical methods imported from the Arabic culture.

This is particularly true in those cases in which the mathematical knowledge to be made readable was conceived of and understood not primarily in verbal form, but by means of geometrical imagination. In the absence of a verbal component that could be directly converted into a text, the solution chosen in the first efforts at readability was to fix in writing a verbal description of the procedure, either in literary or in recipe style. Such descriptions should not in principle be

---

238 Mayer/Hegarty (1996).
239 Jacquemard (2000) p. 113.

interpreted as exclusively aimed at teaching how to do something without understanding it, but should be regarded as efforts to share and communicate the experiences of geometrical imagination which was the basis for understanding both
the mathematical structures involved and eventual philosophical reflections on
them.

When analysing eleventh-century texts devoted to the astrolabe, it is therefore
necessary to make a careful distinction between different criteria of evaluation:

(1) the author's competence in the astronomical, mathematical and philosophical
    field;

(2) the author's interest in and possibility of storing all or part of his knowledge
    in written form. In particular, it is important to ask on which part of his
    mathematical, astronomical and philosophical knowledge the author was fo-
    cussing;

(3) the author's intention and capability of writing a text conforming to con-
    temporary criteria of literary Latin prose on mathematical subjects;

(4) the author's intention and capability of writing a didactical text, i.e. a manual
    capable of transmitting knowledge to a reader otherwise ignorant of the sub-
    ject discussed, with little or no complement from oral teaching, practical ex-
    ample or material aids to visualization.

It is important to distinguish between criteria (3) and (4) which might result in
diverging criteria of readability. In chapters 4 and 5, I shall offer an analysis of
eleventh-century astrolabe texts employing these four criteria. I will argue that
most eleventh-century astrolabe texts were based on a larger amount of knowledge than their (usually poor) didactical value allows them to convey. A significant part of that knowledge was stored and diffused among Latin scholars thanks
to non-verbal and/or non-written methods of knowledge transfer.

Eleventh-century astrolabe texts are evidence of parallel, competitive efforts
at making mathematical knowledge readable. This variety mirrors only in part the
diffusion and assimilations of astrolabe knowledge, and is in large part shaped by
the tensions and transformations in the attitude to the written word which characterized the eleventh and early twelfth centuries.

4 NON-WRITTEN AND NON-VERBAL MODES
OF COMMUNICATION IN THE DIFFUSION
OF ASTROLABE KNOWLEDGE

4.1 MEMORY, NOTES AND GEOMETRICAL PATTERNS
IN HIGH MEDIEVAL MATHEMATICAL ARTS

4.1.1 The main thesis of the present chapter: astrolabe manuscripts as written traces of a primarily non-written transmission

In the followings pages, I shall argue that eleventh-century astrolabe manuscripts are the written traces of a process of transmission and assimilation of knowledge which took place thanks to a combination of written and non-written, verbal and non-verbal strategies of communication. While the written component of the process is in some measure still accessible to us, the contribution of non-written strategies can only be glimpsed indirectly, and reconstructions often have speculative character.

On the other hand, evidence of the employment of non-written strategies of knowledge transfer in medieval times is indisputable, at least in the manual crafts.[240] For example, all efforts notwithstanding, practically no written sources could be located on the features of medieval water mills,[241] on the origin of weight-driven clocks[242] or on the techniques of Romanesque and Gothic architecture.[243] Even though 'craft secrets' are often brought forward as a reason for the lack of technical literature, there are substantial objections to this view, as far as the early and high medieval periods are concerned.[244]

As we shall see, in the mathematical arts, too, there is evidence that knowledge could be transferred and stored with other means than the written word. Disregarding these additional contributions might lead to a bias in interpreting written sources. Apart from oral teaching, the strategies that could be used as alternatives or complements to the written word were: (1) drawings and diagrams combining geometrical figures with text, (2) material devices either embodying astronomical mathematical structures (e.g. a wooden sphere) or tools to be used in producing them (e.g. a compass) and, finally, (3) memory. As in the case of medieval orality and literacy, any clear-cut distinction between different modes of communication is an approximation: what one may expect to reconstruct are hybrid strategies in which different media took part.

---

240 Bischoff (1971), Boehm (1993).
241 Blaine (1976).
242 Dohrn-van Rossum (1992) p. 49–51.
243 Turnbull (2000a) p. 53–87.
244 Shelby (1976). On positive and negative evidence for craft secrecy in premodern times, see: Long (2001) p. 72–101, for high medieval Europe especially p. 78–88.

In the following pages I will review evidence of the employment of hybrid strategies in the medieval mathematical arts and, in particular, in the diffusion of astrolabe knowledge. To understand how effective and epistemologically relevant such strategies could be, it is important to take into account the role played by memory in medieval monastic culture and the way in which the workings of memory were associated with geometrical, mechanical and architectonical structures.

4.1.2. The medieval craft of memory

The term 'memory' is understood today as the ability to passively store and retrieve information without changing it: memory is filled with contents and can retain them, but is not supposed to act on them. Understood in this sense, memory only plays a marginal role in our culture. In ancient, medieval and early modern times, though, the art of memory was not only a reproductive, but also a productive craft and it had a far greater importance than today.[245] At the same time, it would be wrong to consider premodern memory as the same discipline remaining unchanged through the centuries: between ancient and early modern times the art of memory underwent as many transformations as other aspects of European culture.

Frances Yates devoted her work on 'The art of memory' to the changes occurring between the late medieval and the early modern period.[246] Paolo Rossi has argued that mnemotechnics are relevant to the history of science, because in the Baroque period, they became a model for universal languages and encyclopaedic systems.[247] The importance and the particular role of memory in medieval monastic culture have been discussed by Ivan Illich in his study about the changes in literacy that took place in twelfth- century Europe.[248] Mary Carruthers has devoted a series of studies to the investigation of memory in Latin European monastic culture of the Early and High Middle Ages, and the chronological and social frame of her researches matches that of the present study.[249] The expression 'craft of memory' is borrowed from her studies and it serves well to underscore the fact that, as already noted by Yates and Rossi, premodern memory had a productive function: it served not only to store and retrieve information, but also to use it as a starting point to create new statements, experiences, patterns of thought, texts, diagrams or even artefacts. Medieval methods of creative thinking evolved from the techniques of 'inventio' of ancient rhetoric, which taught how to store the ma-

---

245 A collection of papers offering a wide overview on this subject is: Bolzoni/Corsi (1992).
246 Yates (1966).
247 Rossi (1983).
248 Illich (1993) especially p. 29–65.
249 Carruther's main publications on the subject are: Carruthers (1990), Carruthers (2000), Carruthers/Ziolowski (2002). The central points of the thesis are summarized in Carruthers (2000) p. 1–6, Carruthers/Ziolowski (2002) p. 1–31 as well as in the review articles: Hamburger (1999), Green (2000). My discussion is primarily based on Carruthers (2000).

terial for a speech in such an ordered way that the speaker, while retrieving it, would immediately be able to combine it into a speech.[250] This was done by using spatial, and especially architectonical, structures as references to mark the specific path the speaker had to follow in his mind to construct his speech.

Other than ancient rhetoricians, though, medieval monks did not use only well-known, existing buildings as a reference for creative thinking, but imagined their own architectonical or pictorial patterns. These images could be inspired by Biblical text (e.g. the Ark, the Temple) and also, for example, by geometrical or cosmological structures. Patterns of manipulation of numbers, letters or geometrical structures could also be used as a guide for creative thinking. The reliance on vivid mental pictures and on the ability of the students to draw, paint and move pictures in their mind was enhanced in the medieval memory-craft by oriental and Jewish traditions of visionary contemplation.

According to Carruthers, medieval monks devoted great efforts to the task of developing patterns allowing them to retain in memory large amounts of information and providing a guide for using that information as material for creative thinking. This activity was primarily, but not exclusively, aimed at attaining meditative experiences. The methods of memory were learned not so much by means of verbal instructions, but rather through an apprenticeship similar to that used in manual crafts and based on 'orthopraxis', i.e. on constant, systematic imitation and repetition of specific practices, for example, exercise in making mental images.[251]

In her monograph 'The Craft of Thought: Meditation, Rhetoric, and the Making of Images, 400–1200' (1998), Carruthers writes:

> Medieval *memoria* was a universal thinking machine, *machina memorialis* [...] Meditation is a *craft* of thinking. People use it to make things, such as interpretations and ideas, as well as buildings and prayers.[252] [...] Medieval *memoria* thus included, in our terms, 'creative thought', but not thoughts created 'out of nothing'. It built upon remembered structures 'located' in one's mind as patterns, edifices, grids, and – most basically – association-fabricated networks of 'bits' in one's memory that must be 'gathered' into an idea.[253]

4.1.3 The embodiment of mnemonic patterns in images and artefacts

The visual and constructive aspect of the craft of memory is particularly relevant for this study. According to Carruthers, the craft of memory was not only used for meditation or for exegesis and production of texts: the ordered patterns and architectonical structures created as a guide for meditation could also be embodied in artefacts, illuminations, diagrams or architectonical structures. For example, the cloister became a "memory machine and encyclopedia", providing the monks

250 An overview on this subject is given in: Carruthers/Ziolowski (2002) p. 1–31.
251 Carruthers (2000) p. 1–3.
252 Carruthers (2000) p. 4.
253 Carruthers (2000) p. 23.

with constant guidance in their meditations.[254] Frances Yates had already re-marked that no clear-cut border could be drawn between the invisible patterns of memory and the visible patterns of figurative arts, and her arguments have also been taken up by other authors.[255]

It is important to remark that, although the patterns of the high medieval craft of memory could also have a visual character, the visual element was increasingly accentuated in late medieval and especially in early modern times. High medieval memory, on the other hand, depended very much on psychomotoric techniques, in which the cues to memorization were given not only by images but also by ges-tures which, like visual patterns, could either really be performed or simply imag-ined.[256] Even when imagining architectural structures, the medieval monk was supposed to walk around in them, collecting the cues for remembering.[257] This psychomotoric aspect of the craft of memory was particularly important for the medieval mathematical arts in which, as discussed in 2.1.5, the mind's hand was as important as the mind's eye.

Carruthers discusses at length the influence medieval monastic memory had on material culture. In his review of Carruther's work, Jeffrey Hamburger sum-marizes her argument in these words:

> [The meditational mode was] a style of thought and imagination that governed all the monastic arts, not just those genres that we might classify as literature (e.g. prayer, exe-gesis, dream vision, and, of course, meditation itself), but also the visual arts (manu-script illumination and, above all, architecture) and, more important still, the art of visu-alisation.[258]

In her monograph, Carruthers offers no compelling, deductive argumentation, but rather develops the many facets of her central theme by analysing a number of texts, images and buildings focussing on their possible interpretation as patterns produced and used in the memory-craft. It is left to the reader to decide whether she brings the abstract and practical aspects of monastic crafts too near to each other. In his review, Hamburger remarks:

> Carruthers tends to see little difference [...] between an evocation of architecture and an actual building. She admits to being somewhat loose with her terms, a deliberate impre-cision that lends flexibility to her argument, but that also, at times, blurs important boundaries. [...] Buildings [...] were more than ways of structuring meditation. They also served to configure the liturgy, segregate various groups from one another, commemo-rate the saints, shelter relics, and channel pilgrims. They also had to stand up.[259]

---

254 Carruthers (2000) p. 272–276.
255 Yates (1980), Bolzoni (1992), Antoine (1992).
256 The role of gestures in monastic reading and memoria is discussed in: Illich (1993) p. 51–65.
257 The accentuation of the purely visual aspect of the art of memory has been put in relationship with the development of central perspective, that forced those who were looking at a picture to do so by imagining themselves standing still in a particular position chosen by the artist (Antoine (1992)).
258 Hamburger (1999) p. 307.
259 Hamburger (1999) p. 308–309.

In my opinion, what Hamburge perceives as a blurring of boundaries reflects an important feature of the culture which Carruthers is investigating. That is the reason why I believe some of her results may be very helpful in understanding the medieval relationship between mathematical arts on the one side and mechanics and architecture on the other. It is important to evaluate evidence on the modes of communication used in medieval mathematics by taking into account the role of memory in high medieval monastic culture. As I shall argue in chapter 5, the interplay between abstract patterns and their successful material embodiment may have fostered the development of images of knowledge where the construction of artefacts was perceived as a cognitive process of philosophical and theological relevance. In other words, the fact that buildings "had to stand up" may have been seen as more than a material success.

4.1.4 Memory and medieval learning

In discussing the relationship between images, artefacts and the memory-craft, Carruthers focuses on how diagrams, paintings and buildings could be made to serve as patterns for meditation. She also points out that the same material had a pedagogical function not separable from the meditative one.[260] Since Late Antiquity, monastic education had tended to stimulate the ability of pattern recognition even in reading, because texts were written without spacings between the words, and therefore read as collections of individual letters to be ordered into sentences. Mnemonic verses were based on easily recognizable patterns, and were also combined with pictures and diagrams. Advanced products of this craft were the Carolingian 'carmina figurata' and the 'versus rapportati' of the late eleventh-century, both of which combined text and image in a complex pattern. According to Carruthers, diagrams and drawings in medieval manuscripts

> are not 'just' aids to understanding, as we would say, implying their subservient role to language and that they are in some basic way unnecessary to knowing. They are exercises and examples to be studied and remembered as much as are the words.[261]

These considerations refer to diagrams and drawings in general, but correspond very closely to research results on the methods of mathematical teaching and writing used in the tenth and eleventh centuries. In sections 4.3 and 4.4, I shall discuss evidence of how astronomical diagrams could be used as a mode of communication independent from text. This means that the diffusion of new and rather complex mathematical structures, such as those of the astrolabe, could build upon a background of efficient techniques for imagining, memorizing and interpreting non-trivial images, gestures and manual operations with devices.

---

260 The following remarks are based on Carruthers (2000) p. 135–142.
261 Carruthers (2000) p. 142.

4.1.5 Modes of communication in the medieval mathematical arts

In the late tenth and in the eleventh centuries, knowledge in the four mathematical arts (arithmetic, geometry, music and astronomy) could be stored and transmitted in various ways. The written words of Latin authorities preserved a small part of the inheritance of Greek mathematics, often seen through the filter of late-ancient Neoplatonism.[262] In these texts, the mathematical arts were regarded as a key to understanding the divine order of nature, although not necessarily to explain or predict natural phenomena. For astronomy, the main authorities were Ambrosius Theodosius Macrobius' (ca. 360–422) commentary to Cicero's (106–43 b.c.) 'Somnium Scipionis', Calcidius' (fl. ca. 300–350) commentary to Plato's (437–327 b.c.) 'Timaeus' and Martianus Capella's (fl. ca. 410–439) 'De nuptiis Mercurii et Philologiae'. In these works, mathematical concepts were communicated primarily in verbal form and in Roman number notation. Diagrams occasionally illustrated the texts, but very few geometrical constructions were presented. A mathematical formalism that would become very important from the tenth century onward was that of number proportions: it was a system of nomenclatures for ratios of natural numbers originally used to indicate the arithmetical ratios relevant to Pythagorean music theory. The system of proportions was explained in Boethius' 'De aritmetica'.[263] At least from the tenth century onward, written evidence shows that modes of communication different from the written word were employed in teaching the 'quadrivium'. The new methods trained students in imagining and mentally manipulating mathematical structures similar to those we regard as natural numbers, fractions, geometrical figures, musical notes or the homocentric-spheres-model.

A key element of this form of learning was repeated exercise in systematic procedures performed with the aid of material tools.[264] To learn arithmetic, students trained in finger-counting, the abacus and in the game of 'rithmomachia' (i.e. 'battle of numbers'), a board game based on Boethian number proportions.[265] To teach geometry, ninth-century scholars also started collecting passages on geometrical constructions from Roman surveying manuals.[266] In this period, Latin scholars added to the reference texts for the liberal arts the work 'De architectura' by Vitruvius, who was among other things the authority on gnomonics.[267] In mu-

---

262 On the role of the written word in medieval knowledge transmission: McKitterick (2000) p. 13–25; an overview of the textbooks for medieval astronomy is given in McCluskey (1998) p. 114–127.
263 Boethius (1867) p. 42–72 (I, 21–32).
264 All methods are discussed by Beaujouan (1972) and also in Burnett (1997) p. 13–14. On the interaction of the written word with other modes of communication in medieval teaching, see: McKitterick (2000) p. 25–30.
265 On abacus treatises, see: Evans (1976b), Evans (1977b), On 'rithmomachia' as a teaching aid: Evans (1976a), in general (Borst 1986).
266 On the medieval tradition of Roman surveying manuals and on their use as textbooks of geometry, see: Ullman (1964), especially p. 266–273.
267 Schuler (2000), especially p. 320–323.

sic, which in the eleventh century was the leading mathematical discipline, two instruments were particularly important: the Guidonian hand and the monochord.[268] The Guidonian hand was a system invented around the end of the ninth century for representing musical tones on the palm of the left hand. It made it possible to imagine and manipulate elements of music and was a method for memorizing them in such a way as to be able to productively use them. The monochord was a musical instrument with a single chord, on which various tones were produced by letting appropriate portions of the chord vibrate. The monochord provided a means of musical production and representation: the various points at which the chord had to be divided to produce specific tones were indicated in drawings by letters, which came to be used as a musical notation.

Speaking of the monochord, the historian Guy Beaujouan remarks that it was used at various levels: to learn musical intervals, for deciphering, judging and composing melodies and for discussing the principle of music between learned scholars.[269] Thus, like the astrolabe, the monochord had different aspects: it could be a real instrument or a formalism for understanding and discussing music.[270] The same can also be said for other instruments of the 'quadrivium', for example the abacus, whose use could be internalised by thinking of numbers in positional notation. In par 4.1.6, I shall come back to the problems linked to using different number notations and number concepts.

In general, the employment of the mathematical instruments described above both fostered and relied upon the student's capability of seeing and manipulating patterns in his or her mind. As didactic tools instruments did much more than just capture attention and make a complex matter simpler: their effect was to introduce the (usually young) minds of students to specific methods of mathematical thinking. It was in this context that diagrams started being used in astronomical manuscripts as a mode of communication independent from text.

4.1.6 Thinking numbers as geometrical patterns

Before proceeding with the discussion of astrolabe texts, I would like to pause a moment to discuss the problems high medieval mathematicians had to face in a field that today may appear trivial: number notation.

Number-words and Roman numerals were the most diffused written number notation in medieval Europe, but they made computing very cumbersome. Finger-reckoning and especially the abacus, on the other hand, led the students to think of numbers in positional terms (for which no written notation yet existed) and encouraged them to manipulate numbers with the 'oculus mentis' ('eye of the

---

268 On the Guidonian hand as a mnemotechnical device: Berger (2002); on the monochord in the Early Middle Ages: Wantzloeben (1911) p. 35–68. Both instruments are discussed in Beaujouan (1972) p. 640–644 and p. 651–653.
269 Beaujouan (1972) p. 642–644.
270 On the multiple character of the monochord as 'Musikinstrument', 'theoretisches Lehrinstrument' and 'System', see: Wantzloeben (1911) p. 4–34.

mind') or, significantly 'oculus cordis' ('eye of the heart', as in 'learning by heart').[271] The abacus also allowed computations with the Roman uncial system, analogous to today's duodecimal fractions. This could lead to conceptual problems, because, on the one hand, Roman fractions were handled like integers, but on the other they were still perceived as different from integers.[272] Particularly puzzling appeared the fact that multiplying integers with each other made them larger, while multiplying 'unciae' with each other made them smaller.

Finally, playing the game of ritmomachia led to thinking of integers as containing other integers according to specific patterns. Some numbers were associated with others through geometric forms, e.g. 'square', 'triangular' or 'pyramidal' numbers. Square numbers were the same ones as today. One example of a triangular number was 6, because six points can be drawn in a triangle with 3 points on each side. A pyramidal number was 91, because $91 = 1^2+2^2+3^2+4^2+5^2+6^2$ and thus 91 can be thought of as a step-pyramid made up of six squares of side 1, 2, 3, 4, 5 and 6 respectively.[273] The game of 'rithmomachia' was based on Boethian number proportions, which today might be regarded as fractional numbers. However, the game led to thinking of proportions not as numbers, but as operations performed on integers to put them together or take them apart.[274]

In general, the parallel use of different verbal or symbolic notations and operational or visual analogies could lead to confusion, but also to new questions, discussions and philosophical insights. In the late tenth and eleventh centuries, Latin scholars also experimented with new symbolic number notations. One of them was an early version of Arabic numerals, the Ghubar ciphers, which were probably associated with the abacus.[275] Number notations with Greek or Latin letters are also attested, for example, in the astrological texts of the old 'Alchandreana', where David Juste has established their Arabic origin.[276] This kind of notation was also associated with the Latin astrolabe.

On the most ancient Latin astrolabe (**#3042**), numbers are engraved in a letter notation similar to that used by the ancient Greek, as well as by the Arab: ten Latin letters (A, b, C, D, E, V, z, h, T) stood for numbers from 1 to 9, while other

---

271 On this aspect of abacus treatises, see: Evans (1976b). Finger-reckoning fostered positional number notation in that units corresponded to 'digiti' (locations on fingers), while tenths corresponded to 'articuli' (fingers articulations), thus introducing the difference between units and decades. The same terms were taken up by abacus treatises (Evans (1976b) p. 254).

272 On fractions and the abacus: Evans (1976b) p. 258–261.

273 Evans (1976a) p. 266. The associations of numbers with geometrical figures goes back to Boethius' 'de arithmetica', II, 6–31 (Boethius (1867) p. 90–122).

274 Evans (1976a) p. 265.

275 Beaujouan (1972) p. 654–657; on the occurrence of Ghubar cyphers in manuscripts Folkerts (1970) p. 86–87 and tab 1–2.

276 On alphabetical number notation in the 'Alchandreana', see Juste (2007) p. 147-155. Greek letters are used to indicate numbers in a 'ritmomachia' treatise in English manuscripts from ca. 1110 (Evans (1976a) p. 260) and in the 'Liber floridus' from St. Omer, ca. 1120 (King (1995) p. 371, note 29).

letters stood for 10 (I), 20 (K), 30 (L), 40 (M) and so on. [277] For example, the number 12 was written as Ib, because b corresponds to 2 and I to10. It is important to note that this is not a positional notation, because different sets of symbols are used for units and tens (e.g. A is 1 and I is 10). The use of this notation was in all probability of Arabic origin: in the tenth century, non-positional numerals of this kind (the so-called 'abjad' system) were still the most used form of number notation in al-Andalus. In fact, although it is generally recognised that positional Arabic-Hindu numerals reached Latin Europe by way of the Arabic-Islamic culture of al-Andalus, the earliest evidence of their presence in the Western mediterranean area is found in Latin (and not in Arabic) manuscripts.[278] What is particularly interesting for our subject is the fact that the only Latin astrolabe artefact that can be attributed with certainty to the tenth or eleventh century bears traces of experiments with written number notation. Even if this compact notation may have been better suited than Roman numerals for engravings on an astrolabe, the decision to use it indicates an interest in developing new written mathematical forms to match evolution in the mathematical arts.[279]

I have dealt in some detail with the question of number representation to show that even this subject, which to modern eyes may appear rather simple, can become fraught with ambiguities and lead to conceptual and philosophical difficulties. In the course of the eleventh century, as geometrical imagination came to play an increasing role in mathematics, and especially in astronomy, the situation became even more complex. With the rise of geometry, the problem of the relationship between discrete and continuous quantities (i.e. between 'numerus' and 'mensura'), which had often been discussed in Antiquity, became actual again.

A typical effect of confusions at this level was the attempt to use the geometrical features of numbers derived from Boethian proportions to compute the area of polygons. For example, in a letter Gerbert of Aurillac had to explain to his correspondent Adelbold why, when computing the area of an equilateral triangle, one obtained different results according to whether one used (as done today) the geometrical rule (i.e. length of side times height, divided by two) or the arithmetical one, in which the 'area' of the triangle of given integer side length is considered equal to the triangular number with side equal to that integer.[280] With the arithmetical rule, a triangle of side three would have area six:

side: · · ·   'arithmetical' area: · ·
                                                 · · ·

---

277 The letters used in this type of notation by the Greeks, the Arabs and on the astrolabe **#3042** are summarized in Table 1of Kunitzsch (1991/1992), on which the following discussion is largely based. On these engravings see also Beaujouan (1972) p. 656–660, King (1995) p. 371–372. On the astrolabe **#3042** see 3.2.5.

278 Kunitzsch (2003), especially p. 11–12.

279 A particularly interesting case of employment on instruments of a specific kind of number notation in the context of Latin medieval monastic culture is studied in detail in King (2001).

280 Bubnov (1899) p. 43–45. With square numbers the two rules coincide.

In general, for a side N, the 'arithmetical' area would be N(N+1)/2. This subject is relevant for the study of early astrolabe texts, because the attempt of making astrolabe knowledge readable was often coupled with an effort to express and manipulate the geometrical quantities linked to it in the available arithmetical forms, for example, number proportions, as we shall see later on.

4.1.7 Conclusions: quadrivium, memory and manual crafts

The methods used in high medieval mathematical arts resembled those employed in the craft of memory. They also had a similar aim: developing in the student the capability for systematically manipulating information and creatively producing geometrical constructions, numerical computations or musical compositions.

Historians of science have often considered medieval praxis-oriented methods as evidence of a 'practical tradition' in which the mathematical arts were seen as materially useful tools for rather simple everyday tasks.[281] In manuals on the history of mathematics, the medieval quadrivium is dealt with, if at all, in a few pages sandwiched between the description of ancient Greek works and their resurrection in early modern Europe. The standard view is that only with the advent of twelfth-century translations of Greek and Arabic works a real interest in mathematics as theory could develop.[282] However, the beginning of a distinction between 'theoretical' and 'practical' tradition in mathematics cannot be traced further back than the twelfth century: the 'practical' tradition started existing only when, thanks to the new translations, a 'theoretical' one appeared.[283] Even then, the two disciplines remained largely disconnected from each other, and their mutual relationship was very different from that existing today between pure mathematics and its applications.

I believe it would be more appropriate to consider tenth- and eleventh-century mathematics as a type of mathematical thinking deeply linked to non-verbal, non-written modes of communication, but as such not necessarily less speculative, abstract and creative than modern pure mathematics. Seeing the teaching instruments of the quadrivium in the perspective of the medieval memory-craft offers a clue as to how mathematical, philosophical and theological reflections could take

---

281 For example: Mahoney (1987) p. 203–208. An attempt to partially rectify this bias is L'Huillier (1994).

282 For example, in Ivor Grattan-Guinness' history of the mathematical sciences, one chapter is dedicated to: "A quiet millennium: from the early Middle Ages into the European Renaissance". It mostly deals with Arabic and Renaissance mathematics, with one paragraph addressing the question: "European mathematics in the early Middle Ages?" (with question mark!) and the following one devoted to "The awakening of Europe from the twelfth century onward" (Grattan-Guinnes (1997) p. 127–134).

283 On the relationship between 'theory' and 'practice' in medieval and early modern mathematics, see: Beaujouan (1974), L'Huillier (1994).

a outwardly practical form. It may also help explain why high medieval texts often describe procedures that are "unworkable because they are impractical or too imprecise".[284] At the same time, though, the connection between medieval mathematics and experience was not purely formal: the use of non-written and non-verbal modes of communication could bring the mathematical arts nearer to the mechanical ones.[285] There are no reasons to believe that medieval mathematicians never noticed or cared whether a procedure was workable or not, even if written sources do not tell us much about it. After all, as remarked by Jeffrey Hamburger, 'buildings also had to stand', and it is plausible that those patterns that contributed to a successful architectural project could be regarded as having a particular natural philosophical significance.

The methods of the craft of memory provided a context in which not only patterns of thought could find embodiment in artefacts, but in which, at the same time, artefacts or diagrams could become patterns to guide philosophical reflection at all levels. Already in late ancient Neoplatonism, the mathematical arts were understood as a premise to the study of philosophy, though not because of their use in practice, but as a model of axiomatic-deductive knowledge. From the tenth century onward, instead, an interplay between the experience of mathematical practice and philosophical reflection took place and this change was closely linked to the employment of methods for storing and transmitting mathematical knowledge that were radically different from the verbal, axiomatic-deductive formalism of Greek mathematics. While axiomatic-deductive mathematics could be helpful in grasping the neoplatonic idea of emanation of the many from the one, manipulating arithmetical ratios could provide new interpretations of platonic and neoplatonic cosmology, and experimenting with geometrical and architectonical imagination could lead to a conception of the world as a 'machina mundi' and God as its rational artifex.[286] I shall return to this subject in the fourth part of this study. In the following pages, I will offer evidence that non-verbal and non-written methods of knowledge transfer played a role in the transmission of astrolabe knowledge in high medieval Europe.

4.2 MEMORY AND THE ASTROLABE

4.2.1 Memory in astrolabe-related texts

In eleventh-century astrolabe texts, there are often references to memory and they are very explicit: the reader is told to keep well in memory all he has learned or will learn from the text. In the text on astrolabe construction I indicated as **h1**, at the end of some sets of explanations and before going on to another subject, the

---

284 L'Huillier (1994) p. 189–190.
285 Boehm (1993) p. 432–439.
286 Evans (1977b), M. Evans (1980), Molland (1983), Klinkenberg (1986), Beaujouan (1991) p. 481–484, Molland (1996) especially p. 14–15.

author writes "hoc corde fixe retento"[287] or "memorie thesauris efficaciter recon-
ditis premissis descriptiunculis".[288] In the text I indicate as **pt**, the author even
explicitly states that he has to recall from memory the instructions for drawing
altitude circles: "His lucide ut reor designatis, calamum in cordis arcula necesse
est tinguere, si almucantarath aggrediar describere".[289]

The text **J'** on astrolabe use begins by stating that memorizing the structure of
the astrolabe was necessary to whoever wanted to learn: "[...] necesse est ut ista
alzafea, id est tabulam, cognoscat et bene intelligat, et memoriter teneat nomina
laborum vel laboratorum in ipsa.".[290] A little later, before a long, detailed list of
the parts of the astrolabe and of their Arabic names, the author repeats that it is
necessary for the reader to study with attention the names and further things written below and, once they have been studied, to keep them firmly 'in cartula
cordis' ('in the notebook of the heart'), because, once he will have understood this
text and put it firmly in his mind, "the road to finding all previously listed things
[i.e. various astronomical information] shall lay open before him" ("plana erit sibi
via ad invenienda praetaxata").[291] Astrolabe knowledge was, in a way, a new 'ars
inveniendi' to be added to the mental repertory of monastic scholars. At the end of
the same list, we find once more the same advice:

> Hec sunt nomina que necessaria sunt memoriter teneat, ut iam diximus, in capite huius
> textus. Et qui disciplinatus vult fieri in ipsa tabula, debet scire et intelligere ea nomina
> sicut iam explanavimus, et dum ea scierit quanta est utilitas illorum ipse probabit.[292]

I translate this passage as:

> These are all the names that it is necessary for you to remember, as already said at the
> beginning of this text. And whoever wants to be instructed in this flat device ('tabula')
> will have to know and understand these names, as already explained, and once he knows
> them, he shall prove for himself how many their possible uses are.

These references to memory are not just vague invitations not to forget what one
has learned, but specific orders to memorize a long list of complex Arabic terms
and the corresponding elements of the astrolabe.

Admonitions to keep in memory all things said also appear constantly in the
text **J**. There, we also find a reference to the fact that much of what has been said
can only be understood and learned by repeatedly trying it out in first person:

> Sed haec res tectis innexionibus intricata est nec omnibus ad intelligendum enodabilis
> praeter eos, qui usuales in his habent exercitationem. Si quis autem dicta aut dicenda pa-

---

287 "having firmly stored this in memory" (Millás Vallicrosa (1931) p. 298).
288 "having aptly hidden away the previous short descriptions in the treasure chest of memory"
    (Millás Vallicrosa (1931) p. 299).
289 "Now that, as I believe, I have sketched these things in a clear way, it is necessary to dip my
    pen in the ark of memory, if I have to begin describing the altitude cirles" (Millás Vallicrosa
    (1931) p. 323).
290 Millás Vallicrosa (1931) p. 275–276; my translation is "he has to know and understand well
    this alzafea, or flat device ('tabula') and to keep in memory the names of its workings."
291 Millás Vallicrosa (1931) p. 277–276.
292 Millás Vallicrosa (1931) p. 279.

rum ex scriptis intellexerit, ipsam rem probet et procul dubio reperiet satis diligenter omnia digesta esse.[293]

My translation of this passage is:

However, this subject is a complex texture of hidden intricacies and it does not extricate to the understanding of anyone other than those who have familiar exercise in these things. If someone does not understand from the written words what has been or shall be said, he should try it out for himself, and, without doubt, with enough diligence he shall find that all things have been properly digested.

4.2.2 Note-taking as a complement to memory

Arabic star names and their coordinates must have been particularly taxing for the memory of non-Arabic-speaking monks. On the other hand, such information could be quite easily fixed in writing, and it is thus not surprising that the oldest astrolabe manuscripts contain a list of Arabic star names with their coordinates, which was later included in longer texts on astrolabe construction.[294] This table is of a particular kind, classified by Paul Kunitzsch as 'type III'.[295] Star tables of type III have no direct astronomical application, but can be used to draw stars on the rete of a planispheric astrolabe with only the help of a ruler. In other words, such tables would have been useless for all purposes, except drawing and/or constructing the rete of an astrolabe.

It is probable that the star table was one of the first elements of astrolabe knowledge to be fixed in writing and later copied in longer manuscripts. Originally, the names and coordinates of the stars might have been fixed on a scrap of parchment ('schedula') or on wood, as often done in medieval times to aide memory.[296] Similar hypotheses can be made in the case of tables of solar longitudes, climate latitudes or Arabic astronomical terms. In general, note-taking during oral lectures may have played a key role in the production of the earliest Latin astrolabe texts. In high medieval times, the practice of note-taking ('reportationes') was already diffused: notes were taken on wax-tablets in sketchy form during the lecture, in which the teacher spoke at normal speed, and could be later filled in from memory and copied on parchment. Another method for taking notes was to use small scraps of parchment, which the student could then carry around in a leather wallet, like wax tablets. This method was used also to note geometrical drawings and astronomical diagrams and would have been very convenient to store construction drawings of the astrolabe, such as those discussed in section 4.4. I will argue that some of the earliest Latin astrolabe texts describing geometrical constructions might have had as a basis not an Arabic text, but drawings and notes taken from a (possibly bilingual) lecture.

---

293 Bubnov (1899) p. 123.
294 On Latin star tables and their possible Arabic sources, see: Kunitzsch (1966), Samsó (2000).
295 Kunitzsch (1966) p. 23–30. On star tables and astrolabe stars, see also Kunitzsch (2005).
296 The following discussion on note-taking in the Middle Ages is based on: Burnett (1995–96).

4.2.3 Fulbert of Chartres' mnemonic verses and the letters of Radulf of Liège and Regimbold of Cologne

Around the year 1000, the scholar Fulbert was bishop of Chartres and teacher in the cathedral school. Some manuscripts containing his works also contain a list of Arabic names of astrolabe parts, a list of names of astrolabe stars with the zodiac signs they belong to and a short poem providing the same information as the star list in verse form: a mnemonic verse.[297]

It has generally been assumed that Fulbert composed the mnemonic verse using the star list as a basis. Charles Burnett has noted that the names and positions of the stars in the list and in the verse correspond to the information contained in one chapter of the text **J'**.[298] Burnett formulated the theory that that part of **J'** was known to Fulbert, who used it as a basis for the list and the verse. According to Burnett, other astrolabe texts were also known to Fulbert, who used them to compile the list of Arabic terms for astrolabe parts. However, the opposite could also be true: astrolabe knowledge might have reached Chartres by way of memory, mnemonic verse, or notes on scraps of parchment, to be later fixed in manuscripts or eventually used to compose passages in Latin prose. Thus, Fulbert might not have been the author of the mnemonic verse, but rather the one who decided to fix it into writing.

A final piece of evidence that memory and not the written word might have been the main channel of transmission of astrolabe knowledge in the early eleventh century is the much-quoted statement occurring in the correspondence between Regimbold of Cologne and Radulf of Liège. These two scholars, around the year 1025, exchanged letters discussing advanced – and quite abstract – mathematical questions, among other things the meaning of the expression 'external angle' in Euclid's writings and the sum of the internal angles of a triangle.[299] Commenting on their practical approach to mathematics, John Heilbron writes:

> Neither he [i.e. Regimbold] nor Raoul [i.e. Radulf] had the slightest conception of a geometrical proof. Raoul did manage to show that the sum of the interior angles is a straight angle in the special case of a triangle with two equal sides, formed by drawing the diagonal of a square. But the general case eluded him. All he could suggest was either to declare the proposition true by intuition or to draw a triangle on parchment, cut out its angles, and place them together to form a straight angle.[300]

---

297 For an edition and discussion of these texts, as well as for a list of the manuscripts in which they appear, see McVaugh/Behrends (1971). A further edition and translation of the texts can be found in Fulbert (1976).

298 Burnett (1998) p. 335.

299 Tannery (1922), especially p. 240–246.

300 Heilbron (1998) p. 72. Actually, it was Regimbold who proposed to cut the angles out of the triangle: "Quod si adhuc te (latet) nostrae veritas expositionis, cum circini probatione vel proportionali membranam incisione cuncta quae dicimus vera esse poteri comprobare." (Tannery (1922) p. 283), i.e.: "Should the thruhfullness of our explanation not yet be evident to you, you will be able to check that everything we said is true by trying it out with a pair of compasses or by cutting up a piece of parchment according to the [given] proportions."

In one of his letters, Radulf answered a non-extant question of Regimbold concerning an 'astrolabium':

> Astrolabium misissem vobis iudicandum, sed est nobis exemplar ad aliud construendum: cuius de scientia si quid affectatis, ad missam sancti LAN(BERTI) non vos pigeat advenire. Forsitan non penitebit: alioquin videre tantummodo astrolabium non magis iuvabit quam 'lippum pictae tabulae, fomenta podagrum'.[301]

I translate this passage as:

> I would send you the astrolabe so that you can form an opinion about it, but it is our model to make another one: if you are interested in knowing about it, come to the fair of St. Lambert. Maybe you will not regret it: simply seeing an astrolabe would do you no more good than 'a picture to a blind man or a warm compress to someone with gout'.

Whatever Radulf and Regimbold's 'astrolabe' might have been (Arabic artefact? Latin artefact? drawing? device of wood?), one thing is clear: the object itself would not have been enough to transmit knowledge. To really learn something about the astrolabe, explanations and examples would have been necessary. Apparently, those explanations were not available in written form: at least, neither had Regimbold asked for them, nor had Radulf offered them.

4.3 DRAWINGS AS A MODE OF TRANSMISSION OF ASTRONOMICAL KNOWLEDGE

4.3.1 Astronomy in number and/or in measure

Early medieval astronomy, as often noted, was a discipline rather limited in its scopes and methods.[302] In particular, geometrical techniques for performing astronomical computations had been forgotten in the Latin West. Up to the late tenth century, the main tool of quadrivium arts remained the manipulation of (natural) numbers, and astronomy was no exception: arithmetic was the basis of computus, i.e. of the techniques for determining the date of Easter. When Latin scholars started taking interest in the motion of the planets, they had no geometrical models or astronomical tables, and developed the purely arithmetical technique of the 'years of the world'.[303] Arithmetic also dominated the patterns of mathematical thought employed in natural philosophy and theology. In astronomy, following the teachings of Plato's Timaeus and its commentary by Calcidius, Boethian number proportions provided a guide for understanding cosmic harmony.

However, although number manipulation remained a central issue in medieval mathematics for a long time, in the ninth century geometrical structures already started playing a role in mathematical studies, and astronomy largely profited

---

301 Tannery (1922) p. 283. The last words are a quote from the Roman poet Horace.
302 For an overview on early and high medieval astronomy, see: McCluskey (1998), Eastwood (1997), Wiesenbach (1991). On early and high medieval computus Borst (1999) p. 29–79; McCluskey (2003).
303 Juste (2004).

from this tendency. A first indication of this trend was the inclusion in Carolingian computistic manuals of passages taken from the late ancient authorities which dealt with the geometrical structure of the cosmos, describing the celestial spheres, their movements and their dimensions.[304]

4.3.2 Diagrams as an independent mode of communication

In high medieval astronomic compilations, diagrams came to be used with increasing frequency to convey information that complemented and clarified the text.[305] This subject has been studied in detail by Bruce Eastwood, who has analysed the development of astronomical diagrams from the ninth century onward, in particular those related to planetary astronomy.

A central result of Eastwood's investigations has been to show that, already in the ninth century, diagrams could be created or adapted from late ancient models to clarify the text or even provide additional information. An example of this phenomenon were the diagrams traditionally illustrating Calcidius' commentary to Plato's 'Timaeus'.[306] Some diagrams which had been transmitted from the Early Middle Ages in corrupt form were substituted with new pictures correctly conveying the astronomical information provided by the text. Sometimes the scribe also complemented the image with verbal explanations. The new diagram and its new accompanying text could eventually be copied independently from each other, so that the presence of a corrupt drawing did not necessarily imply ignorance on the part of the scribe or of the readers. After discussing a particularly interesting manuscript, Eastwood reaches the following conclusion:

> [This analysis] suggests a process of evaluation and reconsideration that need not have been unique to this manuscript. In other manuscripts with corrupt diagrams [...] the process would simply have taken place outside the manuscript, possibly in the dust on the ground as casual commentators have noted, on a renewable dusted board as Martianus Capella recorded, or perhaps incised on a waxed surface or chalked on a wooden surface.[307]

Thus, new drawings might be transmitted in the form of notes before being finally collected in a manuscript compilation devoted to a particular subject.

In his review of medieval technical literature, Bernhad Bischoff describes one tenth-century manuscript (Schlettstadt 360 (1153bis)) containing, along with excerpts from Vitruvius, also drawings of columns and other architectonic elements which are not described by Vitruvius and which are accompanied by names that

---

304 Eastwood (1997) p. 236–250, McCluskey (1998) p. 114–140.
305 A large number of examples of diagrams used in ancient and medieval learning can be found in: Murdoch (1984). An overview on diagrams as medieval modes of communication: Stevens (1993).
306 The following exposition is based on: Eastwood (1995) p. 225–226, Eastwood (1997) p. 251–253 and Eastwood (1999) p. 178–194.
307 Eastwood (1999) p. 190.

Vitruvius never employed.[308] Bischoff suggests that the drawings were taken from some late ancient source other than Vitruvius: again, a source now lost. Yet in this case, too, the drawings might have been a means of knowledge storage and transfer independent from the written word.

Diagrams could also be stored as psychomotoric mnemotechnical patterns, when one learned to draw them according to a standard procedure: advanced learners might thus have been able to reproduce them when necessary from memory – possibly with errors. For the Late Middle Ages and the Renaissance, there is evidence that techniques employed in architecture and mechanics could be transmitted by means of drawings which, although they had their origin in Euclid's 'Elements', had for a long time been separated from their context and used as a basis for geometrical reflections and experiments aimed at developing new techniques.[309] These figures were stored in the memory of artists, architects and mathematicians and could easily be recalled and referred to thanks to their colourful names, e.g. 'pons asinorum', 'cauda pavonis', 'pax et concordia' or 'figura demonis'. Interestingly, some of these names were of Arabic origin.[310]

In his studies on planetary diagrams, Eastwood has also brought attention to the fact that "exact copying of text was a much more successful art in the 9th to 11th centuries than exact copying of involved linear diagrams".[311] This should not be surprising, given that copying texts (even without understanding them) was a task for which European monks had been trained for centuries, whereas diagrams were a newly developed medium. This may explain why in astrolabe manuscripts sometime spaces are left blank with the obvious intention of filling them up later with drawings: any common scribe could copy the text, but for the diagram a specialist was needed, and was probably not always available. Because of this, the few manuscripts containing drawings of good quality are of particular interest: one of them is BnF lat. 7412 (A.27), which will be studied in some detail in the fourth part of this work.

4.3.3 Stereographic projection in ninth-century astronomic diagrams

As far as the astrolabe is concerned, very significant examples of planetary diagrams are those illustrating the passages from Pliny the Elder's (23/24–79) 'Naturalis historia' describing planetary latitudes, i.e. the varying distance of a planet from the ecliptic (i.e. zodiac) plane.[312] Following Pliny, the zodiac was thought of as a circular band divided into twelve narrow bands, each one degree wide. In the diagrams, these bands are represented as concentric circles, while the orbits of the planets are wavy circles, extending over various degrees as described by Pliny.

308 Bischoff (1971) p. 275–277 and fig. 1–3.
309 On Euclid's figures, their names and their use in late medieval and early modern times: Beaujouan (1974), especially p. 447–458.
310 Kunitzsch (1993b).
311 Eastwood (1995) p. 225.
312 The following discussion is based on: Eastwood (1995) and Eastwood (1997) p. 238–245.

These drawings are particularly interesting for our subject, because they employ some kind of stereographic projection. As we have seen, the tropics become concentric circles when projected onto the equatorial plane: in an analogous way, circles parallel to the zodiac become concentric circles when projected onto the zodiac plane, while the orbits of the planets become wavy circles.

Since these diagrams date back to the ninth century, they cannot be due to the influence of Latin astrolabe literature. Eastwood suggests that knowledge of stereographic projection might have reached Carolingian Europe through "oral tradition and practical instructions", probably thanks to the contacts of the Carolingian court with the Byzantine empire.[313] However, an independent Latin origin of the diagrams is also possible, eventually inspired by remains of Roman astronomical devices employing stereographic projection, in particular anaphoric clocks.[314]

Perhaps, it may not even be necessary to postulate exact knowledge of stereographic projection to explain the structure of the diagrams: rendering the twelve degrees of the zodiac as concentric circles on a plane was a solution not dissimilar, for example, from representing a sighting tube as a tube seen from the side with, at one end, an empty circle, i.e. its aperture as seen from the front.[315] In fact, the twelve circles are usually represented as equidistant, which would not be the case if they were a product of stereographic projection.[316] In a stereographic projection, the circles should grow increasingly distant as they become larger, and it is very interesting to note that this pattern appears only in one manuscript which was written in the eleventh century, i.e. in a time in which the astrolabe had already become known in Latin Europe.[317]

Ninth-century planetary latitude diagrams could be a Latin creation due to the increasing use of geometrical imagination in astronomy. Another type of diagrams testifying to this development are the Carolingian representations of the three continents of earth as portions of a circle.[318] These diagrams are found in two versions, one reversed with respect to the other, and this has been tentatively explained by assuming that astronomers imagined the celestial and the earthly sphere in three dimensions and then, according to the perspective from which they looked at it in their minds, drew on parchment two-dimensional opposite images of the continents.

---

313 Eastwood (1995) p. 222.
314 North (1975). On the anaphoric clock, see 3.1.1.
315 See 5.4.4.
316 Eastwood (1995) fig. 4,5,6,8 and 9.
317 Eastwood (1995) fig. 7.
318 On these diagrams and their possible origin see: Stevens (1993) p. 20–28.

4.3.4 The astronomical diagrams of Abbo of Fleury and Byrtferth of Ramsey; the astronomical devices of Gerbert of Aurillac

According to our sources, two prominent figures of medieval learning took particular interest in developing the geometrical imagination of their students: Abbo of Fleury and Gerbert of Aurillac. Both flourished in the last decades of the tenth century, both played a role in contemporary politics as well as in scholarly life and both were later remembered with great awe by their respective students. Abbo of Fleury used late ancient and Carolingian models to develop new diagrams complementing his astronomic writings. Abbo's example was followed with particular enthusiasm by his English student Byrhtferth of Ramsey, whom Abbo taught during his stay in the English monastery of Ramsey.[319] Among other things, Byrhtferth composed a computistic collection in vernacular which was complemented by a large number of cosmological diagrams. In the computus of Abbo and Byrhtferth, number was at the core of the understanding of nature.[320] It may be added that, although number was the key to understanding nature, the relationship between 'number' and nature was imagined and represented in colourful drawings combining geometric figures and written text. In the analysis of some passages contained in the astrolabe manuscript BnF lat. 7412 (A.27), the influence of Abbo's and Byrtferth's work will be discussed.[321]

Gerbert of Aurillac is remembered among other things for the devices he built and employed in teaching astronomy.[322] The devices are described in his biography, and one of them, the hemisphere with sighting tubes, is also described in one of his letters: later on, I shall discuss drawings of this device which appear in astrolabe manuscripts.[323] Another one of Gerbert's teaching instruments was a celestial sphere provided with a horizon ring: in a recent paper, Marco Zuccato has argued that this instrument, and probably also the other ones, were build by Gerbert following the example of Arabic teaching methods which he had come into contact with during his stay in al-Andalus.[324]

Gerbert himself never wrote treatises describing his astronomical instruments and the way in which he used them. Discussing the role of orality and literacy in medieval teaching, Rosamund McKitterick remarks how Gerbert's letter on the

---

319 On Abbo and Byrtferth: McCluskey (1998) p. 152–157. On Abbo's approach to the mathematical arts: Evans/Peden (1985). On Byrtferth's works, see Hart (2003), in particular Hart (2003) vol. 2 book 2 p. 520–522 for a short sketch of Byrtferth's personality and works.

320 McCluskey (1998) p. 154–157, and fig. 24.

321 See 5.3.1–2, 5.3.4.

322 The devices are described in Richer's 'Historia': Richer (1967 and 1964) vol. 2 p. 50–54. An edition of the text with translation and commentary can be found in Sot (1996), an extended commentary in Poulle (1985). Some interesting remarks on Gerbert's teaching style in relation to the problem of orality and literacy in high medieval Europe can be found in McKitterick (2000) p. 26–28. On Gerbert's mathematical work, see also Lindgren (1976), who offers a very good overview of extant sources.

323 Gerbert's letter is edited in Bubnov (1899) p. 25–28. For the discussion of the drawings, see 5.4.4.

324 Zuccato (2005). I shall discuss the details of Zuccato's thesis in par. 5.3.5.

hemisphere testifies to the difficulties of transferring knowledge "from one medium of explanation to another".[325]

4.4 DRAWINGS AND KNOWLEDGE TRANSMISSION:
THE CONSTRUCTION OF TROPICS AND ZODIAC

4.4.1 The sources: texts **h1**, **h2** and **h3** and drawings **d1** and **d2**

Most of the oldest Latin astrolabe manuscripts contain anonymous texts explaining how to draw astrolabe lines according to the method discussed in 2.2.3. These
texts show how to draw equator, tropics and zodiac and usually also how to construct a chosen horizon and its corresponding altitude circles. In some manuscripts, the two geometric constructions are also shown in two labelled drawings,
which I shall indicate as **d1** and **d2** (fig. 18).

In the following pages, I will compare the procedure used in the drawing **d1**
to construct tropics and zodiac with that described in the three most diffused early
Latin texts on astrolabe construction, which I shall indicate as **h1**, **h2** and **h3**.[326] I
will show that the instructions given in all three texts and the construction shown
in the drawing **d1** follow exactly the same pattern, even in details which could be
freely chosen by the geometer. Moreover, both texts and drawings employ exactly
the same labels to indicate specific points in the drawing.

While the details of the geometrical construction coincide, the strictly verbal
correspondence between the three texts **h1**, **h2** and **h3** is very limited. Historians
have tried to reconstruct from them a chain of transmission and, in this context, **h2**
is usually regarded as the earliest text, possibly a translation from the Arabic. The
text **h1** is seen as its reworking and the text **h3** as its summary. I will instead argue that the only common element between these texts is information which could
have been transmitted by memorizing a particular procedure for constructing labelled drawings such as **d1**. The three texts were in my opinion not directly linked
to each other, but were rather parallel attempts at fixing in writing knowledge that
had spread thanks to non-written and non-verbal strategies (memory, drawings,
models). Later on, in section 4.6, I shall argue that the origin of the knowledge so
transmitted can be sought in the Arabic version of Ptolemy's 'Planisphaerium'.

The texts and drawings to be discussed are described below; the manuscript
tradition of **h2**, **h3**, **d1** and **d2** (not of **h1**) is summarized in table 1;

h1: the text **h1** is the first part of the passage edited by Millás Vallicrosa under
the title **h'**.[327] I have not examined in detail the manuscript tradition of **h1**, but

---

325 On Gerbert's teaching methods: McKitterick (2000) p. 26–28. The quote is from McKitterick
    2000) p. 27. For further references on Gerbert, see 3.1.4.
326 On the manuscript tradition and printed versions of these texts, see app. C and table 1.
327 Millás Vallicrosa (1931) p. 296–300, i.e. from the beginning of **h'** up to (but without) the
    star table.

there are indications that it, too, is made up of originally independent parts which were joined at a later stage.[328] In **h1**, instructions are given to draw equator, tropics, zodiac, horizons and altitude circles.

h2: the text **h2** corresponds to the first part of Millás Vallicrosa's text **h"**.[329] The text **h2** appears in 12 manuscripts, in 6 of which it is associated with the two construction drawings **d1** and **d2**.[330] The text **h2** explains how to draw on the flat sphere the equator, tropics, zodiac, any horizon and its altitude circles. Although **h2** has a much more unitary character than **h1**, it, too, could be of composite nature.[331]

h3: the text **h3** is very short and is sometimes referred to as a 'fragment'. It was printed by Millás Vallicrosa as a variant of **h"**.[332] It occurs in 10 manuscripts and only describes how to draw the equator and tropics.

d1 and **d2**: the drawings **d1** and **d2** are associated to **h2** in 6 manuscripts. They have a very stable form, shown in fig. 18: they always appear with the same structure, labels and even orientation. As far as **d1** is concerned, the only deviation from the norm occurs in the manuscripts BnF lat. 7412 (A.27) and BnF lat. 11248 (A.28) in which an additional line appears. The implications of the presence of this line shall be discussed in 4.5.6.

4.4.2 Relationship between the strictly verbal contents of **h1**, **h2** and **h3**

As far as the strictly verbal content is concerned, the texts **h1**, **h2** and **h3** are not related in any obvious way to each other: on the contrary, they are written in very different styles. The text **h1** is (or tries to be) literarily accomplished and alternates drawing instructions with philosophical considerations. It is usually regarded as a rather poor attempt at teaching how to construct an astrolabe and it is anything but clear.

The low level of readability of **h1** by modern standards is due both to the composite nature of the text and to the author's tendency to philosophise. As discussed in 4.4.6, though, one should not be too fast in dismissing the author or authors of **h1** as mathematically incompetent. The text **h2** leaves very little space to philosophical speculations and concentrates on describing the drawing procedure in a compact but clear style. It lists all labels of the points drawn, even when they

---

328 For example, in Oxf. CCC 283 (A. 25), on f. 81v–85r, **h'** occurs without the long philosophical prologue. The whole structure of **h1** is composite, with instructions sometimes repeated and sometimes interrupted by other considerations.

329 Millás Vallicrosa (1931) from p. 293 line 1 of text III to p. 295 line 45 of the same text, i.e. excluding the final part on the division of the Zodiac.

330 It must be noted that Millás Vallicrosa, in his edition of **h"**, considered **h3** as identical with it, indicating that Ripoll 225 (A.2), Leid. Scal. 38 (A.13) and Vat. Reg. lat. 598 (A.43) contain **h"**, while they in fact contain only **h3** (Millás Vallicrosa (1931) p. 293).

331 For example, sometimes **h2** occurs without its initial lines on the wise philosophers who invented the astrolabe and starts directly with drawing instructions.

332 Millás Vallicrosa (1931) p. 294, note.

are not necessary as references for the construction. In fact, the labels are used to identify elements in the figure: for example, the tropic of Cancer is "circulum inferiorem, i.e. i, k, l, m", while the zodiac is "circulum casarum, id est circulum o, d, i, b".[333] Werner Bergmann notes this habit and considers it a peculiar fixation ('Marotte') of the author, but it is in fact a quite efficient formalism to help the reader follow the construction performed.[334] Finally, **h3**'s verbal component only consists of the few words necessary to sketch the construction procedure.

4.4.3 The construction of tropics and zodiac
in the text **h2** and in the drawing **d1**

All early Latin texts explaining how to draw astrolabe lines describe the procedure summarized in 2.2.15. However, there are in principle a large number of equivalent ways of performing it: one may start by drawing one of the tropics instead of the equator, or choose to cut the arcs defining the circle to be projected in a clockwise or counterclockwise direction. Moreover, even if the drawing procedure is exactly the same, the final drawing can be labelled in different ways and can be rotated at will.

Thus, geometers had in principle much freedom of choice in drawing astrolabe lines. For example, although it is possible to draw both the zodiac and any horizon circle by performing exactly the same steps, the texts **h1** and **h2** and the drawings **d1** and **d2** make use of two different construction procedures: one for the zodiac and the other for the horizon.[335]

As we shall see, the construction procedures described in **h2**, **h3**, **d1** and **d2** coincide in every detail, and those of the text **h1** are very near to them in particularly significant aspects as, for example, the names chosen for labelling the drawings. I will show this explicitly only for the first construction, i.e. that for tropics and zodiac, which is also the only one present in **h3**. I start by discussing **h2**, which is the most readable text by modern standards. The Latin text is:

> Philosophi qui sua sapientia motus siderum cunctaque firmamenti officia inuenere, et inuenta luculentissime et perfectissime scribentes posteris tradidere, quadam similitudinem ex eraminbus construxere per quam facilior existeret intelligentia. Pro qua re astrolapsum inuenere, cuius imaginarias, insequenibus videbis figuras. In quo tres in primis scribebant circulos, quos subtus iacenti aspicis formula. Primum et maiorem circuli capricornii nominantes, II libre et carnarii, III circulum cancri. Quorum medium primitus invenerunt et fecerunt, et inventum per IIII partes dividebant in a, b, c, d, ducentes lineas quas nominabant alcoter, de a usque c et de b usque d, singulis partibus XC donantes et totum ciculum in CCCLX dividebant. Postea de c, b, id est XC partibus, XXIIII abstrahebant, quod est c, e, et

---

333 Millás Vallicrosa (1931) p. 294.
334 Bergmann (1985) p. 113 and p. 115.
335 This can be clearly seen in fig. 18: in **d1**, the points determining the position of the circle to be projected are taken one above and one below the same point of the equator, while in **d2** they are taken on opposite sides of it.

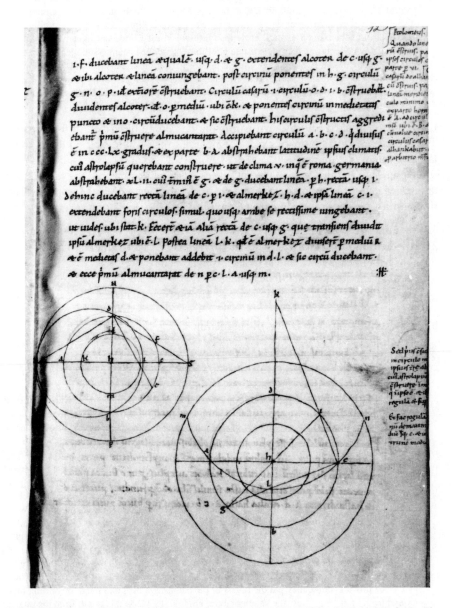

Figure 18: Drawings **d1** (left) and **d2** (right) from the manuscript BnF lat. 7412 (A.27), f. 12r. The image is reproduced with the permission of the BnF from the microfilm stored in the Microfilm Archive of Medieval Scientific manuscripts (Munich University)

de ipso termino, i.e. e aequaliter lineam ducebant, vel filum usque d, que linea transiens truncat alcoter, ex parte c, ubi est i. Quo facto ponebant circinum in h, i, et sic circumdicentes circulum inferiorem, id est i, k, l, m, exarabant. Dehinc ex parte c, d, similiter XXIIII abstrahebant quod est c, f, et de ipso termino id est f ducebant lineam equalem usque d g extendentes alcoter, de c usque g, et ibi alchoter et lineam coniungebant. Post circinum ponentes in h g circulum g, n, o, p, id est exteriorem construebant. Circulum casarum, id est circulum o, d, i, b construebant, dividentes alchoter .i.o. per medium ubi est k, et ponentes circinum in medietatis puncto et in o et in i circuducebant et sic construebant. [336]

I will now summarize in my own words the procedure described in this passage. The accuracy of my summary can be compared to the (more or less) literal translation given in the footnotes.

Describing in third person the actions of 'wise philosophers', the text **h2** tells the reader that three circles have to be drawn: the largest one is the tropic of Capricorn, the second one the equator and the third one the tropic of Cancer:

1) first draw the equator circle;
2) divide it into four quarters, whose extremes are A, B, C and D;
3) join with lines A to C, and B to D;
4) divide each quarter into 90 degrees;
5) starting from C and going towards B, cut on the equator an arc CE of 24 degrees;
6) draw a line from E to D: it will cut the line CA in the point I;

336 Millás Vallicrosa (1931) p. 293–294. My translation is: "The philosophers who, with their wisdom, found out the movements of celestial bodies and all of the duties of the skies, and, by writing down what they had found out in a most splendid and perfect way, passed it on to the following generations, built out of copper a simulacrum through which understanding would come more easily. To this aim, they found out the astrolabe, whose pictures which help imagination ['figuras imaginarias'] you shall see in the following. In the astrolabe they wrote down first of all three circles, which you can see in the geometrical structure ['formula'] below. The first and largest one they named the circle of Capricorn, the second one, the circle of Libra and Aries, the third one the circle of Cancer. Of these circles, they found and drew the second one first, and once they had found it, they divided it into four parts A, B, C, and D, drawing lines which they named 'alcoter' from A to C and from B to D, giving to each quarter 90 parts, and thus dividing the whole circle into 360 parts. After this, from [the arc] C,B, i.e. from 90 parts, they took 24, which is C, E, and from that extreme, i.e. E, they drew a straight line or a rope up to D, a line which, in passing by, on the side of C cuts the 'alcoter' in I. Once this was done, they put the points of the compasses in H and I and, by turning them around from there, they drew the smaller circle, i.e. I, K, L, M. At this point, from the quarter C, D, in a similar way they took 24 [parts], which is C, F, and from that extreme, i.e. F, they drew a straight line up to D, [and then] G, prolongating the 'alcoter' from C to G, and in that point they joined the 'alcoter' to the line. After this, pointing the compasses in H and G, they built the circle G, N, O, P, which is the external one. They built the circle of the houses [i.e. the zodiac], i.e. the circle O, D, I, B by dividing the 'alcoter' I O in the middle where K is, and by pointing the compasses in the middle point and in O and in I, and they turned the compasses around, and so they constructed it."

7) point the compass in H (i.e. the centre of the equator[337]), with its other foot in I, and turn it around: this circle I, K, L. M will be the tropic of Cancer.[338]

8) starting from C and going towards D, cut on the equator an arc CF of 24 degrees;

9) draw a line from F to D and further on to G, where it cuts the prolongation of the line CA;

10) point the compass in H and G, and turn it around, drawing the circle G, N, O, P, which is the tropic of Capricorn;

11) take the segment IO, find its middle point K (!)[339] and point the compass in it, with its other foot reaching I and O. Turn the compass around and you will build the zodiac.

The value of the inclination of the zodiac is 24°, which is very near to Ptolemy's 23°51' 20". It must be noted that, once the tropics are constructed, drawing the zodiac (i.e. step (11)) is not only simple, but even implicit in the previous construction for someone who has understood it.

The construction described by **h2** corresponds both in pattern and labels to that shown in drawing **d1** (fig. 19a–e) and there can be no doubt that the two are very closely linked to each other. How are they related, exactly, though? One might assume that the drawing **d1** was executed following the instructions given in the text **h2**, but it is also possible that the text **h2** was composed to fix in writing the already known procedure employed to construct **d1**. I believe there are strong indications that the latter was the case. The first indication is the attention devoted by the writer of **h2** to quoting all labels, which in my opinion suggests that his source was not so much another text, but rather the memory of the construction procedure, eventually aided by the presence of the drawing **d1** in front of his eyes.

The dominance of construction drawings over text can also be seen in the fact that, when drawings occur associated with **h2**, they have not only the same structure and labels, but also exactly the same orientation, with A B C D distributed counter-clockwise and D on top. These details are not specified in the text and, since many copies of **h2** lack drawings, one might expect variants to occur, if the drawings were reproduced using only information in the text. Yet this did not happen. The scholars who executed the drawings seemed to known how to orient them, even if the information was not given in the text. It is interesting to note that in BnF lat. 11248 (A.28) the text **h2** and the drawings **d1** and **d2** do not occur together, but at a few pages' distance from each other.[340] In the next section, I

---

337 The text uses the label H without having previously defined it.

338 It is not explicitly stated that the compass has to be turned around H, and not I, which suggests that the reader was supposed to know that equator and tropics were concentric.

339 The label K had already been used previously to indicate one of the extremes of the tropic of Cancer. This ambiguity is found in most manuscripts, an exception being the manuscript Avranches 325 (A.1), in which the letter Q is used to indicate the centre of the zodiac, both in the text and in the accompanying drawing.

340 See table 1.

shall discuss a further indication that the first stage of transmission of the construction procedure did not occur via written text, namely the fact that the same construction pattern and labels as in **h2** and **d1** are also found in the texts **h3** and **h1**, even though these texts are never accompanied by drawings.

4.4.4 The construction of the tropics in **h3**: a product of note-taking?

The text **h3** is not easy to understand, when taken on its own. In most manuscripts, the labels are often confused. I give here the version found in the manuscript Leid. Scal. 38 (A.13) on f. 46r, which I consider one of the less erroneous:

> Primus per arbitrium in CCCLX, id est medius, in IIIIor prius diuisus et unaquaque in XC.a. Secundus, id est minimus, sublatis XXIIII de XC.a, de parte C. B. quod est C. E. et de E. usque D., ducitur linea per I ubi fit diuisio linee orientalis et postea de H. usque I. Tercius, id est maximus, sublatis XXIIII. de XC. de parte C. D. quod est F. et extensio linee de C. usque G. et de G. usque D. ducitur linea. Postea de H. usque F. exeunt fit circumductio.

My translation is:

> The first one at will in 360, i.e. the middle one, having it first divided into four and each one into 90. The second one, i.e. the smallest one, taken away 24 from 90, from part CB, which is CE, and from E up to D a line is drawn through I, where it divides the eastern line and after that from H to I. The third one, i.e. the largest one, taken away 24 from 90, from the part CD, which is F, and the extension of the line from C to G and from G to D a line is drawn. After this, a circle is drawn from H to F.

This text sounds very cryptic, but, once the procedure explained in **h2** is known, it is not difficult to recognize it in these few lines. It is exactly the same, both in the construction pattern and in the labels: first, a circle is drawn (A, B, C, D with centre H); after this, two equal arcs of 24 degrees each are cut on both sides of C; finally, the extremities of these two arcs, respectively E and F, are joined to D. The only difficulty is the last line, in which the point F is indicated instead of the point G: this error was present in all the manuscripts whose microfilm copies I consulted.[341]

The close correspondence between **h2** and **h3** cannot be due to chance. Although it could be explained by assuming that **h3** is a summary of **h2**, as usually done, this is not the only possible solution. Since the correspondence only relates to the pattern and labels of the geometrical construction, it could simply be that the same drawing was at the basis of both **h2** and **h3**.

---

341 The manuscripts I consulted are: Leid. Scal 38 (A.13), Lon. BL Old Roy. 15.B.IX (A.15), Oxf. CCC 283 (A.26), Salz. S.P.a.V. 7 (A.40), Vat. Reg. lat 598 (A.43), Vat. Reg. lat. 1661 (A.44). In some of them, additional errors occur, in particular of the kind in which 'ab C' (i.e.: from the point C) becomes 'A B C' (three labels: A B C).

Figure 19a: BnF lat. 11248 (A.28) f. 26r.

Figure 19b: Oxf. CCC 283 (A.26) f. 108r.

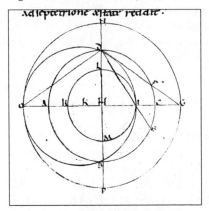

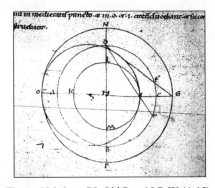

Figure 19d: Lon. BL Old Roy.15.B.IX (A.15) f.75r (© The British Library. All rights reserved).

Figure 19c: BnF lat. 7412 (A.27) f. 12r.

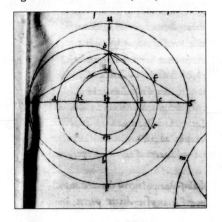

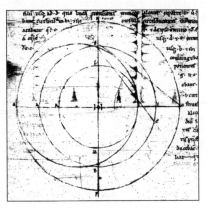

Figure 19e: Salzburg S.P.a.V. 7 (A.40) f. 109r.

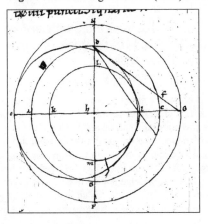

Figure 19a-e: The drawing **d1** as it appears in five different manuscripts. All images are printed from copies in the Microfilm Archive of Medieval Scientific Manuscripts of the University Munich. Permission for the reproduction was given by the relevant institutions: (19a and c) Bibliothèque National de France, Paris; (19b) the President and Fellows of Corpus Christi College, Oxford; (19d) British Library, London; (19e) Library of the Stift St. Peter, Salzburg.

Moreover, the text **h3** can only be understood as notes taken by someone who had already learned the procedure and wanted to fix its salient points in writing. For example, the word 'circle' is always left out, as if taken for granted: the author speaks only of "the first one", "the second one" and so on. Although, in principle, **h3** might be the result of notes taken from a text like **h2**, I believe its compact style rather suggests that it was written down during an oral lecture.

4.4.5 Comparison of the manuscript tradition of **h2** and **h3**

A comparison of the manuscript tradition of **h2** and **h3** (table 1) suggests that they were written independently from each other: in most of the oldest manuscripts, **h3** and **h2** appear separately, suggesting that they had independent origins and were joined together at a later date. Although the diffusion of **h2** is understandably wider than that of **h3**, it is still surprising that such a cryptic text as **h3** was copied so often, and even in the same manuscripts containing **h2**. This could be explained by assuming that it was part of a particularly old and authoritative corpus of Arabic knowledge.

4.4.6 The construction of tropics and zodiac in **h1**

The text **h1** has a more composite nature than **h2** or **h3**, and, for evaluating it in detail, it would be necessary to investigate more closely its manuscript tradition. The passage dealing with the construction of tropics and zodiac is:

> Sit circulus cuiuscumque magnitudinis in plana superficiem descriptus, in quatuorque partes divisus, ab unaquaque in aliam per centrum ductis lineis veluti est A B C D harum unaqueque nonagies seccetur. [...] Sublatis XXIIII de parte C D quod est F ducatur linea ad lineam C que egrediatur usquze G per DF. De hinc circumductio fiat. Quare maior circulus perficitur. Minimus item fit sublatis XXIIII de C B quod sit E ducatur linea E ad D et videatur in quo loco erat per medium linea CA qui locus notetur I, de hinc circino apposito in puncto primo circulus, qui locus notatur H circumscribatur. Quare minimi circuli facta est perspectio.[342]

The instructions given here for constructing equator and tropics are without doubt the same as those in **h2** and **h3**, apart from the fact that, here, the tropic of Capricorn is drawn before the tropic of Cancer, and not vice versa.

---

342 Millás Vallicrosa (1931) p. 29. My translation is: "Let a circle of arbitrary dimensions be drawn on a plane surface, be divided into four parts, and let lines be drawn from each extreme to the other through the centre, or A B C D, of which each shall be subdivided into ninety [...]. Having subtracted 24 from the part CD, which is F, let a line be drawn up to the line C, which arrives up to G through DF. From there, let a circle be drawn. In this way, the largest circle is built. The smallest one is constructed by taking 24 from CB, which is E, and let a line be drawn from E to D, looking where it cuts the line CA, and that point will be called I, from there, having pointed the compasses in the centre of the first circle, which is called H, let a circle be drawn. In this way, the construction of the smallest circle is completed."

Table 1. Comparison between the manuscript traditions of **h2** and **h3**. References to the manuscripts quoted here can be found in appendix A and appendix B under the numbers given in parentheses.

	h3 alone	h3 and h2	h2 alone
11th c.	*Barc. Ripoll 225 (A.2)* **h3**: f. 23r–23v	*Oxford CCC 283 (A.26)* **h3**: f. 89v **h2**: f. 107v–108v **d1**: f. 108r **d2**: f. 108v (**d2** unfinished)	*BnF lat. 11248 (A.28)* **h2**: f. 22v–24v **d1**: f. 26r, **d2**: f. 26v
	Leiden Scal. 38 (A.13) **h3**: f. 46r		*BnF lat.7412 (A.27)* **h2**: f. 11v–12r **d1**: f. 12r **d2**: f. 12r
	Vat. lat. 598 (A.43) **h3**: f. 119v		
11th/12th c.		*Chelt. Phil. 4437 (A.7)* **h3**: f. 53v **h2**: f. 54r–55r	*Zurich C. C 172 (A.46)* **h2**: f. 65v-66v
		Vat. Reg. lat. 1661 (A.44) **h3**: f. 64r **h2**: 79r–79v	
12th c.	*Rostock. philol. 18 (A.38)* **h3**: f. 65r	*BL O. R. 15.B.IX (A.15)* **h3**: f. 70v **h2**: f. 75r–76r **d1**: f. 75r **d2**: f. 75v	*Avranches 235 (A.1)* **h2**: f. 38v–39v **d1**: f. 38v **d2**: f.39v
		Salzburg S.P.a.V. 7 (A.40) **h3**: f.107r **h2**: f. 107r–109r **d1**: f. 109r **d2**: f. 109v	*Göttingen phil. 42 (A.9)* **h2**: f. 13v–14r
			Clm 14763 (A.21) **h2**:f 212r–213r
15t^h/16th c.		*Vat lat. 4539 (B.47)* **h3**: f. 111r **h2**: f. 95v–96v	

Moreover, the same labels for the same points occur in **h1**, **h2** and **h3**: A, B, C, D, H, E, F, I and G. The additional labels found in **h2** and in **d1** do not appear in **h1** and **h3**. As in **h3,** here, too, the writer gives no instructions on how to draw the zodiac after constructing the tropics, and instead goes on directly to the construction of the horizon. This omission is considered by Bergmann a sign of incompetence,[343] but it can be explained by the fact that, as already remarked, once the tropics have been drawn, constructing the zodiac is both conceptually and practically trivial, and may have been taken for granted.

4.4.7 Conclusions: transfer of astrolabe knowledge by means of memory, drawings and notes

There is ample evidence showing that transfer of knowledge on astrolabe construction in the late tenth and early eleventh centuries was inflexibly linked to very specific choices in construction pattern and labelling of geometrical drawings. This inflexibility stands in contrast to the extreme fluidity in wording and style of the verbal component of the texts **h1**, **h2** and **h3**. This discrepancy can in my opinion be explained by assuming a process of knowledge transfer based in large part on non-verbal or non-written strategies (memory, drawings, models), eventually aided by note-taking.

Early in the eleventh century, the inflexibility in construction pattern and labelling would disappear: for example, Ascelin of Augsburg made no use of labels in his very literary construction treatise (**a**). One of the scribes who copied his work, though, felt the need to add a reference to 'point d', using the traditional labelling discussed above: Charles Burnett suggests that the scribe "had available a figured diagram for the construction of the astrolabe very like that in Paris BnF lat. 11248, f. 26r", i.e. our **d1**.[344] Whether the scribe actually had a drawing in front of his eyes or was only working from memory cannot be determined, but in any case his addition to Ascelin's texts testifies to the continued existence of a tradition of astrolabe knowledge linked to a specific drawing and labelling procedure.

In the treatise **h**, Herman of Reichenau used labels, but not the traditional ones: point A was for him the centre of the equator circle and B, C, D, E its four extremes. Moreover, he verbally described the drawing with much more care than the writer of **h2** and specified that the point D had to be at the bottom of the circle, while in the drawings associated with **h2** it was always at the top.[345] By modern standards Hermann's treatise is by far the most readable among all high medieval texts on astrolabe construction, and its explanations can easily be followed even without having a drawing in front of one's eyes. It is hardly surprising that its

343 Bergmann (1985) p. 115.
344 Burnett (1998) p. 345 and p. 346, n. 111.
345 Herman of Reichenau (1931) p. 204–205.

author did not feel any more linked to the conventions of the non-written, non-verbal mode of transmission.

In the course of the twelfth century, the attention devoted to executing drawings seems to decrease: while in older manuscripts (e.g. BnF lat. 7412 (A.27)) all figures are carefully drawn with ruler and compass, in twelfth-century manuscripts (e.g. Salzburg Sankt Peter a. V. 7 (A.40)) the images are much less precisely drawn, sometimes probably even without the help of compass and ruler. In the first stages of transmission of astrolabe knowledge from the Arabic to the Latin world, though, drawing procedures and drawings played an essential role. In the next chapter, I shall argue that the specific procedure used to draw **d1** as well as some of its labels can be traced back to the geometrical constructions discussed in the first chapter of Ptolemy's 'Planisphaerium'.

4.5 THE RELATIONSHIP BETWEEN PTOLEMY'S 'PLANISPHAERIUM' AND EARLY LATIN ASTROLABE TEXTS AND DRAWINGS

4.5.1 Greek, Latin and Arabic version of the 'Planisphaerium'

According to a tenth-century Byzantine lexicographer, the Greek astronomer Ptolemy wrote a work on ἅπλωσις ἐπιφανείας σφαίρας ἐν ἐπιπέδῳ, an expression that can be translated as 'the expansion of the surface of a sphere on a plane'.[346] No copy of the Greek original is extant, but we possess three Arabic manuscripts containing its Arabic translation.[347] This translation was probably completed at the beginning of the tenth century in the Arabic-Islamic East and was later studied and commented by the Andalusi astronomer and mathematician Maslama (Maslama al-Majrīṭī d. 398 H = 1007/1008).[348] Around the year 1143, probably in Southern France, the commented Arabic version was translated into Latin by Herman of Carinthia (fl. 1138–1143). The Greek original of the 'Planisphaerium' is lost and the Arabic version was rediscovered only recently, but Herman's Latin translation was quite popular in late medieval and early modern times and was printed three times before the end of the sixteenth century. In the nineteenth century, it was edited among Ptolemy's works.[349]

346　Toomer (1975) p. 186; Neugebauer (1975) p. 870–871. Neugebauer says that ἅπλωσις actually means 'simplification' and suggests that it should actually be read ἐξάπλωσις. According to the Lexikon zur Byzantinischen Gräzität, though, ἅπλωσις could mean "das Ausbreiten, das sich Ausstrecken über" (Lexikon Gräzität (2001) p. 160).

347　Kunitzsch (1994). A facsimile edition of the version in the Arabic manuscript Istanbul, Ayasofya 2671and its English translation can be found in Anagnostakis (1984). For more details on the Arabic manuscript versions of the 'Planisphaerium', see 4.5.6.

348　Kunitzsch (1995) p. 150–154. Maslama's notes are edited in: Kunitzsch/Lorch (1994). On Maslama's school in Cordoba, see 3.1.4.

349　On the transmission of the Latin version see: Neugebauer (1975) p. 270–273, Jordanus de Nemore (1978) p. 47–48, Sinisgalli/Vastoia (1992) p. 37–41 and p. 62–66 (notes). The Latin version of the 'Planisphaerium' is edited in: Ptolemy (1907).

4.5.2 Relevance of the 'Planisphaerium' for this study

Historians of science have devoted their attention mainly to the Latin version of the 'Planisphaerium', whose literary style is complex and sometimes cryptic.[350] As Otto Neugebauer stated, in the 'Planisphaerium' problems are solved with methods "which we would consider today of 'descriptive geometry'", i.e. with the help of geometrical imagination. In Greek Antiquity, these methods were rather unusual for a work of astronomy, but were often used in treatises dedicated to the mechanical arts.[351]

The 'Planisphaerium' is particularly important for my study, because its Arabic translation is one of the few written sources of Latin astrolabe knowledge that could be identified.[352] Moreover, a Latin star table from the eleventh century seems to bear a relationship to Maslama's works and in particular to his commentary to the 'Planisphaerium'.[353] In some studies dealing with the earliest Latin astrolabe texts, the 'Planisphaerium' is regarded as the model of a theoretical treatise on stereographic projection and, as such, opposed to the allegedly practically oriented Latin texts on astrolabe construction.[354]

The differences between the 'Planisphaerium' and the earliest Latin texts on astrolabe construction are indisputable, both in content and form: most information contained in the 'Planisphaerium' as well as Ptolemy's numerical, geometrical and axiomatic-deductive formalisms find no correspondence in Latin texts. However, I disagree with those historians who see a sharp opposition between the 'theoretician' Ptolemy and the 'practically-minded' Latin. To prepare the ground for my comparison, I shall try to dispel the impression that the 'Planisphaerium' conforms to the modern standards of a theoretical treatise on stereographic projection. To do this, I will offer a short overview of its contents. My discussion is based on the Latin version of the 'Planisphaerium', complemented by Christopher Anagnostakis' English translation of the Arabic version.

---

350 For synopses and discussion of the contents of the 'Planisphaerium', see: Sinisgalli/Vastoia (1992) p. 37–68; Delambre (1817) p. 433–450; Neugebauer (1975) p. 857–868; Jordanus de Nemore (1978) p. 47–52. A German and an Italian translation exist: Drecker (1927–28) and Sinisgalli/Vastoia (1992).

351 Neugebauer (1975) p. 860.

352 The manuscript BnF lat. 7412 (A.27) contains the Latin version of two fragments from the 'Planiphaerium' (Kunitzsch (1993a)).

353 Samsó (2000).

354 This opposition in its extreme form is found in: Conzelmann/Hess (1980–81) p. 59–60. The 'Planisphaerium' is described as a theoretical treatise on stereographic projection also by Borst (1989) p. 16–17 and Poulle (1995) p. 232.

4.5.3 What Ptolemy did not say in the 'Planisphaerium'

What Ptolemy did not say in the 'Planisphaerium' is at least as interesting as what he did state. By modern standards, in a treatise dedicated to the theory of stereographic projection one would expect to find a definition of the projection procedure, such as that given in 2.2.3.2.

However, in the 'Planisphaerium' there is no description of the general rule according to which points on the surface of a sphere are associated to points on a plane. In particular, there is no explicit statement of the role of the pole of projection or of the fact that the plane of the equator is chosen to be the plane of projection. Yet, if the reader were really ignorant of the subject, he would no doubt have difficulties in following the explanations in the treatise, although he might eventually be able to reconstruct the missing background from the exposition itself. The omission is apparently intentional, since the beginning of the treatise is not mutilated. The author of the 'Planisphaerium' assumed his readers to be already familiar with the method of stereographic projection.

Further missing items in the 'Planisphaerium' – again: 'missing' by modern standards – are a general statement and a proof of the circle-into-circle property of stereographic projection, i.e. of the main reason why the method of the flat sphere is so useful for astronomic purposes. Was Ptolemy aware of this property? Could he also prove it? If so, why did he choose not to do it? Various opinions have been expressed on the subject, but in general historians have assumed that Ptolemy was aware of the property as well as of its proof: Olaf Pedersen suggests that he did not state it "because it was well known in his time".[355] Jean Baptiste Joseph Delambre (1749–1822) instead argued that some demonstrations in the 'Planisphaerium' may be understood as proofs of the circle-into-circle property for special cases (maximum circles, circles parallel to the Zodiac) and he doubts whether Ptolemy would have offered such partial proofs if he had been aware that the general property could be demonstrated.[356]

4.5.4 The contents of the 'Planisphaerium'

The Latin 'Planisphaerium' is divided into twenty chapters, usually grouped in a first (ch. 1–13) and a second part (ch. 14–20).[357]

Chapter 1 begins by stating the possibility and necessity of representing the circles of the solid sphere on a plane as if it were flat.[358] It is explained how to draw the equator, tropics and zodiac and then a proof is offered to show that the

---

355   Pedersen (1974) p. 405.
356   Delambre (1817) p. 437–438 p. 454.
357   An extended table of contents for the Arabic version of the 'Planisphaerium' can be found in: Anagnostakis (1984) p. 63–68.
358   Ptolemy (1907) p. 227.

procedure is correct. The contents of this chapter shall be discussed in more detail in 4.5.5.

In chapter 2 and 3, Ptolemy explains that any horizon can be drawn on the plane using the same procedure employed to project the zodiac circle. Using a quite involved deductive argument, partly relying on geometrical imagination, Ptolemy shows that all points on a circle constructed according to the rule explained in his chapter 1 are projections of points lying at opposite ends of spherical diameters.   Historians have offered different interpretations of how far-reaching this proof is.[359] According to Delambre, Ptolemy attempts here to give a proof of the circle-into-circle property for the special case of maximum circles, albeit a proof "pénible et enortillée".[360]

Chapters 4 to 9 describe the geometrical constructions necessary for computing the rising times of the zodiac signs with respect to an observer standing on the earth's equator (i.e. the 'right ascensions' of the zodiac signs). These values are necessary for dividing the zodiac circle on the flat sphere into its twelve signs.[361] In chapters 10 and 11, the results of the previous constructions are used to compute the rising times of the signs with respect to a generic horizon. This is done by computing the portion of equator rising above that horizon at the same time as a given arc of the zodiac. I will discuss this subject more in detail in 5.4.3.

Chapters 14 to 20 are devoted to the construction of the 'arenea', i.e. the rete of some instrument similar to the astrolabe, but not better specified by the author. To position the stars on the rete, circles parallel to the zodiac are drawn on the flat sphere, and then used as reference. In his commentary, Maslama explains that the same construction could be used to draw the parallels to any horizon, i.e. the altitude circles, "which the Arabs call bridges (ar.: al-muqanṭarāt, latinised into almucantarat)".[362] The altitude circles/almucantarat play a very important role in Latin astrolabe texts.

4.5.5 The construction of tropics and zodiac in the 'Planisphaerium'

Both extant versions of the 'Planisphaerium', i.e. the Arabic and the Latin one, show that the instructions for constructing tropics and zodiac given in chapter 1 of that work were very similar to those found in **h1**, **h2**, **h3** and **d1**. I reproduce the relevant texts here, using for the moment the labels given by twelfth-century Latin translators and taken over by both editors, Heiberg and Anagnostakis. The twelfth-century Latin version states:

> Describimus circulum equinoctialem notis A, B, G, D circa centrum E, cuius diametra ortogonaliter se inuicem secantia sint AG et BD. [...] producimus lineam AG utramque in partem sicque de circulo ABGD utraque ex parte G duos arcus aequales resecamus,

359 Sinigalli/Vastoia (1992) p. 5, Anagnostakis (1984) p. 64.
360 Delambre (1817) 435–438.
361 On this subject see 4.6.
362 Kunitzsch/Lorch (1994) p. 22–23 and p. 41–42.

desuper GH, infra GN, continuamusque rectis lineis D cum utrisque notis, ita quidem, ut DH usque in lineam AG perveniat, locumque K signabit, DN vero ubi lineam AG tetigerit, T notabitur. quo facto fixo in E centro ad mensuram EK fiet circulus super diametro KM, sicque non moto consequenter et alter fiet ad mensuram ET lineae super diametrum TL. diuisa deinde TM linea per medium circa diuisionis punctum R describetur circulus ad mensuram medietatis. dico igitur illos duos circulos equidistantes equinoctiali pari utrimque distantia, tertium vero super R centrum decliuem.[363]

Christopher Anagnostakis' English translation of the Arabic version from manuscript Istanbul, Ayasofya 2671 says:

> We consider circle ABGD as the equator, around center E, and we draw in it two diameters intersecting at right angles, namely line AG and line BD. [...] We produce lines AG, BD; we cut off on the circle two equal arcs GZ, GH on the two sides of point G and we join line DTZ and line DHK. We make point E the center and we draw circle TL with radius ET and circle KM with radius EK. I say that these two circles correspond to two circles of the circles on the solid sphere, on the two side of he equator, whose distances from it are equal. Moreover the inclined circle drawn with center (the point) bisecting TM so that it touches these two circles at point T and M, bisects circle ABGD, i.e. it passes through point B and point D.[364]

As we see, the two versions are not identical, but according to both of them Ptolemy gives instructions largely coinciding with those of the high medieval Latin text. However, the Greek author gave the construction without specifying any values of the angles, so that it was valid for the zodiac circle as well as for any horizon. The steps common to Ptolemy and the high medieval Latin texts are:

- draw the equator circle arbitralily, labelling it with the letters A, B, G, D and centre E. Draw its diameters AG and BD;

- cut off on the equator two arcs at equal distance from G;

- join the extremes H and N of the two arcs with point D by straight lines, marking as T and K the points in which the two lines cut the horizontal line AG;

---

363 Ptolemy (1907) p. 227–228. My translation is: "We draw the equator through the points ABGD around the centre E, and let its diameters cutting each other at a right angle be AG and BD. [...] We prolong the line AG on both sides and, on the circle ABGD, on both sides of G we cut two equal arcs, the one above GH, the one below GN. We draw a straight line from D to each of the marked points, so that DH will reach the line AG in a point we shall call K. Where the line DN touches the line AG we shall mark the point T. Once this has been done, draw the circle with centre in E and radius EK on the diameter KM and, after this, without moving [the compasses] draw another one with radius equal to the line ET on the diameter TL. Then, having divided the line TM in the middle, centering on the point of division R draw a circle having as radius the half [of TM]. I now say that those two circles are circles having the same distance from the equator, but on opposite sides. The third one, with centre R, is instead an oblique circle."

364 Anagnostakis (1984) p. 69–70. It remains unclear whether the Arabic version translated by Anagnostakis coincided with the one used and commented by the Andalusi astronomer Maslama, with which Latin scholars in all probability came into contact in the tenth and eleventh centuries (see 3.1.4).

– draw the two circles having E as centre and passing respectively through T and through K. Draw the circle having TM as diameter.

The similarity of the procedure can be appreciated if one considers that it might for example have been possible to choose two equal, diametrically opposite arcs starting one from G and the other from A, or even to draw only one arc, and connect it with two different points on the equator, e.g. B and D.[365]

The pattern of the construction is exactly the same as in **h1**, **h2**, **h3** and **d1**, but this is not all: further evidence of a connection can be found by comparing the high medieval labels with those which were used in the Arabic original. The letters used in **d1** and in the Latin texts **h1**, **h2** and **h3** to label the equator and its centre are: A, B, C, D and H. The letters used as labels for the same points in the Arabic version are: 'alif ‌ا [a], baa' ب [b], jiim ج [j], daal د [d] and haa' ه [h].[366] The first four Arabic letters correspond quite closely to Latin ones: ا ('alif), ب (baa') and د (daal) are pronounced like "a", "b" and "d", while ج (jiim) is a j as in 'joke', and it is plausible that it would have been rendered as c.[367] However, the most interesting point is the use respectively of ه (haa') and H to indicate the centre of the equator, since the ه (haa') is pronounced like the Latin h. While in the Latin texts the choice of A, B, C, D to indicate the first four points drawn is quite obvious and would not require a special explanation; the fact that the letter H was used to indicate the centre of the circle is in my opinion a further indication that the Latin construction procedures were more closely connected to the Arabic version of the 'Planisphaerium' than usually assumed.

4.5.6 The coincidence between **d1** and the drawing illustrating chapter 1 of the 'Planisphaerium'

There is one further element strongly supporting the thesis of a close connection between the Latin knowledge of astrolabe construction and the 'Planisphaerium': in two early manuscripts (BnF lat. 7412 (A.27) and BnF lat. 11248 (A.28)) the construction drawing for the zodiac contains one additional line joining point D with point O (fig. 19a). To understand the origin of that line, we must turn again to the first chapter of the 'Planisphaerium', in which Ptolemy offers a proof that the projection of the zodiac, if constructed according to the recipe given, bisects the equator in the points B and D.[368]

---

365 However, Ptolemy first drew both constuction lines DH and DN, and then both circles, while the Latin constructed one circle at a time.

366 Kunitzsch (1991/1992) p. 8, note 7 and p. 15. The same labels were used by Maslama in his commentary of the 'Planisphaerium' (e.g.: Kunitzsch/Lorch (1994) p. 14–15).

367 Also on the tenth-century astrolabe artefact #**3042**, the letter ج is rendered as c (Kunitzsch (1991/1992), table 1). In twelfth-century Arabic-Latin translations, on the other hand, the standard rendering of ج would be g (Kunitzsch (1991/1992) tables 2, 3 and 4).

368 The statement demonstrated is that "the third circle, the one having centre r and diameter tm and touching the two others in t and m, divides the Equator in two equal parts, cutting it in the points b and d". "Dico igitur [...] tertium [circulum] vero super R centrum declivem quem

During the proof, only one additional line has to be added to the construction already performed: a straight line connecting point D with one of the extremes of the east-west line, which in the drawing **d1** is point O.[369] This is exactly the same additional line appearing in the two manuscripts BnF lat. 7412 (A.27) and BnF lat. 11248 (A.28). In the 'Planisphaerium' manuscript Oxf. Bodl. Auct. F.5.28 (B.27), on f. 88v, and also in the 16th-century edition by Commandino, the text of the first chapter of the 'Planisphaerium' is illustrated by a figure identical to that in the two Paris manuscripts not only in its structure, but even in its orientation, i.e. with the point corresponding to d of the drawing **d1** on top.[370] The labels, though, are different and follow the standard translation for Arabic letters used by twelfth-century Latin scholars.

At this point, it is important to ask: how did the drawing look in the Arabic version of Ptolemy's work? Three manuscripts containing that text have been known to exist, but one of them, which was preserved in Kabul, if it is still in existence, is at present inaccessible.[371] The other two manuscripts are: Istanbul, Ayasofya 2671, published with translation by Anagnostakis in 1984, and the one preserved in Teheran in the private library of Khān Malik Sāsānī. In the Istanbul manuscript, the spaces for the drawings are left blank, but in the Teheran one the drawings are present.

With the help of Prof. Kunitzsch, I was able to see a copy of the drawing corresponding to **d1** (Teheran, private library Khān Malik Sāsānī, p. 3). I reproduce in fig. 20 not the original drawing, but a copy made by myself with the same structure and labels: the lines are identical to those of **d1**, including the additional line, but the whole diagram is rotated by ninety degrees to the right. The labels used for the four extremes of the equator and for its centre are the same as in Maslama's commentary to the 'Planisphaerium', and, if we take the rotation into account, they also correspond to those used in the Latin drawings **d1**.

I believe the presence of the additional line in the oldest manuscripts (particularly in BnF lat. 7412 (A.27), which contains fragments of the 'Planisphaerium'[372]) is evidence that the connection between the earliest Latin texts and drawings on astrolabe construction is closer than usually assumed.

.T.M. linea per aequalia secat quousque utrumque illorum attingat alterum as notam .M. alterum ad punctum .T. equinoctialem per medium secare, quem ad opposita duo puncta .B. et .D. intercipit" (Ptolemy (1907) p. 228). The translations by Drecker (1927–28) p. 256 and Sinigalli/Vastoia (1992) p. 77–79 are misleading here.

369 Ptolemy (1907) p. 228–229, Anagnostakis (1984) p. 70.
370 For the images associated to the Latin version of the 'Planisphaerium', see: Anagnostakis (1984) p. 107–108 and fig. 1a.
371 I owe the information about the diagrams in the Arabic version of the 'Planisphaerium' to the courtesy of Prof. Paul Kunitzsch, who corrected and improved my remarks on this subject. For deatils on the manuscripts, see: Anagnostakis (1984); Kunitzsch (1994).
372 See 5.3.8.

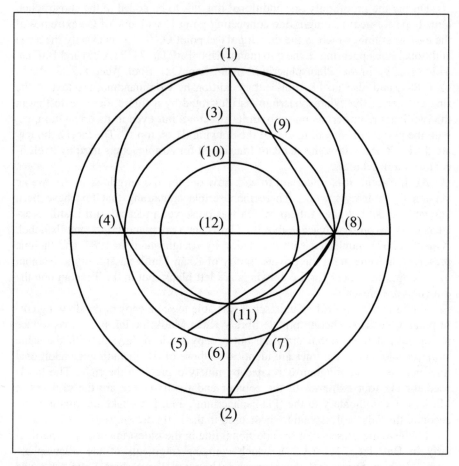

Figure 20: This figure reproduces the drawing associated to chapter 1 of the Arabic version of the 'Planisphaerium' in the manuscript Teheran, private library Khān Malik Sāsānī, p. 3. The following readings of the Arabic labels were kindly provided to me by Prof. Kunitzsch:
On the outer circle: (1) م (miim); (2) ك (kaaf, initial form). On the middle circle: (3) ا ('alif); (4) ب (baa'); (5) ز (zaay); (6) ج (jiim, initial form); (7) ح (haa') (8) د (daal); (9) ن (nuun). On the inner circle: (10) ل (laam); (11) ط (taa'). At the centre: (12) ه (haa').

Since the additional line is not relevant to astrolabe construction, but is necessary for Ptolemy's proof, it would be hard to explain its presence in two out of six diagrams, if not by assuming that Latin scholars somehow came into contact with Ptolemy's proof. The texts **h1**, **h2** and **h3** tell us nothing about this encounter, but some drawings do. This is because the knowledge transfer did not take place through a Latin translation of the 'Planisphaerium', but rather by means of oral teaching, discussions and exercise in which geometrical constructions played an important role. The discussions might have been based on the Arabic version of the 'Planisphaerium', as well as on Maslama's notes to it, probably in a multilingual situation.[373]

The process of knowledge transmission was guided by the interests of Latin scholars, who extracted from Greek-Arabic classics the information they were looking for, exactly as they had been doing with Latin classics such as Plinius, Martianus Capella or Calcidius. Thus, the basic steps needed for the construction of the flat sphere were assimilated and later fixed in writing in various forms (**h1**, **h2**, **h3**), while Ptolemy's deductive proofs were left behind, possibly because they appeared less interesting than the evidence offered by geometrical imagination.

Does the presence of the complete drawing from chapter 1 of the 'Planisphaerium' in Latin manuscripts mean that some Latin scholars of the tenth or eleventh century actually had interest in Ptolemy's deductive proof? I believe this possibility should not be completely ruled out, since that proof relied on geometrical imagination and would have been accessible to Latin mathematicians. No definitive answer is possible, though, partly because the actual scope of Ptolemy's proof remains unclear.[374]

4.6 THE DIVISION OF THE ASTROLABE'S ZODIAC IN ELEVENTH-CENTURY LATIN EUROPE

A subject discussed at length in the 'Planisphaerium' and in Maslama's commentary to it were the methods to divide the zodiac of the flat sphere into its twelve signs. The next sections are devoted to investigating how eleventh-century Latin scholars dealt with this question. I shall analyse both eleventh-century texts and eleventh-century drawings, and I will compare the results with the division of the zodiac on the oldest Latin astrolabe artefact (**#3042**).

Many of the texts to be analysed have been edited by Millás Vallicrosa as parts of longer compositions. However, there are strong indications that some of them originally circulated independently and were later grouped together with other material. Therefore, I have regarded them as independent compositions and,

---

373 On the other, hand, as Prof. Kuntzsch suggested to me, it is not necessary to assume that early Latin scholars had the whole text of the 'Planisphaerium' at their disposition: they might have come to know only the initial parts, as quoted by other Arabic authors.

374 The deductive structure of the proof in question is open to various interpretations: it is usually assumed that the point being proven is simply that the projection of the zodiac bisects the equator (Jordanus de Nemore (1978) p. 49, Sinigalli/Vastoia (1992) p. 57). On the other hand, the proof would seems somehow too elaborate for that aim. In his commentary to the 'Planisphaerium', Delambre argues that the aim of the demonstration is to show that the circle with diameter TM not only bisects the equator, but does so by passing through the points B and D. This fact is very important, because the projection of the zodiac has to pass through those two points, which are common to zodiac and equator and thus project onto themselves. The demonstration would thus prove that the circle with diameter TM fulfils a necessary condition for representing the zodiac on the flat sphere (Delambre (1817) p. 433–434).

when necessary, I have introduced new abbreviations to indicate them.[375] The texts to be analysed are:

- **dz1**, printed by Millás Vallicrosa as the final part of **h"**;[376]
- **dz2**, printed by Millás Vallicrosa as part of **h'**;[377]
- **dz3**, printed by Millás Vallicrosa as part of **h'**;[378]
- **dz4**, printed by Millás Vallicrosa as part of **h'**;[379]
- the passage on the division of the zodiac from **pt**, i.e. an anonymous eleventh-century text on astrolabe construction printed by Millás Vallicrosa;[380]
- the passage on the division of the zodiac in the text **h1**;[381]
- the passage on the division of the zodiac from the text **a**, i.e. the treatise on astrolabe construction by Ascelin of Augsburg;[382]
- the passage on the division of the zodiac from the text **h**, i.e. the treatise on astrolabe construction by Hermann of Reichenau;[383]

The drawings discussed are found in the following manuscripts, listed in appendix A as A.4, A.27 and A.43

- Bern BB 196 f. 2v (ca. 1000)
- BnF lat. 7412 f. 19v (**#4024**) (11th. c.)
- Vat. Reg. lat 598 f. 120r (**#4553**) (11th. c.)

Finally, I shall shortly discuss:

- the division of the zodiac on the oldest Latin astrolabe (**#3042**).[384]

I have chosen to discuss the Latin solutions to the problem of dividing the astrolabe's zodiac for several reasons. First, this is a quite advanced mathematical problem which serves well to address the question of the mathematical competence of Latin scholars. Second, the Latin texts dealing with it show a remarkable variety of literary styles, thus offering a good opportunity of comparing different

---

375 For a general overview on the printed versions of eleventh-century astrolabe texts and on the abbreviations I have introduced for some of them, see app. C.

376 Millás Vallicrosa (1931) p. 295, l. 40–51.

377 Millás Vallicrosa (1931) p. 300, note to lines 137–138 and Herman of Reichenau (1931) p. 214, where the manuscripts in which **dz2** is found are also indicated.

378 Millás Vallicrosa (1931) p. 302 lines 168–169 and Herman of Reichenau (1931) p. 214, where the manuscripts in which **dz3** are found are also indicated.

379 Millás Vallicrosa (1931) p. 302 lines 169–171 and Herman of Reichenau (1931) p. 214, where the manuscripts in which **dz4** is found are also indicated.

380 The text **pt** was printed by Millás Vallicrosa from the twelfth-century manuscript Avranches 235 (A.1), f. 68r–69v (Millás Vallicrosa (1931) p. 323, l. 50–62), but it also occurrs in the eleventh-century manuscript Lon. BL Add. 17808 (A.13) (Burnett (1997) p. 69).

381 Millás Vallicrosa (1931) p. 300. As already explained in par 4.4.1, the text **h1** was printed by Millás Vallicrosa as part of **h'**.

382 Burnett (1998) p. 348 (in ch. [4] of **a**).

383 Herman of Reichenau (1931) p. 207–208 (in ch. 5 of **h**).

384 On this artefact, see 3.2.5.

criteria of readability. In the third place, the division of the zodiac is a well circumscribed subject discussed in short passages, so that problems deriving from text transmission can be reduced to a minimum.

My final reason for focussing on this theme is that the alleged incompetence of Latin scholars in dividing the astrolabe's zodiac has been used as an argument against the possibility of them having really understood how an astrolabe was built. This argument has been used as a main objection to the claim that the earliest extant Latin astrolabe artefact (**#3042**) was made in the tenth or eleventh century. Emmanuel Poulle has argued:

> Les auteurs latins d'astrolabe ont eu du mal à comprendre comment devait être construite la graduation inégale du zodiaque, projection sur le plan de l'équateur de la graduation d'un cercle qui lui est oblique. [...]. Huit textes, aux Xe–XIe siècles, proposent une construction de la graduation du zodiaque de l'araignée. L'une d'entre eux [...] propose un procédé qui revient à exploiter une table des ascensions droites [...], les sept autres textes [...] s'en tiennent tous à une construction consistant à joindre le centre de l'instrument aux divisions égales du limbe, ce qui revient à accorder à tous les signes une même ascension droite de 30°. [...] L'astrolabe de Destombes [i.e. **#3042**] ne connaît pas ces errements: pour avant que on en puisse juger sur des mesures qu'il est difficile de prendre avec précision, sa graduation de l'écliptique est correcte, ce qui l'insère par conséquent très mal dans la première culture astrolabique de l'Occident latin.[385]

As we shall see, Poulle's claim that most early Latin astrolabe texts only contained simplified versions of the zodiac's division is not correct.

4.6.1 The measure of the zodiac

Dividing the astrolabe's zodiac circle into its twelve signs is a non-trivial problem from the mathematical point of view. On the solid sphere, of course, each sign occupies exactly 30° of the zodiac circle.[386] However, since the zodiac is not parallel to the equator, when it is projected onto the flat sphere its signs have different extensions. Methods to divide the astrolabe's zodiac are described in Ptolemy's 'Planisphaerium' and in Maslama's notes to that work. There are various solutions to the problem, but only one interests us here.[387] Its essential idea is to start by considering, for each sign, the two meridians passing respectively through its beginning and its endpoint. These two meridians cut on the equator (and on the tropics) an arc which is usually referred to as the 'right ascension' of the sign in question, and it gives a measure of the time it takes for that sign to rise above the horizon of an observer at the North or South Pole. Since arcs on the equator and on the tropics remain unchanged on the flat sphere, the right ascensions can be directly drawn on the astrolabe plane, and can then used to divide the zodiac as shown in figure fig. 21: First the right ascensions are marked on the outer rim of the flat sphere (i.e. tropic of Capricorn), then lines are drawn connect-

---

385 Poulle (1995) p. 234–235.
386 On the definition of the zodiac signs, see 2.2.2.7.
387 Various solutions are sketched in Poulle (1955).

ing the extremities of these arcs to the centre of the equator: these lines are projections of the meridians considered in the beginning, and they cut the projection of the zodiac in the extremities of the signs. As can be seen in fig. 21, the division of the zodiac on the astrolabe is asymmetric, with the summer signs covering much less than half of the flat sphere's zodiac.

The value of the right ascension is not the same for all signs: the farther a sign is from the equator, the larger its right ascension is. For symmetry reasons, there are only three possible values of zodiacal right ascensions. Those computed by Ptolemy in the 'Planisphaerium' were: 27° 50' (for Aries, Virgo, Libra and Pisces), 29° 54' (Taurus, Leo, Scorpio and Aquarius), 32° 16' (Gemini, Cancer, Sagittarius and Capricorn).[388] These values can be computed once and for all and then used for any astrolabe. In general, it is not at all necessary to understand how the values are derived. As we shall see, Latin astrolabe texts simply provided the reader with a set of values and instructions to use them, without explaining how they had been computed.

In discussing the division of the zodiac, I will use a terminology often employed in Latin texts: on the rete, the two diameters of the equator joining respectively the two equinoctial and the two solsticial points of the zodiac are indicated as 'linea orientalis' ('eastern line', joining the eastern and western equinoctial points) and 'linea meridionalis' ('southern line', joining the northern and southern solsticial points). Sometimes, the half of the eastern line which passes through the autumn equinox is called 'linea occidentalis' ('western line'), and the half of the southern line which passes through the summer solstice is called 'linea septentrionalis' ('northern line').

4.6.2. The division of the zodiac in the text **dz1**

The division of the zodiac is described in the text **dz1** as follows:

> Scimus in omni climate inesse circulos in quorum maximo in alchancabut insignito, in scisiones circuli casarum per metiri possumus. Diuidimus eius quadram in XV incipientes a primo gradu arietis et abstracta XV parte superius iuxta primum gradum cancri quod remanet intra secamus et ecce invenimus primum gradum tauri, et oppositi sui iacente regula in centro et in puncto superius affixo similiter et in aliqua quadra et habemus primum gradum virginis et oppositi sui. Restat videlicet ut primum gradum geminorum et leonis inveniamus quem ita inuestigari oportet. Partimur eandem quadram in III partes et tertiam aponimus priori mensure. Que aposita demonstrat nobis primum gradum geminorum et oppositi sui, et in alia quadra leonis et contra his inventis que restant spacia ad cancrum et capricornum libram et arietem pertinent.[389]

388 Anagnostakis (1984) p. 66.
389 Millás Vallicrosa (1931) p. 295. My translation is: "We know that for all climates there are circles such that, having drawn the largest one of them on the rete, we can properly divide the zodiac. We divide its [the largest circle's] quadrant in fifteen parts, starting from the beginning of Aries and, having subtracted the fifteenth part on top [i.e. at the northernmost point], at the beginning of Cancer, we divide what remains into three parts, and thus we shall have found the beginning of Taurus and also of its opposite sign, by setting the ruler so that it

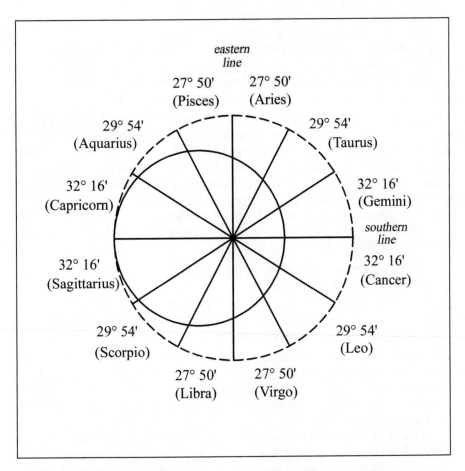

Figure 21: Division of the zodiac circle on the flat sphere according to Ptolemy's values of zodiacal right ascensions. The dashed circle is the tropic of Capricorn, the full one is the zodiac.

passes through the centre [of the equator/tropics]. Fixing [the compasses] on the northernmost point and taking the same interval in the other quadrant, we shall have the beginning of Virgo and of its opposite sign. It only remains for us to find the beginning of Gemini and Leo, which we have to determine as follows. Let us divide the same quadrant in three parts, and let us add one of these arcs to the point which we have found previously. Once we have added it, it shows us the beginning of Gemini and of its opposite sign and, in the other quadrant, [the beginning] of Leo; the spaces remaining on the side opposite of those found correspond to Cancer and Capricorn, Libra and Aries."

The procedure described in the passage is:

    (1) take the quarter of the outer rim of the astrolabe corresponding on the zo-diac to Aries, Gemini and Cancer;

    (2) divide it into fifteen equal parts;

    (3) take the first fourteen parts, starting from the point corresponding to the beginning of Aries.

    (4) divide this portion again into three equal parts:[390] "so we find the first de-gree of Taurus" ("ecce invenimus primum gradum Tauri").

The last step is not clearly described: from the end of the first of the three portions a line must be drawn to the centre of the astrolabe. This line cuts the zodiac in the point corresponding to the beginning of Taurus. After this, the author remarks that, by drawing a straight line from the first degree of Taurus through the centre of the astrolabe up to the other side of the zodiac, the beginning of the sign oppo-site to Taurus (i.e. Scorpio) can be found. Repeating the whole procedure in the adjoining quarter, the beginnings of Virgo and its opposite sign (i.e. Pisces) can be determined. Thus, the reader is explicitly told how to take advantage of the sym-metry of the astrolabe. Now, to the next steps:

    (5) take again the whole quarter of the outer rim, as in step (1);

    (6) divide it into three equal parts;

    (7) take the measure of one of these parts and add it to the arc already found: this arc determines the beginning of Gemini and, thanks to symmetry, also of all remaining signs.

This procedure corresponds to assigning a right ascension of 28° to Aries, Virgo, Libra and Pisces, one of 30° to Taurus, Leo, Scorpio and Aquarius and one of 32° to Gemini, Cancer, Sagittarius and Capricorn.[391] These numbers are not a bad ap-proximations of Ptolemy's values: 27° 50', 29° 54' and 32° 16', as already noted by Emmanuel Poulle.[392]

    Since in the end the estimates for the right ascensions are whole numbers, one might ask why the author chose to express them is such a convoluted way. How-ever, the information conveyed here is much more than the numerical values for zodiacal right ascensions: the author described a practically feasible procedure for drawing on the astrolabe a correct division of the zodiac circle. When performing the construction in reality, dividing a quarter of circle in ninety parts and then tak-ing 28, 30 and 32 of them would in most cases lead to larger errors than dividing it only in fifteen parts, as done here.

    The author of **dz1** knew not only the procedure and the numerical values nec-essary to divide the astrolabe's zodiac, but also how to use them in practice to perform a drawing. The instructions given are quite clear, but some basic steps are not mentioned, as for example the fact that one has to join the points on the outer rim of the astrolabe to the centre and see where these lines cut the zodiac. Some-

---

390 Here I have taken 'intra' to be a corruption of 'in tres (partes)'.
391 The computation is: $90°/15 \times 14/3 = 28°$; $90°/3 = 30°$; $90°-30°-28° = 32°$.
392 Poulle (1995) p. 235.

one already familiar with the general problem and the strategy used to solve it might take this information for granted, but a layman would not: once again, the text appears to have been written rather as a reference manual for advanced students than as a simple instruction booklet for beginners.

4.6.3 The division of the zodiac in the texts dz2, dz3 and dz4

The texts **dz2**, **dz3** and **dz4** are three very short passages – in fact, three sentences – which appear in early astrolabe manuscripts grouped together. I have considered them as independent from each other, though, because each is complete in itself and because each gives different values for the right ascensions. The text **dz2** states:

> Tene ut ad metienda in zodiaco circulo interualla signorum numeres ab utraque parte meridiane linee XXX duos gradus in exteriori umbone ab australi XXVII.[393]

My translation is:

> To divide the zodiac into twelve signs, count on the outer rim of the astrolabe 32 degrees on each side of the southern line and 27 on each side of the eastern line.

This corresponds to assigning right ascensions of 27° to Aries, Virgo, Libra and Pisces, 31° to Taurus, Leo, Scorpio and Aquarius and 32° to Gemini, Cancer, Sagittarius and Capricorn. The values are not completely implausible and the text is very clear, but it only states the numerical values and the sequence in which they had to be assigned: all information relevant to the geometrical construction had to be already known to the reader. Once again, we have a text that looks like a short note taken by an advanced student: it is correct, but would surely have been of no help to a beginner. The text **dz3** says:

> Taurus ab ariete tribus superatur gradibus a tauro Gemini. Sic et in alia parte Leo a Virgine et Cancer a Leone.[394]

My translation is:

> Aries has three degrees more than Taurus, and Taurus three more than Gemini. The same on the other side, where Virgo is larger than Leo and Leo larger than Cancer.

These instructions correspond to right ascensions of 33° (!) for Aries, Virgo, Libra and Pisces, 30° for Taurus, Leo, Scorpio and Aquarius and 27° (!) for Gemini, Cancer, Sagittarius and Capricorn. The numerical values given are a good approximation, but they are given in the wrong sequence, growing instead of decreasing when going from the solsticial (e.g. Gemini) to the equinoctial (e.g. Aries) signs. We shall meet this kind of error again. Finally, the text **dz4** says:

> Primus gradus leonis a XXXta IIIbus gradibus extimi circuli id est umbonis sumitur. Virgo a LXIIIes.[395]

---

393 Millás Vallicrosa (1931) p. 300, note to lines 137–138, Herman of Reichenau (1931) p. 214.
394 Millás Vallicrosa (1931) p. 302, Herman of Reichenau (1931) p. 214.

That is: "the first degree of Leo corresponds to the thirty-third degree of the outer rim of the astrolabe. Virgo starts from the sixty-third degree". This procedure assigns right ascensions of 27° to Aries, Virgo, Libra and Pisces, 30° to Taurus, Leo, Scorpio and Aquarius and 33° to Gemini, Cancer, Sagittarius and Capricorn. We note that in this text, as also in the previous one, not even the aim of the instructions, i.e. the division of the zodiac, is explicitly stated.

4.6.4 The division of the zodiac in the text 'Iubet rex Ptolomeus' (pt)

The text I refer to as **pt** describes in a rather literary style the construction of astrolabe lines, making use of the same construction pattern and labels as the texts discussed in par 4.4.[396] It contains the following instructions to divide the zodiac:

> Ex linea orientali usque meridianam tali discrimine tria signa partire, ut et divisio alia incohata meridiali ut pote signum cancri a proximo ibi signo uidelicet leonis tota et decima parte superetur statu. A tercio autem que in linea finitur orientali tota et sexta parte sui uincatur signi.[397]

That is:

> Divide the three signs comprised between the eastern and the southern line in such a way, that the division starting from the southern line, i.e. the sign of Cancer, will be overcome in extension by the sign nearest to it, i.e. Leo, as by all of itself plus one tenth. By the third sign, instead, i.e. by the one that ends at the eastern line, [Cancer] should be won as it is by all of itself plus one sixth.

In other words: the ratio of the right ascension of Leo to that of Cancer is 11/10 (i.e. Leo is equal to Cancer plus one tenth of Cancer), while the ratio of the right ascension of Virgo to that of Cancer is 7/6 (i.e. Virgo is equal to Cancer plus one sixth of Cancer). Before proceeding with the numerical computation of the right ascensions, we note that here, as already in **dz3**, the division is qualitatively wrong: Cancer should be larger than Leo and Leo should be larger than Virgo, not the opposite!

Let us now compute he numerical values following the instructions. Indicating the three right ascensions as C, L and V (for Cancer, Leo and Virgo), we have:

$$L/C = 11/10 \text{ and } V/C = 7/6,$$

where the values of the right ascensions also have to satisfy the condition:

$$C+L+V = 90°.$$

395 Millás Vallicrosa (1931) p. 302, Herman of Reichenau (1931) p. 214. Drecker adds in the end 'incisio medii circuli in rete LXVI extimi circuli', a sentence thatn in Millás Vallicrosa edition follows to **dz4**, but is in fact independent from it, as can be seen in some manuscripts, where the two are separated by a point (e.g. Oxf. CCC 283 (A.26) on f. 85r).
396 The orientation of the construction, though, is different, as the point d corresponds to the southern line (Millás Vallicrosa (1931) p. 322).
397 Millás Vallicrosa (1931) p. 323.

Combining these equations we obtain:

$$C \times (3+1/10+1/6) = C \times (3+4/15) = 90°$$

and finally:

$$C = 15/49 \times 90° = 27{,}551 = 27° \ 33'$$
$$L = 11/10 \times 15/49 \times 90° = \ 30{,}306 = 30° \ 18'$$
$$V = 7/6 \times 15/49 \times 90° = 32{,}143 = 32° \ 09'.$$

The result is therefore a right ascension of 32° 09' for Aries, Virgo, Libra and Pisces, one of 30° 18' for Taurus, Leo, Scorpio and Aquarius and one of 27° 33' for Gemini, Cancer, Sagittarius and Capricorn. These numerical values are surprisingly accurate. If they were assigned in the opposite order (Virgo–Leo–Cancer instead of Cancer–Leo–Virgo), they would differ from Ptolemy's estimates (27° 50' for Virgo, 29° 54' for Leo and 32° 16' for Cancer) at most by 24'. Such good agreement can hardly be due to chance, so that it may be safely assumed that a very accurate numerical estimate was at the basis of the formula given, but that the order of the signs was inverted, either from the beginning or at some stage of transmission.

Would the text not explicitly quote Cancer and Leo, it would be enough to exchange 'southern line' and 'eastern line' to have a correct result. The right ascensions would then have the same values computed above, but would be assigned in a correct way. Since the signs are explicitly named, though, one might make the hypothesis that the original formula did not state that "Cancer should be overcome ('superetur') by Leo", but that "Cancer should overcome ('superet') Leo" and, similarly, Cancer should win ('vincat') over Virgo and not be won ('vincatur'). However – and here one has to be very careful – this configuration would lead to other values of the right ascensions, determined by the conditions:

$$L/C = 10/11 \text{ and } V/C = 6/7 \text{ with } C+L+V = 90°.$$

This would give:

$$C \times (1+10/11+6/7) = C \times 213/77 = 90°$$

and finally:

$$C = 77/213 \times 90° = 32° \ 32'$$
$$L = 11/10 \times 77/213 \times 90° = \ 29° \ 35'$$
$$V = 7/6 \times 77/213 \times 90° = 27° \ 53'.$$

These values are as good as the previous ones, deviating from Ptolemy's ones by at most 19'. Thus, both corrections are in principle possible, and it will be interesting to compare the two solutions with that offered by the text **h1** discussed in the following paragraph.

4.6.5 The division of the zodiac in the text **h1**

The division of the zodiac is explained in the text **h1** as follows:

> Cuius [i.e. zodiaci] diuisionis ratio, sic dinoscitur dicto. Accepto circino spacium quod continetur inter orientalem lineam et meridianam in tria diuidatur, ita ut diuisio incepta a meridiali linea ad ultimam in sescupla proportione consistat XXX gradibus in uno quoque computatis. Similiter ad secundam comparata decima parte sui superetur, et in alia parte idem sit scilicet inter meridianam et occidentalem lineam. Quibus inventis, a centro linea ducta ad has divisiones videbatur in quo loco alius hemisperii latitudo secetur.[398]

This text has much in common with the previous one, both at the verbal and at the algorithmic level. My translation is:

> The division of the zodiac can be known as follows. Take the compass and divide the space between the eastern and the southern line so, that the division starting from the southern line relates to the last one in a 'sescupla' [6 to 1] proportion, counting 30 degrees in each one. Similarly, when compared to the second one, [the first division] should be smaller by one tenth of itself. The same thing goes for the other part, i.e. between the southern and the occidental line [i.e. the extremity of the eastern line passing through the autumn equinox]. Once these divisions are found, drawing a line from the centre to them one will see where to put the divisions in the other hemisphere.

This text apparently offers an even more corrupt version of the procedure given in **pt**. That it is the same algorithm for the division of the zodiac is clearly shown both by the similar structure of the two passages and by the occurrence of the 11/10 proportion. I shall a offer my reconstruction of a possible original version of this corrupt passage. The first part of the instructions seems hopelessly wrong: the 'sescupla' Boethian ratio (6 to 1) between the right ascensions of Cancer and Virgo leads to completely wrong values (right ascension of ca. 40° for Cancer). A possible correction would be 7 to 6, as in the second reconstruction offered in 4.6.4, a ratio for which the Boethian system also had a name: 'sesquisexta'.[399] Thus, 'sescupla' could be considered as a corruption of 'sesquisexta', which is in my opinion a very plausible hypothesis, given the general similarity of the terms and the fact that 'sescupla' was a much more common term.

It is important to note that, by substituting 'sesquisexta' for 'sescupla', the first part of the instructions becomes fully correct, i.e. Cancer is larger than Virgo and not the opposite, like in **pt**. This means that, if the passage has to make any sense at all, in the second part of the instructions the ratio between Cancer and Leo has to be inverted. This can be done by simply correcting 'superetur' into 'superet'. In this way, Cancer is larger than Leo by 1/10 of its (i.e. Cancer's) value. With both corrections in place, the text **h1** gives the same accurate values of right ascensions computed in the previous paragraph in the second reconstruction: a right ascension of 27° 53' for Aries, Virgo, Libra and Pisces, one of 29° 35' for Taurus, Leo, Scorpio and Aquarius and one of 32° 32' for Gemini, Cancer,

---

398 Millás Vallicrosa (1931) p. 300.
399 Borst (1986) p. 505. It must be noted that many, but not all ratios of integers have a Boethian name. For example, 11/10 has none.

Sagittarius and Capricorn. These results can be achieved, I repeat it, by making only two corrections to the text, i.e. with 'sescupla' becoming 'sesquisexta' and 'superetur' becoming 'superet'.[400]

4.6.6 The language of proportions in **pt** and **h1**

Even if the numerical accuracy of the values of zodiacal right ascensions given in **pt** and in the corrected form of **h1** is rather impressive, one has to ask: did the procedure also allow the medieval astronomer-mathematician to easily draw the division of the zodiac? Even assuming that the instructions were correct, they were not the kind of recipe that it would have been easy to translate into a drawing. Although they might superficially appear similar to those given in **dz1**, these instructions are in fact radically different: there, the divisions were easy to transform into a geometrical construction, here they are not. At least, I do not see any construction that could easily help embody the ratios in a drawing, since two ratios of three unknowns are given coupled to each other.[401] Moreover, such precise values of right ascension would in all probability have been wasted even on a very precise drawing.

The author or authors of the recipe found in **pt** and **h1** mastered quite well arithmetic, possibly less well astronomy, and it would seem that they were not interested in teaching a viable procedure for actually drawing the division of the zodiac. The focus of the passage is rather on transmitting accurate numerical information. This was done by means of the traditional system of Boethian number proportions, which was quite flexible in expressing fractional values. On the other hand, as we have seen, this formalism sometimes proved to be beyond the capabilities of scribes, who in **h1** probably transformed 'sesquisexta' into 'sescupla'. The use of proportions might also be responsible for the other error, i.e. the inversion of right ascensions between equinoctial and solsticial signs. In fact, this error occurs only in texts like **dz3**, **pt** and **h1**, in which the values are expressed by saying that one sign relates to another one according to a certain proportion.

---

400 If this reconstruction of **h1** is correct, then probably the text **pt** should be corrected according to the second reconstruction.

401 Of course, it would have been possible to compute the numerical values (as I did) and then draw them on the graduated rim of the astrolabe, in the improbable case that the rim contained a division in minutes.

4.6.7 The division of the zodiac in the treatises of Ascelin of Augsburg (**a**) and Herman of Reichenau (**h**)

The treatise on astrolabe construction of Ascelin of Augsburg (**a**) is preserved in 5 manuscripts and has been edited first by Werner Bergmann (from ms. Avranches 235 (A.1)) and, later, by Charles Burnett (from all extant manuscripts).[402] In his treatise, Ascelin chose to conform to the models of classic Latin prose. In fact, his introduction shows more interest in the literary than in the mathematical arts: where other authors had explained the importance of the astrolabe to grasp celestial order, Ascelin discusses the power and merits of friendship by referring to Cicero and Symmachus.[403] Ascelin explains how to divide the zodiac with these words:

> Sic repertum, zodiacum partire per duodecim intervalla, dans unicuique sex umbonis maiora spacia, prout ipsorum signorum distenduntur interstitia.[404]

That is: to each sign of the zodiac corresponds one twelfth of the outer rim of the astrolabe. This division corresponds to assigning a right ascension of 30° to each sign. This very simple solution can be regarded either as an error or as a rough approximation which has the advantage of greatly simplifying the construction.

Ascelin's method for dividing the zodiac is the same one proposed by Herman of Reichenau in his treatise **h**.[405] This is particularly interesting because, as we have just seen, it is only in Ascelin's work that this exceedingly simplified method is described: all other texts present more complex solutions. If, as Charles Burnett suggests, Ascelin was indeed Herman's teacher, there could be a direct connection.[406] Herman's treatise is a model of clarity and pedagogical competence: all procedures are explained up to the last detail, and nothing is taken for granted. As discussed in 4.5.3, Herman can hardly be criticized for not explaining the theoretical principles of stereographic projection, since not even Ptolemy did so.[407]

Because of this, I believe that Herman's (and Ascelin's) choice of method to divide the zodiac should not be attributed to ignorance, as often done, but rather to pedagogical competence. In the first place, so many solutions to the problem of dividing the astrolabe's zodiac were known at the time, that it is very improbable that Ascelin or Herman would not have been aware that the one they offered was not strictly correct. On the other hand, because of the many 'inverted solutions' (i.e. **dz3** and **pt**) circulating, and because of the difficulty of securing a correct

---

402 Bergmann (1985) p. 223–225, Burnett (1998) p. 345–349.
403 Burnett (1998) p. 345–346.
404 Burnett (1998) p. 349. Burnett's translation is: 'Divide the zodiac, found in this way, into twelve intervals, giving to each the six larger spaces on the rim, as far as the intervals of the signs themselves are extended' (Burnett (1998) p. 353).
405 Herman of Reichenau (1931) p. 207–208.
406 Burnett (1998) p. 333–334.
407 Hess and Conzelman not only criticize Herman because he did not know "die tieferliegenden theoretischen Voraussetzungen zur Konstruktion des Astrolabs" (Conzelmann/Hess (1980–81) p. 59) and even wonder whether he thought of the earth as a sphere or as a disc (Conzelmann/Hess (1980–81) p. 57).

written transmission of texts containing complex procedures, the simplified solution might have appeared as the best one, since it was almost foolproof against both inattentive students and incompetent scribes. The difference due to the approximation is rather small and was probably negligible for all purposes to which an eleventh-century astrolabe drawing or astrolabe artefact might have been employed.

4.6.8 The division of the zodiac in eleventh-century drawings of astrolabe parts

Three eleventh-century manuscripts contain drawings of the rete of an astrolabe: Bern BB 196 f. 2b (ca. 1000, A.4), Vat. Reg. lat. 598 f. 120r (**#4553**) (11th c., A. 42) and BnF lat. 7412 f. 19v (**#4024**) (11th c., A.27).[408]

The manuscript Bern BB 196 (A.4) was part of the library of the monastery of St. Benedict in Fleury. The initial part (f. 1–8), which was written around the year 1000,[409] contains a collection of astrolabe texts and also drawings of astrolabe parts: a plate (f. 1r), the rete (f. 2v), the back (f. 3v) and the mater (f. 7v), which carry both Latin and Arabic inscriptions.[410] In writing down the numbers, the Arabic letter-notation is rendered in Latin in two forms: (1) by Roman numerals and (2) by writing down in Latin letters the names of the Arabic letters.[411] This notation is therefore different from the one used on the earliest Latin astrolabe(**#3042**).[412] The drawing of the rete is not very accurate, but on the other hand this is the only astrolabe picture in which the zodiac is adorned by artistic representations of the twelve signs. The division of the zodiac, though, lacks any kind of mathematical sense: the author simply divided the zodiac circle into twelve-equal parts, as if it were on the solid sphere and not on the flat one. It is impossible to say whether this was due to error, oversimplification, abstraction or artistic licence.

The manuscript Vat. Reg. lat. 598 (11th c.) was written in Northern France, probably in Fleury or Micy.[413] It contains astrolabe texts and drawings of the back (f.119v)[414] and rete of an astrolabe (f. 120r, fig. 22) which are listed in the Frankfurt catalogue of medieval astronomical instruments as entry **#4553**. This drawing is much more accurate than the one in the Bern manuscript. Using a microfilm

---

408 For bibliographical references on these manuscripts, see app. A, items A.4, A.43 and A.27.
409 This dating has been confirmed to me by Prof. Arno Borst in a private communication. The very early dating (9th–10th century) found in Hagen (1874) p. 246 and Mostert (1989) p. 62 is incorrect as as far as the first folia of the manuscript are concerned.
410 Colour images and a discussion of these folia can be found in: Sezgin/Neubauer (2003) p. 93.
411 I thank Prof. Paul Kunizsch for informing me about the details of the notation used in the inscriptions. See also: Sezgin/ Neubauer (2003) p. 92.
412 See par. 4.1.6.
413 Borst (1989) p. 69, note 121.
414 The image on Vat. Reg. lat. 598 (A.43) f. 119v is reproduced in Millás Vallicrosa (1931) pl. 11.

printout of the image, I have tried to estimate the values of right ascensions used in dividing the zodiac, and the results of my construction are shown in fig. 22.

Even with all uncertainty due to the circumstances, there is no doubt that different values of right ascensions were assigned to different signs: the author of the drawing was conscious that the zodiac should not be divided symmetrically. My attempt to measure the values of right ascension for the six southern signs gave in average, with an error of ± 2°, the result:

Aries and Virgo: 25°; Taurus and Leo: 30°; Gemini and Cancer: 35°.

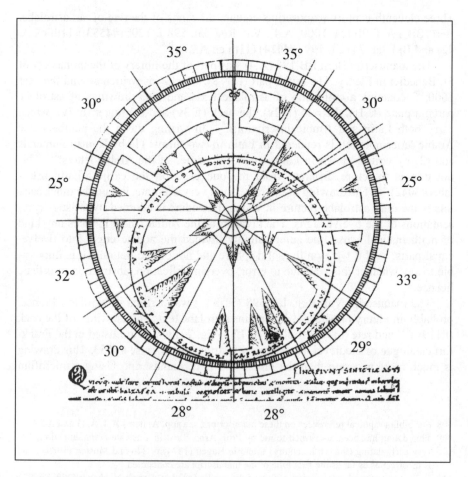

Figure 22: Manuscript. Vat. Reg. lat. 598 f. 120r: rete of an astrolabe (© Biblioteca Apostolica Vaticana (Vaticano)), repoduced from the microfilm stored in the Microfilm Archive of Medieval Scientific manuscripts (Munich University).

These values are poorer approximations than those found in the texts discussed above. However, they at least have the right qualitative structure: the solsticial signs are the largest. The same measure performed on the six northern signs shows instead the wrong distribution of right ascension: Libra (Pisces): 32° (33°) (should be smaller than 30°), Scorpio (Aquarius): 30° (29°) (correct), Sagittarius and Capricorn: 28° (should be greater than 30°). In other words: after approximately but correctly dividing the southern part of the zodiac, the scribe made a mistake in dividing the northern half, assigning the largest right ascensions to the signs nearest to the eastern line rather than to those nearest to the southern one. The asymmetry appears very clearly in the drawing if one tries to prolong the division lines to the other side of the zodiac (dashed lines in fig. 22): the prolonged line should coincide with the division of the opposite sign, but is instead far off it.

Apparently, the geometer was aware that the problem of dividing of the zodiac is not trivial, and also had some numerical values of right ascensions at his disposition. However, he made a similar – but not the same – error as in **dz3** and **pt**. I believe this error can hardly be ascribed to carelessness or gross ignorance, but is rather due to a systematic misconception of the geometrical construction involved in dividing the astrolabe's zodiac. As seen in 4.6.1, the values of right ascension are completely determined by the solid spherical model and are free from distortions due to stereographic projection. Therefore, they are symmetric with respect to the southern as well as to the eastern line. This fact was not clear to our artist, who respected the east-west symmetry, but broke the north-south one, which is the one broken by the projection onto the plane. Even though his division of the zodiac was wrong, it was no simplistic approximation, but rather the result of (misguided) reflection on the problem. Since this type of error was probably produced by Latin scholars rather than taken over from an Arabic source, the fact that it appears in the drawing in Vat. Reg. lat. 598 shows that this figure was not(directly or indirectly) copied from an Arabic original, but independently drawn by Latin geometers.

The drawings of astrolabe parts in the manuscript BnF lat. 7412 (11th c., A.27) f. 19v–23v are very accurately executed and are listed in the Frankfurt catalogue of medieval astronomical instruments as entry **#4024**. The drawings are discussed more in general in 5.6 and here I will only comment on the division of the zodiac in the drawing of the rete, which occurs on f. 19v. In fig. 23 I reproduce he drawing together with my estimates for the values of right ascensions. Averaging among the values which should in principle be identical, I have obtained (with an error of ± 1°): Aries, Virgo, Libra and Pisces 32°, Taurus, Leo, Scorpio and Aquarius 30°; Gemini, Cancer, Sagittarius and Capricorn 28°.

The values are quite good, although the accuracy is hardly surprising since, as the Arabic inscriptions show, the drawing was either executed by some Arabic-speaking scholar or was copied from an Arabic original.

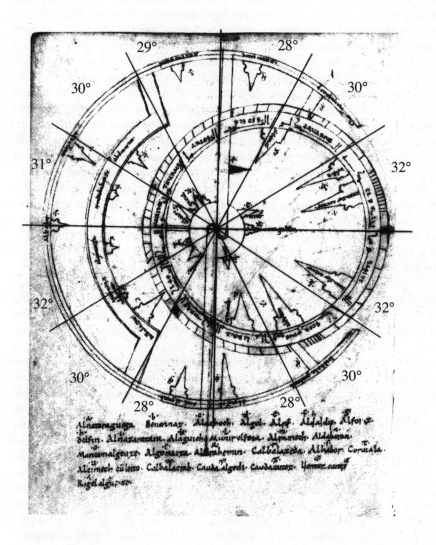

Figure 23: Ms. BnF lat. 7412 (A.27) f. 19v: rete of an astrolabe. Image is reproduced with the permission of the BnF from the microfilm stored in the Microfilm Archive of Medieval Scientific manuscripts (Munich University).

Table 2: Values of zodiacal right ascensions found in 11th.-c. astrolabe texts and in Ptolemy's 'Planisphaerium'.

	Gemini, Cancer, Sagittarius, Capricorn	Taurus, Leo, Scorpio, Aquarius	Aries, Virgo, Libra, Pisces
dz1	32°	30°	28°
dz2	32°	31°	27°
dz3	27° (!)	30°	33° (!)
dz4	33°	30°	27°
pt	27° 33' (!)	30° 18'	32° 09' (!)
pt (corr.)	32° 09'	30° 18'	27° 33'
h1(corr.)	32° 32'	29° 35'	27° 53'
a	30°	30°	30°
h	30	30°	30°
Planisph.	32° 16'	29° 54'	27° 50'

4.6.9 Conclusions: the problem of dividing the astrolabe's zodiac as a focus of natural philosophical discussion in high medieval astrolabe studies

The analysis in the previous paragraphs provides examples of the many facets of the process of assimilation of astrolabe knowledge in eleventh-century Latin Europe. There is indisputable evidence that Latin scholars of the tenth and eleventh century were well aware that the problem of dividing the astrolabe's zodiac needed a non-trivial solution. A variety of numerical values for the right ascensions of the zodiac signs circulated: they are summarized in table 2 and compared once more with Ptolemy's results. I have included in this table the corrected (i.e. inverted) values from **pt**, as well as my reconstruction of the original version in **h1**.

If we accept the corrections in **h1**, no text gives completely wrong numbers. The error occurring most often is not an oversimplification, as claimed by Poulle, but the fact that numerically correct values are attributed to the signs in inverted sequence. This mistake looks rather typical of advanced students more experienced in dealing with number ratios than with geometrical structures. Moreover, trying to grasp the division of the flat sphere's zodiac by geometrical imagination is very tricky, because one has to cope with two distorting factors at the same time: the difference in right ascension due to the inclination of the zodiac and the additional north-south distortion caused by stereographic projection.

The great variety of values found in the texts analysed suggests that Latin scholars were not only aware of the problem of dividing the zodiac, but also actively experimented, producing different solutions for it. We have no hint as to the methods they might have used to perform the computations or geometrical constructions. The very accurate values given by the text **pt** and, according to my reconstruction, also by the text **h1** are evidence of the high level of competence reached. Even assuming that the geometrical constructions on which those values were based were passively taken from Arabic sources, there can be no doubt that the task of expressing the results in number ratios was performed by Latin mathematicians. That might have appeared as the best solution to express degree values which were not integer.

The fact that the texts not only present different values of zodiacal right ascensions, but also different literary styles, is evidence of diverging choices in attaining readability. These choices were guided by the intellectual focus of the author. For example, the use of Boethian notation was fitting when arithmetical accuracy and philosophical literacy were the aim, but it was not very helpful if the aim of the text was to explain a viable procedure for dividing the zodiac on an astrolabe drawing.

Ascelin's and Herman's solution was an optimal compromise between didactical clarity and a geometrical accuracy appropriate to a drawing or artefact. The mathematically more advanced aspects of the problem, on the other hand, were left aside: Herman chose to make his treatise 'readable' in the sense of 'capable of transmitting knowledge exclusively through the written word, unaided by other means'. To achieve this aim, he left out those subjects that he felt could not be properly treated in writing, for example a correct division of the zodiac. Herman's choice surely contributed to the success his treatise enjoyed as a handbook for astrolabe drawing, but it does not imply that the knowledge that could not be entrusted to the written word was lost.

4.6.10 The zodiac circle on the astrolabe #3042

A single Latin astrolabe artefact can be dated to the High Middle Ages, probably the late tenth century: the astrolabe **#3042**, also known as 'Destombes' astrolabe' or 'Carolingian astrolabe'.[415] In this and in the following paragraph, I will discuss the evidence on the division of the zodiac on its rete and compare it with the results of the analysis given in the previous paragraphs. Even within the limits of this subject, the discussion will offer examples of the problems encountered when studying early European astrolabe artefacts.

Paleographical studies have shown that the rete of the astrolabe **#3042** is inscribed in Gothic letters datable to ca. 1200: much later than the inscriptions on the mater and plates of the artefact.[416] Yet there are no indications that the rete itself is a replacement: it is therefore assumed that it was produced at the same time as the rest of the astrolabe, but was engraved later.[417] The first difficult question is therefore: when was the zodiac on the rete of **#3042** divided?

---

415 For a short introduction and bibliography to this artefact, see 3.2.5.
416 Mundó (1995) p. 303.
417 G. L'E. Turner (1995) especially p. 423–427, King (1995) p. 306.

The signs on the rete are not only divided from each other, but also subdivided into smaller parts. The latter graduation is very imprecise and was probably made at the same time as the Gothic inscriptions.[418] Was the zodiac divided into signs at the same time as it was (incompetently) graduated and inscribed with the names of the signs, i.e. after 1200? Or may we assume that the division of the zodiac into signs was performed earlier? In his discussion of the craftsmanship of **#3042**, Gerard Turner remarks:

> The worst marking is on the ecliptic circle in the rete. The lines indicating the single degrees slant in various ways, but most are radial to the centre of the ecliptic circle, and not to the centre of the astrolabe. The dividing lines between the signs of the zodiac are, however, aligned with the centre of the instrument.[419]

Thus, the division of the zodiac (i.e. ecliptic) somehow differs from its graduation: it is possible that it was carried out in a first stage, taking advantage of symmetry, as suggested by most Latin texts. Moreover, a rete lacking a division of the zodiac would have made it very difficult to locate the actual position of the sun on it and would thus have made it hard to use the artefact at all. It is therefore possible to assume that the zodiac was divided when the rete was made, and I shall do so. I also assume that the engravings were done by Latin craftsmen, and not by Arabic ones.

4.6.11 The division of the zodiac on the astrolabe **#3042**

The division of the zodiac on **#3024** was studied in detail by Raymond d'Hollander as part of a general comparison of **#3042** with a Western Arabic astrolabe built by Abū-Bakr ibn Yūsuf in Marrakesch in 1216–1217 (**#1090**).[420] In his paper, d'Hollander measured on the artefact **#3042** the portions of astrolabe rim corresponding to the six signs on the left part of the zodiac and also checked that the division is symmetric with respect to the north-south line.[421] His conclusion was

---

418 "The rete [of **#3042**] originally bore no astronomical markings beyond the star-pointers and no inscriptions. The absurdly inaccurate divisions of the ecliptic scale were added by an incompetent. And the names of the signs were added by an incompetent, probably the same person. It seems beyond doubt that these were added during the period 1200–1400" (King (1995) p. 366). "Some of the degree divisions are so variable that they are twice the width of others [...] The number of divisions in the signs is anomalous, as Libra to Pisces have 31, except for Capricorn which has 32, and Aries to Virgo have 16 each, presumably two unit divisions because of the cramped spacing in that part of the ecliptic." (G. L'E. Turner (1995) . p. 425).

419 G. L'E. Turner (1995) p. 425.

420 The following description is taken from d'Hollander (1995), especially p. 416–417.

421 d'Hollander's aim was to compare these values with theoretical ones computed using as input the value of the inclination of the zodiac previously deduced from a direct measurement on **#3042**. He then compared these results with those obtained by performing the same procedure on the Arabic artefact **#1090**.

that the zodiac of the Latin astrolabe **#3042** was divided correctly, even if with less precision than that of the Arabic artefact **#1090**.[422]

d'Hollander's opinion coincides with that of Emmanuel Poulle who, as already mentioned, expressed the view that no tenth- or eleventh-century Latin European would have been capable of dividing the zodiac so accurately. However, as we have seen, enough information was available in eleventh-century Latin manuscripts to allow a Latin mathematician to divide the astrolabe's zodiac correctly. Let us look at the values measured by d'Hollander on **#3042** and compare them with those taken from Latin astrolabe manuscripts.[423] The values of right ascensions on the astrolabe artefact **#3042** as given by d'Hollander are:

Cancer	31° 36 '	Libra	27° 24'
Leo	30° 24'	Scorpio	30°
Virgo	28°	Capricorn	32° 36'

Table 3: Values of zodiacal right ascensions found in eleventh-century astrolabe texts and on the astrolabe artefact **#3042**.

	Gemini, Cancer, Sagittarius, Capricorn	Taurus, Leo, Scorpio, Aquarius	Aries, Virgo, Libra, Pisces
astrolabe **#3042**	32°00' ± 30'	30° 00' ± 30'	27° 30' ± 30'
dz1	32°	30°	28°
dz2	32°	31°	27°
dz3	27° (!)	30°	33° (!)
dz4	33°	30°	27°
pt	27° 33' (!)	30° 18'	32° 09' (!)
pt (corr.)	32° 09'	30° 18'	27° 33'
h1(corr.)	32° 32'	29° 35'	27° 53'
Planisph.	32° 16'	29° 54'	27° 50'

As we see, values that should be equal (Cancer/Capricorn, Leo/Scorpio, Virgo/Libra) are instead different, but d'Hollander provides no estimate of the measurement error. Since it seems safe to assume that the division of the zodiac

---

422 "La mise en place des extrémités des signes du zodiaque est correcte pour l'astrolabe dit 'carolingien' [i.e. **#3042**], bien qu'on not un écart maximal de 0,7°; la mise en place est excellent pour l'astrolabe d'Abū-Bakr" (d'Hollander (1995) p. 417).

423 The following values are taken from d'Hollander (1995) p. 417, but converted into the same form which I have been using up to now, i.e.. the width of the arc of equator corresponding to each sign measured in sexagesimal degrees and minutes. d'Hollander instead gives in decimal notation the coordinates of the division points measured on the astrolabe rim assigning 0° to the equinoctial point and respectively +90° and –90° to the solsticial ones.

was carried out by using the same three values of right ascension for all quarters of the zodiac, I will take the mean values and round up the results to the next half degree with an error of 30' (the values add up to 90° within the error). In table 3, these results are compared to the values given in the texts **dz1**, **dz2**, **dz3**, **dz4**, **pt**, **pt** (corr.), **h1** (corr.) and in the 'Planisphaerium'. The entries indicated with (corr.) are corrected as in table 2.

The division of the zodiac on the Latin astrolabe **#3042** could have been performed using the instructions contained in Latin manuscripts. The correspondence with **pt** (corr.) looks very good, but given the imprecision of the artefact it is doubtful whether such precise values could actually have been used. In general, the errors are too large to make any decisive statements, but the agreement with **dz1** seems better than with **dz2**, **dz3** (which is anyway wrong) or **dz4**. This is interesting because we remember that **dz1** did not directly provide numerical values for the right ascension, but described a construction procedure for performing the division.

In his analysis of the craftsmanship of the astrolabe **#3042**, Gerard Turner concluded that the artefact had been made rather by a scholar than by an expert in metalworking:

> This astrolabe shows an intelligence behind it, but one lacking in craft skills. The brass could be acquired fairly readily. The mathematical lines are drawn in the same manner as would be done on vellum or paper.[424]

This fact suggests that the maker of the astrolabe had experience in drawing mathematical constructions, but not in engraving them, as could be expected by students who had learned the mathematical arts not only from texts but also from exercise. Turner continues:

> A pair of compasses has been used to draw out the lines. The centres of circles are revealed by pronounced dots that would have located the leg of the compass firmly. Many very short arcs show how positions were stepped by the compass.[425]

If the maker of **#3042** was skilled in using compass and ruler, he might have used the geometrical construction described in **dz1**. On the other hand, he also graduated the rim of the astrolabe in single degrees and could have used them to locate the zodiac signs. It would be interesting to know whether the artefact bears signs of how its maker actually performed the division.

---

424 G. L'E. Turner (1995) p. 425.
425 G. L'E. Turner (1995) p. 424.

5 THE PHILOSOPHICAL ASPECTS OF THE ASTROLABE

5.1 WHY DID MEDIEVAL LATIN SCHOLARS STUDY THE ASTROLABE?

5.1.1 The astrolabe as a new mathematical and philosophical instrument of the mathematical arts

The present chapter is devoted to investigating the motivations that might have led Latin scholars to take an interest in the astrolabe. In historical studies, it has often been taken for granted that tenth- and eleventh-century astrolabes could be useful for time-keeping or that they might be interesting as didactic tools.[426] It has also often been denied that high medieval astrolabe studies had anything to do with astrology, although in the last decades the importance of this aspect of the astrolabe has been increasingly recognized. Charles Burnett has pointed out that the diffusion of astrolabe texts in eleventh-century Europe can be linked to that of a group of astrological works which David Juste has recently edited and commented: the 'Alchandreana'.[427]

Since the astrolabe is a material device, it has usually been assumed that it was interesting not for philosophical, but mostly for practical purposes like time-keeping or eventually as a tool for astrological predictions. However, I believe that Latin astrolabe studies might have been motivated by philosophical interests which, although they had to do with the measure of time and with astrology, were not identical with time-keeping or astrological fortune-telling.

The astrolabe was philosophically interesting because, like the other instruments of the mathematical arts (abacus, compass, monochord), it provided a way to both rationally understand sensible phenomena and relate to them in practice. It was a means to understand, reconstruct or predict them according to mathematical patterns. As such, the astrolabe was also an astrological tool. I believe there can be no doubt that Latin interest in the astrolabe was at the same time an interest in astrology, since astrology was at the time also a rational worldview in which the patterns regulating cosmic order could be, at least to some extent, grasped with

---

426 Lindgren (1976) p. 33, Bergmann (1985) p. 57 and p. 216–217, Borst (1989) p. 77–84, Mc-Cluskey (1998) p. 170–173. Marianne Hess and Peter Conzelmann have discussed this question in the specific case of Herman of Reichenau, but their treatment is so biased by the assumed opposition between understanding the theory of stereographic projection and using the astrolabe for practical purposes that they inevitably conclude "daß er [i.e. Herman] die Materie der Astronomie für die das Astrolab betreffenden Belange nicht vollständig beherrschte, da ihm zu seiner Zeit und in seiner persönlichen Lage die Voraussetzungen nicht gegeben waren, daß er aber über die Grenzen einer Anwendung der Astronomie allein für die praktische Berechnung des Computus hinausstrebte." (Conzelmann/Hess (1980–81) p. 63).

427 Burnett (1998), especially p. 338–343. The same opinion is expressed in Sturlese (1993) p. 54–65. The edition of the texts is Juste (2007).

mathematical methods.[428] From the middle of the ninth century onward, the Magi of St. Matthew, who had been condemned as 'malefici' by Augustine, began to be rehabilitated, even as astrologers.[429] Medieval astrology only became a forbidden discipline when the extent to which cosmic order was thought of as rational and mathematical became a menace for the free will of Man or, even worse, for that of God.

My views on the astrological component of the philosophical relevance of the astrolabe are supported by the results of David Juste's study and edition of the 'Alchandreana'.[430] According to Juste's analysis, the 'Alchandreana', which never mention the astrolabe, are older than the oldest Latin astrolabe texts, and were probably written even before the introduction of the instrument in Catalonia.[431] However, the 'Alchandreana' texts share with astrolabe literature not only part of their manuscript traditon, but also some terminology, while astrolabe manuscripts provide plenty astrologically relevant information. What is even more important, Juste explicitly notes how no distinction can be made between astrological and astronomical matters:

> L'examen des manuscrits anciens met en évidence un fait suffisamment récurrent pour qu'il ne puisse être passé sous silence: tous ces manuscrits sont des manuscrits scientifiques, et des manuscrits scientifiques d'excellent tenue. [...] Quoi qu'il en soit, ces manuscrits montrent bien que, dans l'esprit des savants de la fin du Xe au début du XIIe siècle, les 'Alchandreana' font partie des disciplines du quadrivium.[432]

It was precisely in such a context that the astrolabe could acquire a high philosophical significance.

The reason why the astrolabe, and not some Greek-Arabic astronomical text, first attracted the attention of Latin scholars must be sought in the specific features of high medieval mathematical thought. As shown chapter 4, mathematical knowledge circulating in tenth- and eleventh-century Europe was stored and transmitted not only by means of the written word, but also by using non-verbal and non-written modes of communication. Within this cultural context, Arabic knowledge about the astrolabe, its design and its possible uses could appeal more to Latin mathematicians than the Arabic versions of Euclid's or Ptolemy's masterworks. Then again, non-verbal and non-written astrolabe knowledge was complemented by written words in the form of excerpts from Ptolemy or al-Khwārizmī.[433]

For medieval Latin scholars, the astrolabe was more an answer than a question: it was a new instrument to acquire astronomical knowledge needed to address problems which had already interested Latin mathematicians in Carolingian times. In high medieval mathematics, practice and reflection did not

428 Garin (1976), North (1987), Grafton (2000) especially p. 82, Kieckhefer (1994) p.128–146.
429 Flint (1991) p. 364–375.
430 Juste's edition has appeared during the final revision of the present work, and therefore I was unable to take advantag of it beyond a few remarks in sections 5.4.3 and 5.5.1.
431 Juste (2007) p 246-249.
432 Juste (2007) p. 265.
433 See 3.3.9 (al-Khwārizmī) and 4.5.2 (Ptolemy).

stand side by side, but largely coincided. The astrolabe, like the abacus or the monochord, was at the same time a material device, an abstract pattern of thought and a symbolic representation. This (for us) ambiguous status was not peculiar only to the instruments of the arts of quadrivium: also artefacts used in everyday life had the same character, for example tally sticks and standards for measures.[434] As discussed at length in 4.1.3–5, the blurring of the border between an abstract pattern and one perceived by the senses was a characteristic feature of the monastic craft of memory.

The astrolabe provided a means to conceive space and especially time as measurable, i.e. as something that could be put in relationship with geometrical constructions or numbers. For medieval times, this was a great conceptual innovation, and the subject of the measure of time and motion would provide material of discussion at least up to late medieval times.[435] As I shall endeavour to show with my analysis of BnF lat. 7412 (A.27), the focus of interest of eleventh-century astrolabe manuscripts were those mathematical or rational patterns that through sensory, bodily experience and through the study of literary authorities could be recognized as ordering Creation and in particular as structuring space and time. For example, the pattern of equal and unequal hours or that of the seven climates

One caution: the idea of searching for mathematical patterns ruling phenomena might at first sight appear in some sense 'scientific', but any comparison with the modern experimental method should be avoided, since the basic premises for such a comparison are lacking. On the one side, it is impossible to conceive the experimental method of modern science without the distinction theory-practice which, as we saw, was absent in high medieval natural philosophy.[436] On the other hand, trying to separate medieval natural philosophy from theology or literary studies could only lead to artificial oversimplifications.

5.1.2 The main thesis of the present chapter: the astrolabe as 'architectonica ratio'

Different forms of mathematical communication and thought can be linked to differences in the image of knowledge in which the mathematical arts are embedded, as already discussed in 4.1.7. The medieval mathematical arts worked in practice, for example in surveying, gnomonics, architecture or mechanics, and they could thus be regarded as a form of rational knowledge leading to effects perceivable to the senses and not only to the mind. For example, the act of using the compass could be both a way to think of a circle and an act of embodying the abstract geometrical structure in a real drawing. Because of this close interplay between thought and action, the construction and use of geometrical structures or mechani-

---

434 Witthöft (1983), Kuchenbuch (1999).
435 Sarnowsky (1983).
436 Various kinds of distinctions between 'theoretical' and 'practical' knowledge were present in medieval thought, but they did not correspond with the modern conceptions. On this subject, see 4.1.7.

cal devices could be regarded as a method not only for pursuing a material aim, but also for attaining knowledge of cosmic order, an order that was constantly thought of as being at the same time rational, natural and divine.

I believe that in the context of Latin medieval astrolabe studies a concept of 'rationality' ('ratio') could prevail in which material experience could play a role in determining whether ideas or methods were 'rational' or not.[437] When patterns could be employed to generate predictable effects and to relate a broad range of phenomena to quantities in a uniform way, this fact could be regarded as validating their rationality. These rational patterns were thought of not so much as abstract structures (e.g. a circle, a two-dimensional geometric model), but as methods of construction and/or use of material devices (e.g. using a compass to draw a circle, making and using an astrolabe).

As I will try to show with my analysis of BnF lat. 7412 (A.27), I believe this is the reason why astrolabe manuscripts group together literarily elaborated texts of abstract philosophical content (5.2), inelegant, recipe-like passages explaining how to measure or count the parts of space and time (5.3), sketches illustrating various devices to be built and used to the same aim (5.4, 5.5, 5.6), geographical excerpts from the Latin classics (5.3.4) and, at times, also descriptions of experiments impossible to perform (5.3.7). As I will argue in 5.2.2, the idea that, in some cases, practising mechanical crafts might be a path to philosophical knowledge could find expression in the frequent use of Vitruvius' term 'architectonica ratio', which could also become a 'mechanica ratio'. In astrolabe texts, the cosmos was a 'machina mundi' not just because it was mathematically ordered, but also because that order could eventually be investigated and understood thanks to 'machinationes'.

In telling the hour of the day, in correctly predicting seasonal changes of daylight, or in estimating the depth of a body of water, mortals could be regarded as showing their 'similitudo dei' because they proved to have grasped the 'ratio' ordering God's creation. The rational patterns according to which human artifices worked - if and when they worked - were analogous (or even the same?) to those employed by God in creating the world. Symbolic representations of this concept of rationality were the miniatures of 'deus geometra' with compass and scales. Although the best known images of this kind are those from thirteenth-century Bible Moralisée, the earliest ones are from the eleventh century, for example the one illustrating the act of Creation in the Eadwin Gospels (ca. 1025) as well as in other illuminated bibles from the middle of the eleventh century.[438] In the twelfth century, stone angels holding astrolabes in their hands would represent cosmic ratio on the Royal Portal of Chartres cathedral (ca. 1150).[439]

---

437 'Ratio' was a central concept in the images of knowledge of high medieval Latin culture and it was then, like 'rationality' today, a highly problematic concept. On the problematic aspects of the concept of rationality, and especially on its variety, see: Turnbull (2000b) especially p. 45–48.

438 On the earliest images and description of 'deus geometra' see Block Frieman (1974), for 11th.-century images: fig.1–3. On the thirteenth-century images: Tachau (1998).

439 Houvet (1925) p. 67.

In the context of this image of knowledge, the mechanical arts could come to be valued as much as the liberal ones, a trend that was present in the monastic culture of the tenth and eleventh century, as testified also by the fact that many prominent figures of the time occasionally devoted themselves to mechanical arts such as architecture or metal-working.[440] This trend would be excluded from the literary culture of twelfth-century cathedral schools and thirteenth-century universities, but would remain alive outside of them. When discussing the history of the astrolabe in the Late Middle Ages, I will suggest a similarity between the epistemology traditionally linked the astrolabe and the 'artisanal epistemology' of the Renaissance described by Pamela H. Smith in her work 'The body of the artisan' (2004).[441] However, in the present chapter I shall argue for my thesis exclusively on the basis of contemporary, i.e. high medieval, sources.

5.1.3 Was there actually an interplay between the abstract and the material side in astrolabe studies?

According to my thesis, in some eleventh-century Latin circles, the act of constructing and successfully using devices such as astrolabes or sundials could become epistemologically relevant. While in surveying, architecture or mechanics an interplay between abstract and material mathematical construction can hardly be doubted, one may wonder whether a practical feed-back from astrolabe studies actually existed. First of all, let me clearly state that this question can only be asked in the case of a relatively small circle of learned persons: in the eleventh century, astrolabe studies cannot be expected to have generated and received practical feedback at a broader social level. Only from the late eleventh century onward, due to the growing importance of astrology as a political factor, astrolabes and other astronomical devices may have aroused the interest of more than a few scholars.

Eleventh-century astrolabe texts have normative character, in that they describe how the astrolabe should be used: descriptive texts on how astrolabes were actually used to some practical aim are much rarer. No written source tells us that astrolabes were ever actually used to tell time. As I shall discuss in 5.2.5, 5.3.7 and 5.4.5, I believe that, while the mathematical aspects of the astrolabe are essential for relating experienced time flow to the geometrical patterns of celestial movement, the actual quantitative estimates necessary to quantify duration (i.e. altitude of sun or stars) might have been performed in practice with other instruments (quadrant, gnomon, geometric square). There is only one text describing an instance of actual use of an astrolabe in the eleventh century: it is a passage in which Walcher (d. 1135), monk from Lotharingia and prior of the monastery of

---

440 On the value assigned to the 'artes mechanicae' in the Middle Ages, see: Alessio (1965), Sternagel (1966), Nobis (1969) esp. p. 41–47, Conant (1971), White (1978b), Skubiszewski (1990) esp. p. 285–308, Boehm (1993), Galloni (1998), Hiscock (2000) p. 158–166.
441 P. H. Smith (2004), see 6.6.

Great Malvern in England, described how he used an astrolabe to measure the exact moment of a lunar eclipse in the year 1092:

> Mox enim apprehenso astrolapsu horam qua totam nigredo caliginosa lunam absorbuerat diligenter inspexi, et XIa noctis agebatur hora .iii. puncto peracto.[442]

The exact intepretation of this passage has been object of discussion, and it is not sure whether it should count as evidence that astrolabes were actually used to perform altitude measurements. In any case, even counting Walcher's testimony, as already noted in 2.3.4., the astrolabe was in all probability very rarely used for measurements: in the field of astronomy, immediate connection with the phenomena was rather provided by naked-eye observations, eventually aided by sighting tubes, quadrants or sundials. Observing the movements of sun and stars could provide confirmation of the validity of the 'ratio' of the astrolabe, but the most fruitful activity from the mathematical and philosophical point of view was the construction of 'horologia', which is constantly mentioned as a reason for studying the astrolabe. The term 'horologium' could indicate a quadrant, an astrolabe or a table of length of shadows, but more often than not it referred to sundials.[443]

In the twelfth century, as we have seen, the most widely diffused collection of astrolabe texts was made up of a treatise on astrolabe construction, a treatise on astrolabe use and a text devoted to a vertical sundial that could be built for any latitude thanks to the help of the astrolabe. The sundial described there is often taken to be an invention of Herman of Reichenau, to whom the text is attributed, but a recent archaeological discovery has shown that similar, if not identical, dials already existed in Roman times.[444] As we shall see later on, in the manuscript BnF lat. 7412 (A.27) a particular interesting drawing of a sundial occurs.[445] I believe that the construction and use of sundials might have provided an important component of astrolabic and, in general, astronomical studies. In the next paragraph I shall therefore sketch the (little) evidence we possess on medieval sundials.

---

442 Haskins (1924) p. 115. My translation is: "Immediately taking an astrolabe, I determined by careful inspection the hour in which the cloudy darkness had absorbed the whole of the moon, and it was the ninth hour of the night, in the fourth quarter". For Walcher, a 'punctum' corresponded to a quarter of an hour. Walcher's text and its interpretation are discussed in: Haskins (1924) p. 114–118 and in McCluskey (1998) p. 180–184.

443 On the term 'horologium', see McCluskey (2003) p. 215, n. 23.

444 Arnaldi/Schaldach (1997) especially p. 114–115.

445 See 5.5.

5.1.4 Evidence on medieval sundials

A comprehensive history of Latin medieval sundials still has to be written, and no systematic documentation of material and written sources on this subject exists.[446] A large amount of information was collected and published by Ernst Zinner, and more is possibly still hidden in his archive, which is today located at Frankfurt University.[447]

Sundials were built in medieval Europe from the seventh century onwards and they were different from the standard horizontal Greek-Roman ones: most extant medieval stone dials are vertical and have the form of a half-circle divided into a number of equal parts.[448] They are usually found on the South walls of churches. In the tenth and eleventh century, interest in building sundials in new forms apparently increased, especially in England, although interpretations of the form and possible uses of extant dials remain tentative.[449] It is not even clear whether high medieval sundials on churches were really supposed to show the time of prayer or whether they had a representational function.[450] In some instances, they might even have been elements of earlier sun-worship which had been assimilated by Christendom.[451] At the present state of research, material evidence cannot tell us much about the motives for building and using gnomons and sundials in the Early and High Middle Ages.

It is interesting for our subject to note that also in tenth- and eleventh-century al-Andalus interest in building sundials seems to have increased. Extant sundials are mostly of the classical Greek-Roman kind, with shadow lines in the form of hyperbolae, but there is evidence also of another kind of dials, where a circle was divided into twelve equal parts.[452] As far as written sources are concerned, the classic authorities on gnomonics and sundials were Pliny and Vitruvius. In particular, Vitruvius devoted much attention to the description of rather complex geometrical constructions and devices and, as we shall see, he was quoted as an authority on the astrolabe, too.[453] In medieval times, new texts on sundials were composed, but they have not yet been properly investigated. For example, a short text circulated which described how a gnomon can be used to estimate the time by

---

446 For a recent review of the actual state of documentation and research on the history of gnomonics and sundials, see: King (2004). A very recent publication with images and descriptions and images of sundials in Mecklenburg-Vorpommern from the Middle Ages until ca. 1800 is: Hamel (2007). A short overview on the history of gnomonics and sundials is: A. J. Turner (1989). On vertical sundials: Rau/Schaldach (1994).

447 Zinner (1964), Zinner (1967). On the Zinner Archive: Ackermann (2001).

448 Gunther (1923) p. 97–101, Zinner (1964) p. 1–4, Zinner (1967) p. 46–52, Rau/Schaldach (1994) p. 273–276.

449 Rau/Schaldach (1994) p. 286–288.

450 The half-circle dial divided into twelve equal parts is rather imprecise at higher latitudes (Zinner (1967) p. 46–47).

451 Rau/Schaldach (1994) p. 277–281.

452 Samsó (1991) p. 5–6, Samsó (1992) p. 98–105. Julio Samsó suggests that these dials might have had some connection with the ones found in English churches.

453 Schuler (2000) p. 321–322.

the length of its shadow. It was attributed to Beda Venerabilis (673/674–735) and was sometimes associated to astrolabe texts.[454] In one of his letters, Gerbert of Aurillac gave indications on how to build a sundial taking into account local latitude as suggested by Martianus Capella.[455] The texts on sundials that can tell us more about their possible epistemological relevance, though, are passages in treatises on 'computus' where experiences with 'horologia' are mentioned.

5.1.5 Evidence on gnomons and sundials as devices for astronomical observations

Passages on astronomical observations in computistic literature seem at first sight to suggest a growing interest in sundials and gnomons, but it is rather difficult to estimate how much they tell us about actual observational practice and its philosophical relevance.[456] In discussing the determination of the correct date of the vernal equinox (21st vs. 25th of March) to be used in computing Easter, Beda claimed that the Nicean date (21st March) could be proven correct also by means of 'horologica inspectio'. In Bede's time, though, the 21st of March did not correspond any more to the observable equinox, which had moved to the 16th or 17th of the same month. Not enough details on Bede's 'horologica inspectio' are known to evaluate it more precisely. Perhaps the measurement precision was only enough to tell that the equinox had moved backwards, or maybe the experiences mentioned were never actually performed.

More detailed instructions on how to observe the equinox were given in the computistic treatise of Helperic of Auxerre (9th c.): solstices and equinoxes could be determined by observing the varying direction of the rays from the rising sun hitting the West wall of a room. Helperic's description is very detailed and the author explicitly states how, by performing the experience with great care ('studiosissime', 'vigilantissime'), the Nicenian date for the winter solstice could be observed – even if today's computations tell us that it could not. However, it is difficult to pass judgement on Helperic's account without at least testing the feasibility and sensitivity of his experiment with a reconstruction. Joachim Wiesenbach attributes the result to autosuggestion, while Stephen McCluskey believes that the description had a 'didactic focus', i.e. was rather aimed at illustrating astronomical concepts and confirming the accepted system than at testing it.[457] Thus, the description may have to be taken as an exemplary thought experiment. However, the mere fact that such a detailed description of an 'experimentum' was introduced into a computistical treatise is evidence of an image of knowledge in which systematic, rational practice could at least in principle be expected to de-

---

454 Ps.-Beda (1904b), Wiesenbach (1991) p. 128.
455 Bubnov (1899) p. 38–41.
456 On the relevant passages and the problems of interpretation: Wiesenbach (1991) p. 115–124, McCluskey (2003) p. 205–211, on which the following discussion is based.
457 Wiesenbach (1991) p. 122, McCluskey (2003) p. 205.

liver epistemologically relevant results – even if the final answer could be safely be assumed to agree with traditional authorities.

5.1.6 The analysis of ms. BnF lat. 7412 as evidence on the motives behind for Latin astrolabe studies

In the following chapters, I will look more closely at the contents and structure of one particular eleventh-century manuscript: BnF lat. 7412 (A.27).[458] As is the case with most eleventh-century astrolabe manuscripts, this one, too, contains an apparently inhomogeneous collection of material only loosely bound under the heading 'mathematical arts': abstract philosophical reflections expressed in a rather elaborate Latin style, recipe-like texts describing how to build and use surveying tools or how to draw astrolabe lines and also carefully executed drawings of devices which are accompanied at most by a couple of explanatory verses.

I will analyse and offer an interpretation of some of this material, searching for a line of thought connecting the elements of this apparently disparate collection of texts and images. A special effort is devoted to understanding the texts and drawings which may at first appear more devoid of sense (e.g. chapter **J 8** or the drawings scattered on f. 14v). Occasionally, I have complemented the analysis of BnF lat. 7412 (A.27) with references to material taken from other contemporary sources.

Apart from the fragments of Ptolemy's 'Planisphaerium', there are no texts which are only peculiar to this manuscript: all material contained can be found in many other contemporary parchments, and the same is valid for most of the drawings. Because of this, the results of my analysis provide a general indication on the interest behind the composition of eleventh-century astrolabe manuscripts. The thesis that I hope to support with my analysis is the one stated in 5.1.1. and 5.1.2. I do not aim at offering a cogent argument in its favour, but I trust that, even if the reader will not find all of my interpretations convincing, enough indications will remain to make my arguments at least plausible.

5.2 THE ASTROLABE AS 'ARCHITECTONICA RATIO' OF THE 'MACHINA MUNDI'

5.2.1 The first chapter of the text J: the 'machina mundi' and its 'architectonica ratio'

The first nine folia of the manuscript BnF lat. 7412 (A.27) contain twenty out of twenty-one chapters of the text **J** on astrolabe use. Apart from **J 19**, the chapters are in the same form and order as in Bubnov's edition. Chapter **J 1** is a general

---

458 For references on this manuscript, see app. A, item A.27.

introduction and contains very clear statements on the motives for learning the uses of the astrolabe:

> Quicunque astronomicae discere peritiam disciplinae et coelestium sphaerarum geomet-ricaliumque mensurarum altiorem scientiam diligenti veritatis inquisitione altius rimari conatur, et certissimas horologiorum quorumlibet climatum rationes, et quaelibet ad haec pertinentia industrius discriminare nititur, hanc Walzagoram, id est planam sphaeram Ptolomaei seu astrolapsum,sollerti indagatione perquirat et discat.[459]

These introductory words put into focus central elements of the image of knowledge governing the whole text: the aim to be attained was higher rational knowledge of the geometrical measure of the heavens, which was at the same time knowledge of the 'certissimas rationes' for building specific artefacts. Those who wished to attain this aim were advised to become well acquainted ('perquirat et discat') with the astrolabe, i.e. with "hanc Walzagoram, id est planam spaeram Ptolomaei seu astrolapsum".[460] This phrase can be translated as "this walzagora, that is Ptolemy's flat sphere, or the astrolabe", and in fact the three terms ('walzagora', 'plana sphaera Ptolomaei' and 'astrolapsus/astrolabium') are mostly used as synonyms to indicate the astrolabe. However, the term 'walzagora' (also: 'wazzalcora', 'waztalchora') was clearly the Latin transcription of an Arabic expression, and there are indications that this expression might have been the Arabic title of Ptolemy's 'Planisphaerium'. In Arabic, Ptolemy's treatise was called the book "on the expansion of the surface of a sphere on a plane" (fī tasṭīḥ basīṭ al-kura), an expression whose final part, i.e. 'basīṭ al-kura' meaning 'the surface o the sphere', is pronounced more or less 'basit alkura', which sounds similar, albeit not identical to 'watzalgora'.[461]

Thus, it is possible that the term 'walzagora' was meant here to refer to Ptolemy's work which, as discussed in 4.5, was very closely linked to the earliest Latin texts on astrolabe construction and has also left many traces in the manuscript BnF lat. 7412 (A.27).[462] The term 'astrolapsus', on the other hand, clearly refers to the device described in the text. However, I would suggest that, as in the case of the monochord, the term did not necessarily mean a material artefact.

After the opening passage, chapter **J 1** lists the astronomical information that could be obtained by means of the astrolabe, showing that the interest in it also had a practical side. Thanks to the astrolabe, one could determine:[463]

---

459 Bubnov (1899) p. 114. My translation is: "All those who strive to become experts in astronomy and, thanks to a diligent search for truth, reach a higher degree of knowledge of celestial spheres and geometrical measures, and desire to actively discern the surest rational patterns of sundials for any climate and all possible things relating to them, they should become well acquainted with this walzagora, i.e. the flat sphere of Ptolemy, or astrolabe."
460 Bubnov (1899) p. 114–115.
461 Kunitzsch/Lorch (1994) p. 13, n. 1. The identification of the two terms with each other in Kunitzsch (1992) p. 517–518 had been based on an older, incorrect reading of a manuscript. On examining personally the manuscript, Kunitzsch revised his opinion to a more careful one.
462 See 5.4.3.
463 The following list is a summary of the text printed in: Bubnov (1899) p. 116.

- times of rising and setting and positions of stars and zodiac signs,
- the position of the sun on the zodiac,
- the altitude above the horizon of the sun and stars,
- the circumference of the earth sphere,
- the hours of night and day, both natural and artificial ('naturales sive artificiales'), as well as their increase and decrease;
- the changes in sundials ('horologia') depending on the climate.

None of these data are immediately relevant for everyday monastic timekeeping, i.e. for fixing the time of prayer. The only exception are temporal hours ('horae artificiales'). In fact, after mentioning them, the author adds that knowledge of the hours is necessary for celebrating services and for eliminating 'false time-keepers' ('pseudohorologia'). However, as already noted by Bubnov, this remark seems introduced rather to prevent objections to the piety of astrolabe studies than to explain the main reason to study and use the astrolabe.[464]

The list quoted above shows that the 'utilitates' of the astrolabe were the possibilities it offered to reconstruct observable natural phenomena with geometrical certainty. The astrolabe could do this thanks to the information inscribed on it ("ejus perigraphiis"), that could be compared to an "architectonica seu mechanica ratione".[465] I suggest that this expression can be translated as "architectonical or mechanical rationality", i.e. a form of rationality expressing itself in (and recognizable from) material constructions which work, e.g. a building that stands or a sundial that correctly tells the time of day. After having mentioned the "architectonica seu mechanica ratio" in the introductory chapter, in the following one the author uses the verb 'machinare' various times to describe the most ingenious details of the astrolabe, such as the altitude circles on the horizon plates, which are "artificiosa industria machinati".[466]

5.2.2 The Prologue 'Ad intimas' (pa) and its possible relation to Vitruvius

In other eleventh-century astrolabe texts the word 'machina' is used often in the sense of a 'machina mundi'. This is the case in a text which is usually referred to as 'Ad intimas' and regarded as a prologue to a (lost) astrolabe treatise. This text, which I shall indicate as pa, was edited by Bubnov among Gerbert's 'opera incerta' and later also by Millás Vallicrosa, and is sometimes attributed to Lupitus

---

464 Bubnov (1899) p. 116, n. 7.
465 Bubnov (1899) p. 116–117.
466 Bubnov (1899) p. 119, 'machinare' found also on p. 121 and 122. Markus Popplow states that in medieval times "'Machina' betonte [...] den Aspekt eines zusammengefügten, in sich stabiles Gebildes" (Popplow (1993) p. 14). In his essay, Popplow focuses on the stability of the medieval 'machina', as opposed to the mobility of the Renaissance one. However, his analsis shows how the idea that the stability resulted from a raional construction was just as important.

to Barcelona.[467] The text **pa** occurs in BnF lat. 7412 (A.27) f. 16v–18r. I shall discuss it here briefly, because it has much in common with the first chapter of **J**.

The text **pa** is longer than the chapter **J 1** and its Latin style is more elaborate, but its message is very similar and the references to the importance of using the astrolabe for connecting the rational pattern of mathematics to the evidence of the senses are very explicit. The author explains that astronomy deals with the divine and rational order connecting heavenly and earthly phenomena, and knowledge of this discipline can be attained by actively investigating natural phenomena and by reaching up from there, with the help of reason, to understand divine order "quasi ab activis ad contemplativa".[468]

> Quod pulchrius documentum, quam corporis animeque intentione celum subire totamque illam supernam machinam indagabili ratione rimari et quedam velata archana divina ratione percipere et mysticis notitiis discere et per visibilis sphere contemplationem ad invisibilium attingere confinium.[469]

Astronomy is a particularly worthy discipline, because its subject matter is the heavenly machine, which works not because of human activity, but thanks to a necessary, eternal rule:

> Digna est tenenda astronomica disciplina, cuius substantia superna est machina. Non enim hec actu consistit humano: habet enim suum non utrumlibet, sed naturalem immutabilitatis effectum, normam necessaria perpetuitate dispositam.[470]

The text **pa** contains a long overview on the phenomena in whose variations, thanks to astronomy, order can be recognized: the day-night cycle and its variation during the year, the movements of the stars, the seasons, the rhythms of plant-life, the moon phases and tides, the flow of humors in earthly bodies.[471]

After discussing these matters, which have little or nothing to do with liturgical time-keeping, the author makes an explicit reference to the Bible: the nativity star and the eclipse at Christ's Passion prove that it pleases God to use astronomical phenomena to confer a 'showy' side to his miracles ("Deus [...] miracula sue divinitatis arbitrio superventura, per hec ostentaria stellarum novitate vult insignire").[472] However, he adds that all 'Chaldaic' doctrines which fully subordinate human will to the stars are to be rejected, and then states the importance of astronomy for correctly computing the Easter date and for performing liturgical of-

467 Bubnov (1899) p. 370–375, Millás Vallicrosa (1931) p. 271–275.
468 Millás Vallicrosa (1931) p. 271.
469 Millás Vallicrosa (1931) p. 274. My translation of the passage is: "What more beautiful proof, than nearing the sky by intension of body and soul, exploring the whole machine above us through a rationality that can be investigated, and perceiving hidden, high secrets through divine rationality and learning them with mystical concepts, and, by contemplating the visible sphere, reaching the border to the invisible".
470 Millás Vallicrosa (1931) p. 272. My translation is: "The discipline of astronomy must be regarded as having a special dignity, because its subject is the machine above us: this machine does not exist and persist out of a human act: it has its natural, unchanging function and rule fixed not arbitrarily, one way or the other, but by eternal necessary permanence".
471 Millás Vallicrosa (1931) p. 272–273.
472 Millás Vallicrosa (1931) p. 273.

fices at the appropriate time. The author claims that astronomy and astrology are not generally condemned by the Holy Scriptures, and refers to the authority of Abraham, who had introduced the mathematical science of the stars into Egypt.[473] The association of Abraham with astrology, based on the authority of Flavius Josephus (ca. 37-97), was not uncommon in medieval times, as it could be found in the 'Etymologiae' of Isidor of Sevilla (ca. 570–636).[474] As we shall see in 5.6.3, Abraham could be associated not only with astrology, but also with the astrolabe itself.

The liturgical utility of astronomy does not occupy a central position among the motivations for learning this discipline. It would rather seem that the subject is quoted to support the argument that the science of the stars is fully compatible with Christian religion. The main focus of the text **pa** remains the search for order behind phenomena with the help of the 'architectonica ratio' of the astrolabe, i.e. Ptolemy's flat sphere:

> Nam inter cetera huius artis insignia ab ipso [i.e. Ptolomeo] subministrata adiumenta quoddam instrumentum et utillimum discentibus et magnum miraculum considerantibus adinuenerit. Quo quidem inter omnia inuenta nil prestantius ad intimas doctrinarum indagationes et matheseos artes nihilque utilius ad totam illam supernam machinam investigandam et ad omnia astronomica studia atque geomericalem scientiam. Est autem Wazzalcora divina mente comparata, quod latine sonat plana spera, que etiam alio nomine astrolapsus Ptolomei. In qua Wazzalcora secundum celi rotunditatem formata naturali ratione tota celestis sphere describitur forma et omnia ritu celestium figurationum architectonica ratione notantur.[475]

In the prologue **pa,** neoplatonic influences are quite evident and one may ask whether the references to the investigation of natural phenomena really express an interest for observation and experiment or whether they are only to be understood as a literary topos. At the same time, though, not a few expression in this text are taken from the Latin author who devoted most attention to real 'machinae': Vitruvius. First of all, we note that in the last lines of the prologue Vitruvius is quoted

473 "Cuius quidem propter celebrem sue dignitatis cultum et multiplicem utilitatis fructum et si plurimi essent auctores, silicet ut Abraham, qui ut Iosephus refert, arithmeticm et astronomiam primum Aegyptiis tradidit, tamen inter omnes precipue Ptolomeus hac claruit disciplina." (Millás Vallicrosa (1931) p. 274).

474 The statement is found in Isidor of Sevilla's Etymologiae, book III, 25 (Isidor of Sevilla (1911)) who refers to Flavius Josephus's Antiquitates judaicae, I, 7–8.

475 Millás Vallicrosa (1931) p. 274. My translation is: "Among the additional instruments given by the same [Ptolemy] to this art, he invented a certain instrument which is very useful for those who are learning and is object of great admiration to those who reflect. Among all discoveries there is none which is more efficient for an intimate exploration of the doctrines and for the mathematical arts, and none which is more useful to investigate the whole of the machine above us, as well as for all astronomical studies and for geometrical knowledge. The wazzalcora is similar to the divine mind, its name in Latin sounds 'flat sphere', but it is also called astrolabe of Ptolemy. In the wazzalcora, there is a description of the whole form of the heavenly sphere, a form determined by natural rationality in accordance with the roundness of the sky. In the wazzalcora, all rules of celestial configurations are represented with architectonical rationality."

as a reference along with (and at the same level as) the 'canones Ptolomaei'.[476]

Even more important, there are some clear analogies between the language of the text **pa** and that of Vitruvius' chapters devoted to gnomonics ('De architectura', book 9). At the beginning of those chapters, Vitruvius introduces the 'analemma' (an astronomical and geometric construction used for drawing sundials) with these words:

> Ea autem sunt divina mente comparata habentque admirationem magnam considerantibus, quod umbra gnomonis aequinoctialis alia magnitudine erat Athenis, alia Alexandriae, alia Romae, non eadem Placentiae ceterisque orbis terrarum locis. [...] *Analemma* est ratio conquisita solis cursu et umbrae crescentis ad brumam observatione inventa, e qua per rationes architectonicas circinique descriptiones est inventus effectus in mundo.[477]

We note here in particular the expressions 'divina mente comparata' and 'architectonica ratio', which are also found in the prologue **pa**. In Vitruvius, though, the first expression means "constructed by a divine mind", with possibly a reference to Ptolemy. In the medieval text, instead, I believe that the term 'divina mente' is to be understood as 'the mind of God', and the term 'comparata' as 'similar to'. Vitruvius was in fact the ideal reference for a philosopher who was not only interested in abstract mathematical-cosmological reflections in neoplatonic style, but also in using the compass to re-construct the material effects of the rational structures of nature.[478] In the beginning of his work, Vitruvius offered his well-known definition of architecture as a combination of 'fabrica' and 'ratiocinatio':

> Architecti est scientia pluribus disciplinis et variis eruditionibus ornata [...]. Opera ea nascitur et fabrica et ratiocinatione. Fabrica est continuata ac trita usus meditatio, quae manibus perficitur e materia cuiuscumque generis opus est ad propositum deformationis. Ratiocinatio autem est quae res fabricatas sollertiae ac rationis proportione demonstrare atque explicare potest. Itaque architecti, qui sine litteris contenderant, ut manibus essent exercitati, non potuerunt efficere, ut haberent pro laboribus auctoritatem; qui autem ratiocinationibus et litteris solis confisi fuerunt, umbram non rem persecuti videntur. At qui utrumque perdidicerunt, uti omnibus armis ornati citius cum auctoritate, quod fuit propositum, sunt adsecuti.[479]

---

476 Millás Vallicrosa (1931) p. 275. On the canones Ptolomaei (late ancient astronomical tables) see Pingree (1990).

477 Vitruvius, 'De architectura' 9, 1,1. "It is ordained by the divine spirit and inspires great wonder in those who consider it, that the shadow of the gnomon at the equinox is of one magnitude at Athens, another at Alexandria, another at Rome, is different at Piacenza and in other parts of the world. [...] The analemma is an exact contrivance invented by observing the course of the sun and the lengthening of the shadow towards the winter, by means of which through achitectural calculations and the use of the compass, the action of the sun in the universe is discovered."(Vitruvius (1931 and 1934), vol. 2 p. 210–213).

478 Toulze (1996) p. 40–51.

479 Vitruvius, 'De architectura' 1,1,1–2."The science of the architect depends upon many disciplines and various apprenticeships which are carried out in other arts. [...] Craftsmanship is continued and familiar practice, which is carried out by the hands in such material as is necessary for the purpose of a design. Technology sets forth and explains things wrought in accordance with technical skill and method. So architects who without culture aim at manual skill cannot gain a prestige corresponding to their labours, while those who trust to theory and lit-

According to Vitruvius, knowledge in architecture was to be attained through a combination of non-verbal, non-written experience and of the written word.

Even though the context and intentions of Vitruvius' words were completely different from those of high medieval mathematics, what is important for my argument is the way those words could be interpreted by Latin scholars. Since in Latin Christian Europe the written word was a source of knowledge about God and the order of the world, the same might apply to 'fabrica', particularly thanks to the connection between meditation and manual crafts provided by the craft of memory.

In this context, it is interesting to note the contents of a philosophical dispute held by Gerbert of Aurillac against the scholar Otric (d. 981) in front of emperor Otto II (973–983) and his court in Italy (Pavia or Ravenna) in the year 980, according to the 'Historia' of the monk Richer (written ca. 991–998).[480] Richer relates that Otric had falsely spread the word that Gerbert, in dividing philosophy into its branches, had subordinated physics ('physica') to 'mathematics' ('mathematica').[481] Asked by the emperor to explain or correct this opinion, Gerbert replied that mathematics, physics and theology were equally subordinated to the same genus ("Dico itaque mathematicam, phisicam et theologicam aequaevas eidem generi subesse").[482] When prompted to offer more details on his ideas concerning the division of philosophy, Gerbert referred both to Boethius and to Vitruvius, arguing that philosophy could be divided into two species: theoretical and practical ("[...] secundum Vitruvii atque Boethii divisionem discere non pigebit. Est enim philosophia genus cuius species sunt practice et theoretice"), and that theoretical philosophy comprised "philosophia naturalis, mathematica intellegibilis ac theologia intellectibilis".[483] Yet again, Gerbert said, it was not unreasonable to place mathematics under physics ("Rursusque mathematicam sub physicam non praeter rationem collocamus").[484]

Thus, according to Gerbert, mathematics, as a part of philosophy, was on the same footing as physics and theology, but could in a sense also be seen as subor-

erature obviously follow a shadow and not reality. But those who have mastered both, like men equipped in full armour, soon acquire influence and attain their purpose."(Vitruvius (1931 and 1934), vol. 1 p. 6–7).

480 On Richer and his 'Historia' see the preface of Latouche's edition: Richer (1967 and 1964) vol. 1 p. I–XIV. On the date and place of the dispute, see: Richer (1967 and 1964) vol. 2 p. 67, n. 1 (Pavia) and Beumann (1997) p. 117–118 (Ravenna). Schuler (2000) erroneously refers to the emperor as Otto III (994–1002).

481 "Etenim cum mathematicae phisica par atque coaeva a Gerberto posita fuisset, ab hoc mathematicae eadem phisica ut generi species subdita est; incertumque utrum industria an errore id factum sit" (Richer (1967 and 1964) vol. 2 p. 66 (b. 3, ch. 56)). On Gerbert, see 3.1.4 and 4.3.4.

482 Richer (1967 and 1964) vol. 2 p. 72 (b. 3, ch. 59).

483 Here I quote "Vitruvii" as in: Richer (1839) p. 620, l. 18, also reproduced in Migne's *Patrologia Latina* 138 (1853) c. 107c, and quoted in Schuler (2000) p. 323. In Richer (1967 and 1964) vol. 2 p. 72 (b. 3, ch. 60), "Vitruvii" is corrected in "Victorini", as explained at the end of the present paragraph.

484 Richer (1967 and 1964) vol. 2 p. 72 (b. 3, ch. 60).

dinate to physics. According to Richer, Otric had instead accused Gerbert of making the opposite error.

Although Richer's account is too sketchy to allow an analysis of the philosophical questions at issue, two elements in this episode deserve to be noted: first, the reference to Vitruvius and, second, the focus on the problematic relationship between 'physica' and 'mathematica'. One is tempted to understand Gerbert's position as an attempt to harmonize, on the one side, Boethius's neoplatonic conception of the role of mathematics as a path from 'intellegibilia' (i.e. that which can be known with sense and reason) to 'intellectibilia' (i.e. that which can only be known through reason excluding the senses) and, on the other, Vitruvius' idea of knowledge as a symbiosis of 'fabrica' and 'ratiocinatio', in which mathematics is not only an abstraction, but also a basic component of the practical, sense-dependent study of nature and, as such, in a way subordinate to physics. Significantly, the reference to Vitruvius, which appears in the only manuscript in which Richer's work is preserved, has been disputed by some authors and corrected into a reference to Victorinus, the translator of Porphyrius 'Isagoge', another work of neoplatonic character.[485] However, as Stefan Schuler argues, the appearance of Vitruvius' name would fit in the context of tenth-century mathematics and philosophy.[486]

5.2.3 The 'architectonica ratio' of the astrolabe as a forbidden art

As we have seen, the authors of astrolabe texts took pains to distance themselves from forbidden astrological views. Astrolabe studies were dangerous because, in eleventh-century Arabic and Byzantine culture, the astrolabe belonged to the tools of fortune-tellers, who were becoming more and more influential.[487] Yet it was also because of their rationality that astrolabe studies were problematic. Rationality was in the eleventh century a highly controversial subject: an excessive faith in the rational methods of logic, grammar or mathematics was object of strong critique by figures like Petrus Damiani (1007–1072).[488] Damiani expressed himself very strongly in his 'De sancta simplicitate scientiae inflanti anteponenda', where he showed the creative power of grammar by letting the Serpent teach Man the plural of 'god':

> Ecce, frater, vis grammaticam discere? disce Deum pluraliter declinare. Artifex enim doctor [i.e. the Serpent], dum artem inoboedientiae noviter condit, ad colendos etiam plurimos deos inauditam mundo declinationis regulam introducit.[489]

485 Richer (1967 and 1964) vol. 2 p. 72–73 (esp. note 1). On the one manuscript of Richer's 'Historia': Richer (1967 and 1964) vol. 1 p. XII–XIV.
486 Schuler (2000) p. 323–324 and note 17.
487 See 3.1.2.
488 On Petrus Damiani: Fumagalli Beonio Brocchieri/Parodi (1998) p. 131–135.
489 Pier Damiani (1943) p. 166, (De Sancta simplicitate, 1). My translation is: "Here, brother, do you want to learn grammar? Learn to decline God in plural. Thus the doctor and art-maker

In the same work, Damiani offered one of the very few eleventh-century mentions of an astrolabe outside of astrolabe-texts: Ugo, a cleric from Parma, was so ambitious in his study of the arts, "ut astrolabium sibi de clarissimo provideret argento" ("that he provided himself with an astrolabe made out of the purest silver").[490] Ugo became chaplain of the emperor Conrad II and had great hopes for his future career, but was murdered by bandits on his way home.

Thus, authors writing on the astrolabe had to be particularly careful in defending themselves and their subject matter from the accusations of wanting to become fortune-tellers. As we have seen, Abraham could be referred to as the authoritative founder of the Christian science of the stars. In 5.6.3 we shall see how the patriarch could not only be quoted as an authority on Christian astrology, but also represented as a user, or possibly the inventor, of the astrolabe.

5.2.4 Chapters J 2–7: determining temporal hours with the astrolabe

After having discussed **J 1** and other texts in our manuscript expounding the reasons for studying the astrolabe, let us go back to the first folios of BnF lat. 7412 (A.27), which contain the text **J**. Chapters **J 2–7** offer a description of an astrolabe artefact, with the advice to keep it well in memory (**J 2**), explain the basic principles for finding the position of the Sun on the zodiac and for individuating it on the rete (**J 3** and **J 4**), and explain in a simple and clear way how to estimate the current temporal hour with the astrolabe, by day (**J 5**) and by night (**J 6**).[491] Here the author remarks that this function of the astrolabe is very important for liturgical aims. This is the last time that the subject of practical monastic timekeeping is mentioned in **J**.

To determine the current temporal hour, the altitude of the sun or of a star has to be measured. The astrolabe can then be set onto the present celestial configuration and the position of the sun (or of its nadir) will indicate the actual temporal hour. In chapter **J 7**, the twelve temporal hours of the day are associated to three of the four directions: hours I–IV to east, V–VIII to south and IX–XII to west. This is a system which, given space orientation (e.g. the knowledge of the direction south), easily allows orientation in time by estimating the position of the sun in the sky. After this, two folios written only on one side are bound into the manuscript: the first one bears a list of Arabic star names and coordinates on it (though

---

[i.e. the Serpent], when he founded anew the art of disobedience, introduced in the world a rule of declination unheard of before, so that it would be possible to worship many gods".

490 Pier Damiani (1943) p. 184–186, (De Sancta simplicitate, 6). The passage is usually interpreted as saying that Ugo gave or wanted to give the astrolabe to the Emperor as a gift, but the text does not explicitly say so.

491 All procedures are those described in 2.3.3.3–4.

not the usual star table of type III), the other one a table of solar longitudes and a short chapter from the text **J'**.[492]

5.2.5 Was the gesture of raising the astrolabe a practical recipe or a symbolic act?

According to the instructions given in **J 5** and **J 6**, the altitude of the sun or of a star has to be measured using the alidade as a sighting instrument.[493] The description of this procedure is very detailed and it is almost literarily taken from al-Khwārizmī's treatise on the use of the astrolabe: the left shoulder is opposed to the sun, the astrolabe hangs from the right hand, the alidade faces 'the eyes of the sun' and has to be moved up and down until a ray of light passes through both holes at the same time.[494]

Given the problems inherent to the determination of solar and stellar altitude with a small astrolabe artefact,[495] one may ask how far this description, whose literary origin has been traced, was actually a recipe to be enacted by readers of the text. Were author and readers unaware of how inaccurate the result of such a measurement might be? Is this passage evidence that Latin scholars never really tried to use an astrolabe in practice?

Possibly, the passage taken from al-Khwārizmī can be compared to high medieval descriptions of crowning ceremonies or other rituals with a symbolic content.[496] It has been recognized that not all gestures and objects featuring in the literary description of rituals are to be taken literally, as if the text were an eyewitness account. The description was in many cases aimed at conveying the symbolic content of the ritual enacted by describing it 'as it should be'. This attitude did not exclude a faithful description of events, but left the possibility open for introducing details of particular significance, especially when a literary example could be found.

In the present case, the original Arabic description might have provided an appropriate representation for the act of taking the altitude of a celestial body, even if Latin astronomers were aware that the procedure used in practice could also be another one, for example involving a gnomon or a geometrical square.[497] In fact, in the manuscript BnF lat. 7412 (A.27), on f. 13v, it is also explained how to build a geometric square out of wood or copper, and it is explicitly stated that "the larger the better" ("quanto maior tanto melior").

---

492 On the star table in BnF lat. 7412 (A.27), see: Kunitzsch (2000a) p. 395–396; Samsó (2000) p. 512.
493 On the problems relevant to this use of the astrolabe, see 2.3.4.
494 The same description with a slightly different wording is found in the treatise **J'** and has been identified by Paul Kunitzsch as a translation from al-Khwārizmī's treatise (see above 3.3.9).
495 See 2.3.2.1 and 2.3.4.
496 Hageman (1999).
497 See 2.3.2.2.

Finally, the short description from al-Khwārizmī is not only elegant, but also conveys a very clear visual impression, which could be a precious help for memory. In the twelfth century, most images of astrolabes represent astronomers (or the personification of Astronomy) in the act of taking the height of a celestial body by using the alidade of an   astrolabe. These images have contributed to spread the idea that the astrolabe was primarily an observational instrument, although this was not the case.

5.3 THE ASTROLABE AND THE MEASURE OF TIME AND SPACE

5.3.1 The measure of time as a philosophical problem

The chapters **J 8–13** deal with a new subject: estimating the quantity of hours. This is one of the main focuses of interest of the whole manuscript. I believe it is right to say that, for Latin scholars, the astrolabe was primarily a device for measuring time, yet in the sense that it was a device to let experienced time be grasped as a measurable quantity. In the astrolabe, the heavenly rotation became a reference frame for quantitatively estimating experienced time duration. This was something different from regulating the rhythm of everyday life by using observable celestial phenomena such as day and night, lunar phases or the yearly movement of the sun through the zodiac.

The idea of experienced time as something qualitatively homogeneous, that could be divided in numerable parts was not traditionally part of medieval culture.[498] In general, the whole modern concept of 'measure' does not apply to measures and measuring in medieval times. As the historian of metrology Harald Witthöft wrote:

> Die abstrakten Einheitsnormen des metrischen Systems verstellen heute den unmittelbaren Zugang zum Verständnis von Maß und Gewicht des Mittelalters. [...] Das mittelalterliche Maß war qualitative und quantitative Größe in einem, mit deren Hilfe Zustände und Verläufe geordnet, ihr Sinn erfaßt, beschrieben, handhabbar gemacht wurde – das Maß war zugleich Realität, Metapher, Symbol innerhalb des mittelalterlichen ordo.[499]

Thus, the idea of determining the number of parts in a temporal hour using the 360 degrees of the celestial rotation as a reference – which is what was done in the text **J**, as we shall see – was conceptually far less trivial than might appear at first sight, and should not be interpreted immediately as a 'measure of time' in modern sense. In a way, it was a novel philosophical method, which still appeared problematic to late medieval philosophers like Jean Buridan (d. ca. 1360), who

---

498 Some examples of the complexity of the medieval concept(s) of time can be found in: Lecoq (1992) (on the representation of time in medieval world-maps), Bourin (1992) (on time-awareness in medieval witness reports), Telesko (2000) (on 11th-c. changes in the concept of time and its representaion).

499 Witthöft (1983) p. 235.

discussed the philosophical questions surrounding the concept of measure in general and of measure of time in particular.[500]

One high medieval example of reflections on the quantity of time occurs in Abbo of Fleury's commentary to the 'Calculus' of Victorinus of Aquitaine.[501] In that work, Abbo discussed 'weight, number and measure' in a context influenced by late ancient Neoplatonism. His aim was to show how to proceed from knowledge of the visible to knowledge of the invisible – just like in astrolabe texts. Abbo, too, paid much attention to phenomena, and discussed the question of how not only to perceive, but also to divide incorporeal things like 'temporal quantity'.[502] In order to do this, he showed how the temporal quantity called 'day' varied according to the time of the year and could be divided into a number of equinoctial hours. To this aim, Abbo suggested a method found in Macrobius: a water-clock was calibrated to equinoctial hours by watching the movement of the zodiac signs: according to Macrobius, each sign rises above the horizon in two equinoctial hours. This statement is, in fact, wrong, because the rising times of zodiac signs depend upon local latitude and also upon the zodiac sign, but we shall find a similar suggestion also in **J**.[503] In any case, Abbo's discussion of the division of incorporeals shows that, at the time, temporal quantity and its division could be a non-trivial philosophical example. Abbo's text also shows how a mechanical device (the water-clock) could play a central role in solving an abstract philosophical problem (dividing the incorporeal), even if only in an astronomically incorrect thought experiment.

5.3.2 Chapter **J 8**: the astrolabe and the parts of the hours

Chapter **J 8** begins by stating that, once the temporal hours ('horae artificiales') have been 'digested', the author shall explain about equinoctial and unequal hours ('horae aequinoctiales' and 'horae inaequales') This statement is somehow confusing, because seems to suggest that three kinds of hours existed: temporal, equinoctial and unequal ones. In fact, though, unequal hours are temporal hours or, to call them with their Arabic name, 'horae Ezemeniae'.[504]

About equinoctial hours, the author says: "in the zodiac circle, that is in 360 parts" ("in toto circulo zodiaco, id est CCCLX partibus"), there are 24 equal hours, each of which has 15 parts.[505] The erroneous reference to the zodiac instead that to the equator may be attributed to the influence of Macrobius and Abbo of Fleury. The author goes on to explain how many unequal hours there are in those same 360 parts: one has to give 19 parts to each unequal hour, and will find 19 of

---

500 On measure in general: Zimmermann (1986),on the measure of time: Sarnowsky (1983), especially p. 159–160.
501 On Abbo of Fleury, see 4.3.4. On the commentary: Evans/Peden (1985).
502 The following summary is taken from Evans/Peden (1985) p. 119–120.
503 See 5.3.2.
504 Bubnov (1899) p. 131.
505 Bubnov (1899) p. 132.

them in the whole circle. Since 19 times 19 is 361, this is not a bad approximation for 'the whole circle'. Still, it is not clear what this computation is supposed to mean: what are 'unequal' hours of 19 parts?

After this puzzling beginning, though, the text becomes again clear: hours are 'equinoctial' only twice a year, while for all other days they vary, alternatively increasing and decreasing. However, the shortest unequal hour is never less than 11 parts and the longest one never more than 19. These are well known remarks about the varying length of daylight and night-time, to be found already in traditional Latin authors. However, the conclusion of the author is peculiar: "Unde XIX partes sibi attribuuntur, quae tamen aut crescendo, aut decrescendo, ad aequinoctialis horae partes, id est XV, recurrunt"[506] That is: to each unequal hour are somehow (virtually?) attributed 19 parts, which decrease down to 11 and then increase to 19 again, passing through the 15 parts (i.e. degrees) of equinoctial hours. With this last remark to explain the initial statement, chapter **J 8** ends.

5.3.3 The assimilation of temporal and equinoctial hours

I shall now attempt to interpret the rather puzzling remarks in **J 8** and their philosophical significance. The system of temporal hours had never been aimed at providing a quantification of experienced time duration. In fact, it could hardly ever have done so, since its variable hours cannot be used as measurement units. Once temporal hours had been represented on the astrolabe, though, this device could serve as a mediator to relate temporal hours to the division of celestial, circular motion into 24 equal ones.

When assimilated to equinoctial hours, the temporal ones would become 'unequal' hours as opposed to 'equal' ones: as the beginning of chapter **J 8** suggested, the subject of equal and unequal hours was not exactly the same thing as that of 'artificial hours' ('horae artificiales'). Here, the use of the term 'artificial' might indicate that temporal hours were now perceived as being less natural the equinoctial ones. Abbo of Fleury, following Macrobius, had proposed to use a water-clock to mediate between celestial revolutions and earthly duration, but now the astrolabe provided an infinitely more practical and precise solution. What's more important, one did not need to look at the sky anymore: the revolving celestial circles were represented on the device itself and, thanks to the astrolabe structure, the flow of earthly and celestial time could be imagined and manipulated in the mind. The visual representation of the experienced flow of time was a new element in medieval Latin culture, too, where the duration of the daily activities, as far as it was not determined by the activity itself, was usually regulated by auditory signals, for example bells.[507]

---

506 Bubnov (1899) p. 132. My translation is: "Therefore [unequal hours] give themselves 19 parts and, either increasing or decreasing, they converge to the parts of euinoctial hourse, i.e. 15. "
507 Dohrn-van Rossum (1992) p. 44–45 and p. 185–201.

Thanks to the astrolabe, temporal hours could be thought of as corresponding to a certain number of the 360 parts of a circle, albeit a changing number, and could be accordingly manipulated. Observable phenomena came in again, though, to determine that the parts of an unequal hour would never be less than 11 or more than 19. It is not possible to say if these numbers are taken from observation or from literature, but some manuscripts put the maximum length at 20, others at 18 parts.[508]

At this point, comes a step which, to modern readers, appears particularly devoid of sense: unequal hours are (apparently arbitrarily) assigned 19 parts each and then, on this premise, the author asks how many unequal hours are contained in the circle made up of 24 equal hours. The (approximate) answer is 19 (19 x 19 = 361). At the latest at this point, unequal hours are no temporal hours anymore, since for temporal hours the answer would have been, by construction, twice twelve.

The context in which this passage appears is a clear, reliable composition, competently expounding non-trivial mathematical and astronomical matters. If it were not so, the apparent confusion surrounding unequal hours could be attributed to ignorance. Yet this seems to me quite improbable. Moreover, the same subject (i.e. 19 hours of 19 parts each) is dealt with also in **J' 10–11**.[509] I believe the discussion in **J 8** should be considered as a sort of abstract speculations on the 'number of parts' of experienced duration.

Once more, it is important to underscore the role of the astrolabe as mediator between the traditional monastic hours and the hours of astronomy and mathematics, corresponding to 15 out of 360 parts of the celestial revolution. It was only thanks to the astrolabe – both as a geometrical structure and as a tool to be manipulated – that the two could be represented as analogous and eventually assimilated to each other. In this process, both kinds of hours were changed: equal hours became units of experienced duration, and not only of celestial movement, while temporal hours took up a quantitative aspect. However, unequal hours did not become abstract mathematical entities: the limits to their length were those imposed by the maximum and minimum duration of daylight in the inhabited portions of the earth.

After having introduced the subject of equal and unequal hours, the text **J** offers a series of examples of how to use the astrolabe for telling the number of parts of experienced duration: counting how many parts (degrees) are in a temporal hour on a specific day (**J 9**) or night (**J 10**); estimating the quantity of the periods of daylight and night (**J 9** and **J 12**); computing the number of equinoctial hours in a day or night, and the pattern according to which those numbers increase and decrease (**J 13**); provide a quantitative estimate of the duration of dawn (**J 14**).

---

508 Bubnov (1899) p. 132, n. VIII,3.
509 "quomodo turnas horas rectas in horas tortas" and "ut de horis tortis facias rectas" (Millás Vallicrosa (1931) p. 285). I am not aware of other mentions of 19-hours-systems in high medieval literature.

5.3.4 The order of the sky and its projection on earth: the seven climates

Chapters **J 15–17** are devoted to observational astronomy and astrology: reconstructing the current celestial configuration and, in particular, which star or zodiac sign is rising or setting, i.e. determining the astrological ascendent and descent, a very important information (**J 15**);[510] given a star represented on the astrolabe, locating its position with respect to the zodiac signs (**J 16**); learning the Latin and Arabic names of stars and constellations and recognizing the form of the constellations (**J 17**)[511]. Chapters **J 18** and **J 19** are instead devoted to a system relating earthly space to the celestial sphere : the seven climates.[512]

Chapter **J 18** explains what is local latitude and how to estimate it, pointing out that knowing latitude is necessary to build 'horologia' (sundials).[513] It provides a table of the latitudes of the seven climates and of the length of their longest days expressed in equinoctial hours. Each line of the table is given first in Arabic language written in Latin letters and then translated into Latin. The author also refers to traditional authorities: Ptolemy, Erathostenes and Martianus Capella.

Chapter **J 19** deals with the climates, too, but in a qualitative way, offering an overview of cities, lands and peoples belonging to each climate.[514] In fact, while in Bubnov's edition all seven climates are discussed together, in the manuscript BnF lat. 7412 (A.27) on f. 8v–9r only the description of the seventh climate occurs. This is the climate in which France, England and Germany are located. The descriptions of the remaining six climates come later on (f. 18r–18v): a further indication of the composite nature of the text, that apparently grew together according to the specific interests of Latin scholars. The contents of the climate descriptions were studied by Uta Lindgren, who could not trace them back to any of the Latin authorities on the subject.[515] The description of the seventh climate from **J 19** is followed by two more texts devoted to the climates. The first one (f. 10r–10v) is an extract from Byrhtferth of Ramsey's 'Glossae' to Bede's 'De temporum ratione' and it is in large part a summary of Pliny the Elder's 'Naturalis historia', book VI, 65–72.[516] The second text (f. 11v) is taken from Martianus Capella's 'De nuptiis Philologiae et Mercurii', book VIII, 876–877.[517]

---

510 Juste (2007) p. 248. On the use of the astrolabe to compute the ascendent, see 5.5.1 and 5.6.3.
511 Chapter **J 17** and its manuscript tradition have been studied in great detail in Bergmann (1985) p. 76–79 and p. 221–222 and in Kunitzsch (2000b).
512 On the seven climates, see 2.2.2.9.
513 Bubnov (1899) p. 138–142.
514 Chapter **J 19** is edited in Bubnov (1899) p. 142–149. An edition, French translation and commentary of the version of **J 19** occurring in the manuscript Paris BnF lat. 14065, f. 51 can be found in Gautier Dalché (1996).
515 Lindgren (1985). Lindgren suggests that the author (Gerbert?) drew his description from a map he had in front of his eyes (Lindgren (1985) p. 636–637). For further references on Gerbert see 3.1.4.
516 The occurrence of Byrhtferth's text in this manuscript could be interpreted as a sign that it was related to Fleury. On Byrhtferth of Ramsey, see 4.3.4, on his commentary on Beda's work: Hart (2003) vol. 1 book 1 p. 219–238. The text is edited as a complement to Beda's

In both texts, the number of climates is eight instead of seven. They both offer a broad panoramic of the lands and cities to be found in each climate, as well as the measure of their longest and shortest days. The climates, with their quantitative and qualitative features, summarized and ordered the whole quantitative and qualitative variety of the inhabited world. It is worth remembering Witthöft's statement on the quantitative and qualitative, material and symbolic character of measures in medieval times: the climates can be seen as a measure unit for ordering and understanding the whole of inhabited earth, both in its qualities and in its quantities. In this attempt at unification, universality was more important than coherence. The qualitative texts devoted to the climates offer different, sometimes incompatible views on the same subject in a universal overview. The new knowledge from the Arabs follows the newly written Latin text **J**, and is followed by the slightly older text by Byrhtferth, commenting on Beda Venerabilis and quoting from Pliny. The late ancient author Martianus Capella comes at the end. The panoramic attempts to be universal not only in space, but also in historic memory, unifying different languages as well as eras, even though this task may not appear very rational to a modern reader.

5.3.5 The three-dimensional sphere and the measure of time (**J'a**)

On f. 9r–10r of BnF lat. 7412 (A.27), we find seven of the chapters published by Millás Vallicrosa under the title: 'De horologio secundum alkoram id est speram rotundam (**J'a**)'.[518] Although Millás Vallicrosa described these passages as dealing with a spherical astrolabe, the instructions given here, at least as far as they can be understood, apply just as well to the use of a simple celestial globe, i.e. a celestial sphere resting on a circular support representing the horizon, eventually equipped with a sighting tube for determining the altitude of the sun. The terms 'alkoram' (i.e. 'sphere') and 'spera rotunda' can mean both a globe and a spherical astrolabe.[519] As we shall see, this same conclusion has also been reached by the historian Marco Zuccato in his recent analysis of this same text.

In the case of **J'a,** the attempt at making astronomical knowledge readable was particularly unsuccessful: the text is very unclear, both in its grammar and in its terminology, which is a mixture of Latin and transliterated Arabic. I will not attempt to reconstruct its contents in detail, and shall only offer as an example my tentative interpretation of the first chapter of **J'a**, which purports to explain how to compute the (equinoctial?) hours of the day. In Millás Vallicrosa's edition as well as in our manuscript the text reads:

'De temporum ratione' in: Byrtferth of Ramsey (1904) c. 445–446 and 431–432 among the 'Glossae et scholia', indicated as 'Brid. Rames. Glossae'.
517 Martianus Capella (1987). On the role of excerpts from Martianus Capella in astronomical compilations, see: Abry (2000).
518 Millás Vallicrosa (1931) p. 289–292, l. 28.
519 Poulle (1994) p. 224–225.

In primis de horis diei. Quando quaeris scire horas diei, ponebis gradum solis subtus circulum medie terre, ubi libet sic ut iaceat altitudo terre super alcotob, et videbis ubi iacet angulus, ubi se coniungunt III circuli ubi est oriens, et pone ibi signum, postea alzabis ipsam alkoram quousque sol inest per foramina super gradum solis, postea computabis quantum est de angulo II loco usque ad primum locum unde se movebat, vel primo loco usque ad II locum ubi tunc est, et tot hore sunt.[520]

Since no description of the instrument is extant, in order to interpret the text it is necessary to guess at the same time both the shape of the parts of the device and the actions performed with them. Some of the terms have a standard meaning in astrolabe texts: 'gradum solis' indicates the point of the zodiac in which the sun is on a particular day of the year, known today usually as solar longitude; the word 'alcotob' corresponds to the Arabic 'al-quṭb', meaning 'pole, axis'.[521] My interpretation of the passage, given in form of a loose translation, is the following:

First of all, about the [equinoctial] hours of the day. When you want to know the [equinoctial] hours of the day, position the celestial sphere on its hemispherical stand so, that (1) the actual position of the sun on the zodiac lies on the rim of the circular stand, and (2) the local latitude coincides with the vertical axis of the sphere. [That is: the direction of the local zenith must be aligned to the verical, so that the horizontal rim of the circular stand corresponds to the local horizon.] Now note the point where the three circles meet in the East [i.e.: the point corresponding to the eastern intersection of horizon and zodiac, through which necessarily also a meridian will pass. This is the point the sun was occupying that same day at sunrise.] Note the angular position of that particular meridian. After this, rotate the sphere around its North-South axis [which does not coincide with the vertical!], until the sun, passing through the holes [of the sighting device], hits the point of the zodiac where the sun should be ('gradum solis').[522] Then measure how large is the angle [covered by the meridian] from the second position to the first one, from which it moved, or alternatively from the first position, from which it moved, to the second one, where it now is, and so many are the hours.

I leave it up to the reader to decide whether my intrepretation is convincing or not. Further passages of the texts are even harder to understand, and I shall not attempt to do it. However, even though the text hardly allows safe conclusions, I believe that it describe how to use an instrument similar to the globes built and used by Gerbert of Aurillac and his correspondents, as well as by the monks of the cloister

---

520 Millás Vallicrosa (1931) p. 288; BnF lat. 7412 (A.27) f.9r. The only difference between the two texts is that where Millás Vallicrosa has 'alkoram', our manuscript has 'anchoram'.

521 Kunitzsch (1982) p. 545–546.

522 I believe that the sphere was equipped with a sighting device similar to the alidade of an astrolabe: two small rings connected in a straight line. When a ray of sun passed through both rings, the direction of the sighting tube corresponded to that of the sun. If the sighitng device was connected to the sphere so that it would always point to its centre, the place where the sunray hit the surface indicated on the sphere the altitude of the sun above the horizon at that precise moment. I assume that the sphere was standing so, that its horizon ring was actually horizontal. The inerpretation of the expression 'foramina' ('holes') as a sighting device is supported by the second passage of **J'a**, where the author refers to 'the holes of the alidade' ('foramina alhidade') as a moveable part of the whole device (Millás Vallicrosa (1931) p. 289).

St. Gallen in the early eleventh century.[523] This tool may have been particularly useful for introducing students to the concept of equal hours and their parts through visualization and manipulation of the three-dimensional sphere. The use of the astrolabe would have been the second, more advanced stage.

In a recent paper, Marco Zuccato has analysed both old and new evidence relating to Gerbert's stay in al-Andalus in the years 967–970: his results are in agreement with my interpretation of **J'a**, and I shall briefly summarize them.[524] Zuccato argues that **J'a** describes one of Gerbert's instruments, namely a celestial globe with horizon ring. Gerbert had come to know this kind of globe in al-Andalus, in particular thanks to a treatise on the subject written by the Jewish philosopher Dunāsh ibn Tamīm ibn Ya'qūb al-Isrā'īlī al-Qarawī, who was active at the Fatimid court in Egypt in the years 925–960. Dunāsh had sent his treatise to another Jewish scholar, Abū Yusuf Ḥasdāy ben Isḥāq ben Shaprūṭ. The latter was a prominent cultural and political figure at the court of Cordoba, and the Latin scholar John of Gorze had met him when he visited al-Andalus around the middle of the tenth century.[525] Gerbert might in turn have been able to access the treatise in Catalonia thanks to the mediation of Gotmar, who in 940 went to Cordoba as ambassador of the Count of Barcelona, and in 944 was elected bishop of Gerona. The text **J'a** would then be a Latin version of some parts of Dunāsh's treatise. Unfortunately, the Arabic treatise is not extant: we only have its title and a brief overview of its contents.

I thoroughly agree with both Zuccato's interpretation of the instrument and with his thesis that Gerbert had come to know it in al-Andalus thanks to Dunāsh treatise and some Jewish intermediaries. However, I would tend to interpret the extant Latin text **J'a** not so much as a translated excerpt from an Arabic original, but rather as a written trace of the teaching Gerbert or other Latin mathematicians received from Arabic-Islamic or Arabic-Jewish astronomers.

5.3.6 Abstract patterns and their embodiment in astronomical devices and surveying methods

On folios 11r–14v of BnF lat. 7412 (A.27) we find some fragments of Ptolemy's 'Planisphaerium', the text **h2**, chapters 20–25 of the **GIA III** and the short text **dpp**.[526] All texts offer examples of how to relate celestial and terrestrial motion, time and space. In this manuscript the texts are accompanied by carefully executed geometrical drawings which complement the information given verbally and testify to the attention of the author not only to the generic geometrical principles, but also to what we today might call technical details: the mathematical

---

523 Wiesenbach (1994) p. 386.
524 The following remarks are based on: Zuccato (2005). On Gerbert's instruments, see 4.3.4.
525 See 3.1.4.
526 On the **GIA**, see above (3.4). The texts are edited in: Bubnov (1899) p. 331–334, p. 365 and figs. 58–64 (**GIA III 20–25**) and Jacquemard (2000) p. 105 (**dpp**).

artist was at the same time philosopher and craftsman. I will not discuss in detail all this material, but will only focus on those aspects most closely linked to astrolabic studies. In paragraph 5.3.7, I will analyse the contents of the text **dpp**, in which an astrolabe is used as a tool to measure time duration, and discuss its possible philosophical significance. In paragraph 5.3.8 I shall summarize the various links of this part of the manuscript with Ptolemy's 'Planisphaerium'.[527]

5.3.7 The sinker and the astrolabe (**dpp**)

The text **dpp**, which appears on f.12v of BnF lat. 7412 (A.27), is the one whose manuscript tradition has been discussed in section 3.4.[528] This text describes how to estimate the depth of a body of water by measuring the time needed for a loaded sinker to reach its bottom and then come up again after having discharged its load. Catherine Jacquemard and Alain Haire have analysed this passage as far as the sources and practical feasibility of the procedure described are concerned.[529] Their results offer a solid basis to suggest an interpretation of the role played by the astrolabe in this context and I shall briefly summarize them.

The text **dpp** begins by giving instructions on how to build a sinker of copper and iron with a ballast that is discharged once the sinker hits the bottom. I shall not describe here the details of the construction, but only note that the method was tested and found workable.[530] Similar devices were described in the Late Middle Ages and Renaissance. The key factor making the procedure feasable is the fact that, because of the water's viscosity, the sinker, going down, almost immediately reaches a limit velocity which then remains constant until the weight reaches the bottom. The same happens when the floater travels back up to the surface: it moves at constant speed, albeit not the same speed as before. In short, the movement of the sinker in the water offered a rare example of uniform motion, which might have been of great interest to those looking for mathematical, rational order in nature.

Once the sinker has been built, the Latin author instructs his reader to lay it in the water and, as soon as it starts to go under, to "take the height of the sun in the astrolabe and see what hour it is".[531] When the sinker comes up again, the reader should "again take the hour by means of the astrolabe ('astrolapsus') and he shall see how much it is from the beginning of the immersion up to the return: whether

---

527 For a short description of the contents of **GIA III 20–25**, see Jacquemard (2000) p. 108–109.

528 The text is edited in Jacquemard (2000) p. 105. This passage and its relevance for understanding the natural philosophical significance of the astrolabe in high medieval Latin Europe are discussed in more detail in: Borrelli (2006).

529 Haire/Jacquemard (2000a).

530 Haire/Jacquemard (2000a) p. 228–229; Haire/Jacquemard (2000c).

531 "tu eadem hora accipias altitudinem solis in astrolapsu et vide quae hora sit" (Jacquemard (2000) p. 105).

one hour or two or whatever".[532] The 'astrolapsus' quoted here is clearly a device comprising two functions: one based on the shadow triangle (for taking the height of the sun) and one on the flat sphere (for showing how many hours have elapsed). The hours mentioned here have to be equinoctial hours, i.e. constant units of duration, and cannot be understood as temporal hours.

The author continues: after having performed the whole operation, the reader has to take a rod ('hasta') or some other measure ('mensura') and use it to estimate the depth of water. Thus, he will know how deep the water is in 'pedes', 'cubitos' or 'status' and "how many hours there are in its (i.e. the water's) immersion and emersion".[533] All this, explains the author at this point, has to be done in shallow water, so that later the procedure can be repeated in deep water, "giving so many 'pedes' or 'cubitos' for so many hours".[534] In modern terms, one can say that what is being described here is the calibration of a measuring instrument. More precisely, it is shown how the depth of a body of water can be put in proportional relationship not only to a predefined length, i.e. that of a measuring rod, but also to a predefined time duration, i.e. that of an equinoctial hour. As already noted, the element making the correspondence possible is a physical object moving of uniform motion: the sinker in the water.

In principle – or, shall we say, in theory – the procedure is correct and clearly described. The construction and use of the sinker are also realistic. The only fully unrealistic point in the whole account is the use of the astrolabe to estimate elapsed time. Whether or not astrolabe artefacts could be used to measure time in general, they surely could not estimate such short durations (a few minutes) as those involved in this measuring procedure, not to mention the difficulties involved in using an astrolabe while sitting in a boat, as would be the case when surveying a large body of water.[535]

In this respect, then, the whole operation seems to resemble more a thought experiment than a surveying method. The Latin author, though, is somehow aware of this problem: at the end of the chapter he suggests an alternative method to estimate the elapsed time: a special kind of water-clock. It is a vase of clay with a hole in its bottom ("vas tellureum subtus perforatum"[536]), which must be laid on the water surface in the moment in which the sinker goes down. When the sinker comes up again, the quantity of water in the vase shall provide the means to relate the unknown depth to a known one, just as the astrolabe did before. This method surely looks more practical than using an astrolabe, and the author even offers possible quantitative relationships between the quantity of water that has flowed into the vase and the depth measured: a quantity of water of four 'argenteos' corresponds to ten times the height of an average man.

---

532 "tu iterum accipe horam per astrolapsum et videbis quantum est ab hoc quod coepit immergi usque regressum fuerit, si est una hora vel duae vel quotlibet." (Jacquemard (2000) p. 105).
533 "et mensurabis quot pedes vel cubitos vel status ipsa aqua in profunditate eius et quot horas habeat in immersione et emesione eius" (Jacquemard (2000) p. 105).
534 "tot pedes vel cubitos dabis ad tot horas" (Jacquemard (2000) p. 105).
535 Haire/Jacquemard (2000a) p. 228–229.
536 Jacquemard (2000) p. 105.

Jacquemard and Haire have devoted much attention to studying this device, the 'sinking bowl', and its possible oriental antecedents.[537] There are plenty of indications that the sinking bowl can be used to measure short time periods: it was apparently unknown in Greek-Roman Antiquity, but had been widely used in India since the fifth century. Later, it became known in the Arabic world, although there it never supplanted the traditional form of water-clock, where the water flows out and not in.[538] The passage studied here is the earliest evidence that the sinking bowl was known to Latin scholars.

Jacquemard and Haire express the opinion that the method described in the passage was composed of various elements, all of them of oriental origin: the loaded sinker, the astrolabe and the sinking bowl. They argue convincingly that the procedure originally involved the sinker and the sinking bowl, while the astrolabe was an unrealistic addition probably due to Latin scholars.[539] They conclude:

> Le chapitre consacré à la mesure des grandes profondeurs paraît bien être le résultat latin d'un 'melting pot' surprenant d'expérimentations empiriques et de traditions géométriques qu'on aurait pu croire a priori complètement étrangères.[540]

The authors realize very well the extraordinary novelty of the ideas implicit in the text and in the procedure. In another article devoted to the same subject, they state:

> Il nous faut insister particulièrment sur le fait que les notions de mesures de vitesse, de temp, d'espace et de masse sont parfaitement maîtrisée, autant dans l'invention elle-même que dans le texte qui la décrit. [...] Il semble que l'idée courante d'une science de la dynamique inexistante au XIe siècle doive être revue à la lumière de ce texte.[541]

Even though, as stated in 5.1.1, I believe one should avoid making direct parallels to modern science, I share the opinion that the text **dpp** testifies to an awareness of and interest in those mathematical structures that could provide a universal, rational formalisation of procedures which could be employed to observe and quantify motion phenomena, both earthly and celestial ones. It is because of this, that the (apparently absurd) use of the astrolabe in the procedure becomes particularly significant.

Jacquemard and Haire state that the use of the astrolabe in the procedure described in **dpp** is "difficile à interpréter".[542] Considering that this passage occurs in the context of a consistent body of astrolabe knowledge, it seems rather improbable that writers and readers would not realize the unfeasibility of the surveying procedure. Because of the practical impossibility of using the astrolabe, Emmanuel Poulle even proposes that the term 'astrolabium' should be understood

---

537 Haire/Jacquemard (2000a) p. 230–232 on the history of the device, p. 232–234 on is technial feasibility.
538 Dohrn-van Rossum (1992) p. 28–30.
539 Haire/Jacquemard (2000a) p. 225.
540 Haire/Jacquemard (2000a) p. 235.
541 Haire/Jacquemard (2000c) p. 261.
542 Haire/Jacquemard (2000a) p. 224 and Haire/Jacquemard (2000b) p. 245, note 8.

here as 'water-clock': "d'ou je propose de tire l'équation astrolabium = clepsy-dre".[543] I do not believe this to be the right solution to the problem.

I would like to suggest that it was because of natural philosophical reasons that here the astrolabe played the main role as a time-measuring tool. When read in the context of astrolabe literature, this short chapter conveys the important natural philosophical message that, by means of a specific procedure involving quite complex tools (sinker, astrolabe), a mathematical relationship between earthly and celestial movement could be established. The focus of attention here is on neither abstract mathematical structures nor material measuring procedures alone, but rather their mutual relationship as mediated by technical devices. It is because of this focus that it had to be the astrolabe, and not the sinking bowl, the tool used to estimate how long the sinker was underway. Only the astrolabe, thanks to its combined functions of shadow triangle and flat sphere, could mediate between heavenly and earthly movement. The mathematical structures embodied both in the procedure and in the tools let earthly and celestial phenomena appear numerable in the same units of space length and time duration. The sinking bowl, even if probably more fitting as time-measuring tool for the task at hand, could not relate the movement of the sinker to that of the celestial sphere like the astro-labe did.

5.3.8 The traces of the 'Planisphaerium' in ms. BnF lat. 7412

As already discussed in 4.5, Latin knowledge on astrolabe construction can be traced back to the Arabic version of the 'Planisphaerium'. The evidence of a connection between BnF lat. 7412 (A.27) and the 'Planisphaerium' is:

- f. 11r-11v: a short Latin text which Paul Kunitzsch has identified as the translation of the first lines respectively of chapter 2 and chapter 3 of the 'Planisphaerium', both dealing with the construction and properties of horizon circles;[544]
- f. 11v-12r: the text **h2** and the construction drawings **d1** and **d2** (f. 12r), where **d1** contains the additional line making it identical to the construction of chapter 1 of the 'Planisphaerium';[545]
- f. 14v: drawing of a geometrical construction taken from chapter 10 of the 'Planisphaerium'. I shall discuss this image in 5.4.3.

It is very interesting to note that the text **h2** does not occur here in the same form in which it was published by Millás Vallicrosa : on f. 11v, just after the fragments of the 'Planisphaerium', the last part of **h2** occurs, i.e. the passage explaining how to draw the altitude circles. After that, the beginning of **h2** follows. This might be simply taken to be a copying error, if it were not for the fact that the passage on

---

543 Poulle (2000) p. 447.
544 Kunitzsch (1993a).
545 See 4.5.6.

the altitude circles addresses the reader in the second person singular, while the rest of **h2** describes the construction in the third person plural, as actions performed by 'wise philosophers'. This discrepancy strongly suggests that **h2** might also be a composite text, and that the manuscript BnF lat. 7412 (A.27) testifies to a very early stage of its transmission and compilation.

5.4 EUCLID'S AND PTOLEMY'S GEOMETRIC CONSTRUCTIONS

5.4.1 The drawings in ms. BnF lat. 7412 and the knowledge they conveyed

The manuscript BnF lat. 7412 (A.27) not only testifies to the literary aspects of the Latin medieval astrolabe, but also provides evidence of the more practical side of astrolabe studies. The knowledge linked to these practical aspects was the most difficult to store and transmit in verbal form, but writing on parchment offered one possibility of capturing at least in part the philosophical crafts of cosmometry: drawings. Our manuscript contains a large number of carefully executed drawings: they are geometrical constructions, illustrations of surveying procedures and pictures of devices that could be used in astronomical observations and computations.

Some of the drawings are accompanied by texts, others stand isolated. In the latter case, it is only possible to formulate hypotheses on the knowledge they may or may not have conveyed. For example, on f. 14v, many geometrical figures appear which are not associated with any text: they are discussed in 5.4.2–3. The most immediate evidence of the connection of astrolabe studies to experiments involving the construction and use of rather complex devices are the images of a sphere with sighting tubes and of a polar sighting tube (f.15r), of a quadrant (f. 16r), of a sundial (f. 19r) and of the various parts of an astrolabe (19v–23v). Apart perhaps from the sphere, these devices did not receive as much attention as the astrolabe in high medieval literature. However, since they are all relatively simple from the technical point of view, it seems probable that they were not only drawn, but also actually built and used. In them, we have an example of knowledge transfer taking place primarily through strategies different form the written word.

5.4.2 Drawings from Euclid's 'Elements' (f. 14v)

Folio 14v of our manuscript contains a number of different, apparently unrelated texts and images (fig. 24). In the upper part of the page there are the final lines of **GIA III 25**, together with the corresponding drawing. The long figure on the left side seems to represent an alidade, while the larger, rather complex drawing near the centre of the page is a construction taken from Ptolemy's 'Planisphaerium', as we shall see in 5.4.3. The rest of the page is occupied by a number of geometrical figures, all unlabelled. Ten of them are numbered from I to X, and they are the figures corresponding to the first ten propositions of book I of Euclid's 'Elements'.[546] They might have been reproduced from an Arabic manuscript, but also from a Latin source, since those propositions were among the ones contained in the text known as the 'Geometry II' circulating in high medieval Europe as a work of Boethius (which it was not).[547] The figures might also have been a good copy of notes taken during a Latin or Arabic lecture based on Euclidean material. The numbering of the last two images is exchanged: the one indicated with VIIII corresponds to Elements I.10, while the one labelled X illustrates Elements I.9.herefore, the last proposition represented is I.9, on the bisection of an angle. The distribution of the ten Euclidian figures, which from top to bottom of the page fill up the spaces between the other images, strongly suggests that they were the last ones to be drawn. Why were they reproduced in this manuscript?

A possible answer suggested itself to me when I read a passage in John L. Heilbron's 'Geometry civilized' (1998), a monograph devoted to the contents and historical significance of Euclid's 'Elements'. Proposition I.9, corresponding to the figure labelled X in BnF lat. 7412 (A.27), explains how to bisect an angle with ruler and compass. It is discussed in a paragraph of Heilbron's monograph entitled 'Graduating an arc'.[548] The discussion is associated to images of a quadrant and an astrolabe because, by repeating this procedure many times, an arc of circle can be graduated and used to measure the height of celestial bodies above the horizon. Thus, the presence of this figure in BnF lat. 7412 (A.27) might be evidence not only of a generic interest in Euclid, but of a specific effort to study a construction procedure whose correct performance was crucial for building an efficient tool for the measurement of altitudes.

Heilbron adds that "this graduation will make your circle a useful quadrant [...], when you paste to it on cardboard and attach a pivoted ruler, or 'alidade', at its centre".[549] This remark helps us to see a direct connection between the images from Euclid and the alidade sketched on the left of the page.

5.4.3 Drawing from the 'Planisphaerium' (f. 14v)

The large drawing appearing near the centre of f. 14v of the manuscript BnF lat. 7412 (A.27) illustrates chapter 10 of Ptolemy's 'Planisphaerium'.[550] Given its rather complex structure, it can be identified without doubt even if it is not perfectly drawn, in that the two intersecting circles are equal, while they should represent the flat sphere's equator and its zodiac, and should therefore be different.

---

546 Euclid (1883) vol. 1. p. 10–31.
547 This work is edited in: Folkerts (1970). The same figures appearing on f. 14v of our manuscript can be found in that work on p. 221, fig. 47–53 and 55–56. See also the notes on p. 233–234.
548 Heilbron (1998) p. 90–97.
549 Heilbron (1998) p. 90.
550 Drecker (1927–28) p. 266. The figure is clearly described in the text and can be found in the early printed editions of the work.

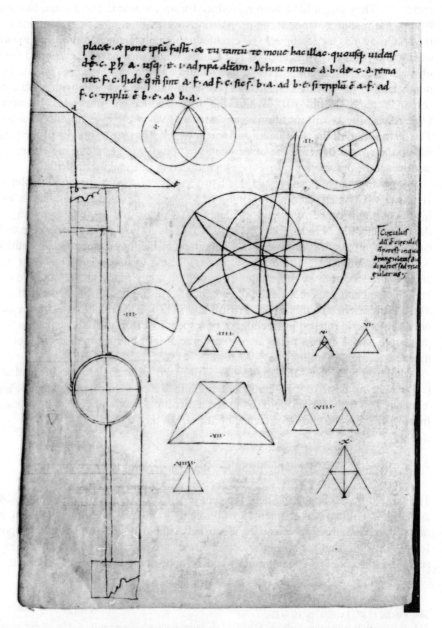

Figure 24: Ms. BnF lat. 7412 (A.27) f. 14v: final lines of **GIA III 25** with figure (top and top left); an alidade (left); geometrical figures from Euclid's 'Elements' I.1-10 and Ptolemy's Planisphaerium' ch. 10 (centre and right). Image reproduced with the permission of the BnF from the microfilm stored in the Microfilm Archive of Medieval Scientific manuscripts (Munich University).

To understand the meaning of the image, it is necessary to shortly discuss the contents of chapter 10 of the 'Planisphaerium'. The question addressed in that chapter is the following: given a horizon and a specific arc of the zodiac circle, determine which arc of the equator rises above that horizon at the same time as that portion of the zodiac.[551] For example: which arc of the equator rises in Rhodos at the same time as the sign of Pisces? In general, the answer is different for different latitudes, as well as for different zodiac signs. This apparently abstract question has a great practical significance, because equal arcs of the equator always rise in the same time at all latitudes and at all times of the year, namely: an arc of 15° of the equator rises in one equinoctial hour. Therefore, determining the arc of equator rising together with a particular zodiac sign for a certain latitude is equivalent to computing the rising time of that sign for that latitude at all times of the year. This is helpful not only for estimating duration at night, but also to compute the ascendent, i.e. the zodiac sign which is (was or shall be) rising at the horizon of a specific place at a certain moment time. The ascendent sign and its significance were discussed in the 'Alchandreana', the high medieval Latin corpus of astrological texts.[552]

The question of the rising times of zodiac signs had received some attention on the part of late ancient Latin authorities, but had not always been satisfactorily handled. As already discussed in 5.3.1, in Macrobius' late ancient commentary to Cicero's 'Somnium Scipionis' – a highly influential reference for high medieval astronomy – it is stated that each zodiac sign rises in two hours independently from latitude.[553] This astronomically incorrect statement was taken up by Abbo of Fleury. Another late ancient, authoritative source on the subject was Martianus Capella who gave values of rising and setting times of zodiac signs computed for the latitude of Carthage or Alexandria.[554] These passages were copied in some medieval manuscripts containing astrolabe texts and/or parts of the 'Alchandreana', showing how this question interested Latin scholars both astronomically and astrologically. In the 'Alchandreana', a method was described to assign a zodiac sign to each temporal hour of the day and night: the first two temporal hours of the day belonged to the sign in which the sun was in that period, the following two hours to the next sign, and so on.[555] This method can be used as an approximation for computing the rising times of the zodiac signs taking at least partly into account the local latitude, and as such it is a better approximation for computing the ascendent than the methods of Macrobius/Abbo and of Martianus Capella.[556]

---

551 Ptolemy (1907) p. 241–244.

552 Juste (2007) p. 175-184. On the 'Alchandreana' see 5.1.1, on the ascendent and the astrolabe see also 5.5.1.

553 McCluskey (1998) p. 117–119.

554 Abry (2000) p. 198-199; Juste (2007) p. 140 n. 104. The relevant passages are chapters 844-845 of book 8 of Capella's 'De nuptiis'.

555 Juste (2007) p. 140-141.

556 Juste remarks that this method does not take into account latitude, but this is not correct: it keeps into account latitude in that it uses latitude-dependent temporal hours. However, Juste

However, the construction in chapter 10 of the 'Planisphaerium' offers the chance to compute the rising (and setting) times of the signs exactly, and also makes it possible to grasp the geometrical mechanism through which a regularly uniform, circular motion of the celestial sphere is translated into a regularly variable, seasonal phenomenon observable on earth.

In conclusion, one reason for Latin scholars to take interest in chapter 10 of the 'Planisphaerium' might have been that it allowed them to geometrically grasp why the rising times of zodiac signs vary both with the sign and with latitude, correcting the statements of Macrobius, Martianus Capella, Abbo and the 'Al-chandreana', and becoming able to use the astrolabe for computing the ascendent in a more precise way.

This is not all. Using the construction on f. 14v, Ptolemy not only showed that different parts of the zodiac rise in different times, but also remarked that it was because of this fact that, for a specific latitude, the sun remained above the horizon for a longer or shorter period depending on the position it occupied on the zodiac. This subject, too, was of interest for Latin astronomers. Martianus Capella had explained that the varying rising times of zodiac signs determined the changing duration of days and nights, but had offered no clear explanation of this statement.[557] The construction associated to the drawing on f. 14v helped understand Capella's claim, because it also allowed the representation of the durations of the longest and of the shortest day of the year for a specific latitude as arcs of the equator circle. The proportion between the extension of each arc and the full circle corrisponded to that existing between the duration represented by the arc and the twenty-four hours.

In conclusion, the knowledge that might have been transferred thanks to this geometrical construction would have offered an additional angle on the same subjects discussed in the text **J**.

5.4.4 Drawings of a sphere with sighting tubes and of a polar sighting tube (f. 15r)

Some of the drawings in BnF lat. 7412 (A.27) are not geometrical figures, but representations of existing devices or sketches for building and using them. On f. 15r, we find the sketch of a hemisphere equipped with sighting tubes and that of a polar sighting tube (fig. 25). Similar drawings can be found in other Latin manuscripts: for example, the sphere is represented also in London BL Old Roy. 15.B.IX (A.15) f. 77r and in Vat. Reg. lat. 1661 (A.44) f. 60 and f. 77v.[558] The hemisphere looks similar to the device described by Gerbert of Aurillac in his

---

correctly notes that other effects are neglected, as for example, the fact that rising and setting times of the same sign on the same day are not equal (Juste (2007) p. 140, n. 103).

557 McCluskey (1998) p. 120–122.

558 For the drawing in the manuscript London BL Old Roy. 15.B.IX (A.15), see Wiesenbach (1991) p. 140.

letters, which was a celestial globe that, thanks to the sighting tubes, could be aligned to the north-south direction and used to identify the position of celestial equator and tropics. [559] However, Gerbert's sphere was a vertically cut hemisphere, not an horizontal one.

The second, smaller drawing on f. 15r can be interpreted as the image of a polar sighting tube thanks to the researches of Joachim Wiesenbach.[560] Wiesenbach has been able to reconstruct the structure and function of this device, which is often depicted in Latin manuscripts, but never verbally described. Wiesenbach has convincingly argued that the polar sighting tube was a device consisting of an empty tube with a graduated, movable disc at one end. It was used to locate in the night the celestial north pole, observe the rotation of the sky around it and use this observation to estimate temporal hours. According to him, the tube was developed in Latin Europe in the ninth century, but it is also possible that it was introduced from the Arabic world, where it is was also present from the ninth century onward.[561] In any case, using the sighting tube to detect the celestial north pole was probably done only in Latin Europe.

In order to find the celestial pole with the sighting tube, one first points it in the direction of the North Star, i.e. the last star in the handle of the Little Dipper. Today, this star is very near the celestial north pole, but in the year 1000 it was about 6° off it: using the graduated disc, which was provided with sighting holes, one could use the North Star as a reference to locate the position of the north pole, and then point the sighting tube at it.[562]

The method is schematically shown on f.15r, where the graduated disc and the Small Dipper are clearly recognizable. The black triangle indicates the North Star, called here (as also in other manuscripts) 'computatrix'. The position of the pole is indicated by another star lying inside the disc and labelled 'polus'.[563] At the bottom, the eyes of the observer are shown, one fixed on the North Star, the other one looking at the north pole through the tube.

It is worth noting how the tube itself is represented as if lying inside the graduated disc, and is complete with two eyes looking into it. Not only BnF lat. 7412 (A.27), but also other manuscripts represent the observer inside the circle, and even more in detail. In the ms. Chartres 214 (A.6), which was destroyed in 1944, the head, arms and shoulders of a figure holding the tube can be seen.[564]

559 Bubnov (1899) p. 24–28. On Gerbert's sphere see 4.3.4, on Gerbert in general see 3.1.4.
560 Wiesenbach (1994). This article contains many images of the sighting tube in Latin manuscripts. See also Eisler (1949).
561 Morelon (1996) p. 9–10.
562 Wiesenbach (1994) p. 379–380.
563 According to Wiesenbach, this is a star of the Giraffe constellation (32 H Camolopardalis), which at the time stood very near the Pole (Wiesenbach (1994) p. 381).
564 The drawing in ms. Chartres 314 is reproduced in Wiesenbach (1994) p. 380, fig. 5 and in Michel (1954) p. 176.

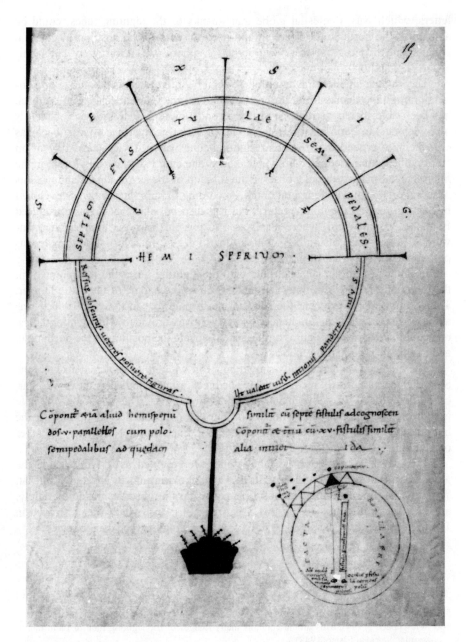

Figure 25: Ms. BnF lat. 7412 (A.27), f. 15r: a hemisphere with seven sighting tubes (above) and a polar sighting tube (below to the right). The image is reproduced with the permission of the BnF from the microfilm stored in the Microfilm Archive of Medieval Scientific manuscripts (Munich University).

On f. 32v of the manuscript Avranches 235 (A.1), a whole human figure is shown sprawled in a very uncomfortable position inside the circle, so that his eyes have a position corresponding to those drawn in BnF lat. 7412 (A.27).[565] What was represented here was not simply the instrument as an object, but the whole experience of using it. The polar sighting tube was a useful device not only for estimating temporal hours in the night, but also for finding the direction of the north-south axis, i.e. for determining local latitude. Whether it had been imported from the Arabic world in the ninth century or not, in the late tenth century the tube belonged to an older Latin tradition of observational astronomy, which in some astrolabe manuscripts was associated with knowledge more recently imported from the Arabic culture.

5.4.5 The universal horary quadrant: text and drawing (f. 16r and f. 18v–19r)

The passage on f. 16r of our manuscript describes how to use a device of Arabic origin known as 'universal horary quadrant' or 'quadrans vetus', which was used to measure the current temporal hour and to determine local latitude. Texts on f. 18v–19r describe two more methods to find the local latitude using it.[566] The quadrant is clearly identifiable thanks to the drawing, although the accompanying text does not describe very clearly its features and functions (fig. 26).[567] The essential feature of this device is a cursor sliding on a plumb-line fixed to the corner of the quadrant. The position of the cursor has to be adjusted according to the local latitude and to the day of the year. In this way, the current temporal hour can be estimated – though not converted into equinoctial hours – and the local latitude can be found by measuring the altitude of the sun at midday and using a relevant scale on the quadrant. The principle on which the quadrant works is an approximate formula that the Arabs had taken over from Indian sources.

Once the reader knows how the quadrant works, the text on f.16r appears not only understandable, but even detailed and correct. Once again, in judging a high medieval mathematical text, a clear distinction has to be made between the competence of the author in mathematics, astronomy or instrument-making and his capability to translate that knowledge into words or drawings. The quadrant can also be used for measuring the altitude of the sun and has in principle a better precision than the astrolabe, because its radius is usually larger. Moreover, not the whole device, but only its plumb-line has to hang perpendicular to the ground, making operations easier. The presence of a text devoted to the quadrant in this astrolabe manuscript as well as in other ones suggests that the Latins were aware

---

565 The drawing in Avranches 235 (A.1) is reproduced in Holtz (2000), in the third page of the appendix.

566 My description of the universal horary quadrant is based on King (2002).

567 These are explained in King (2002) p. 240–242, with a particularly useful fig. 4(b) on p. 241 and I shall not attempt to summarize them here.

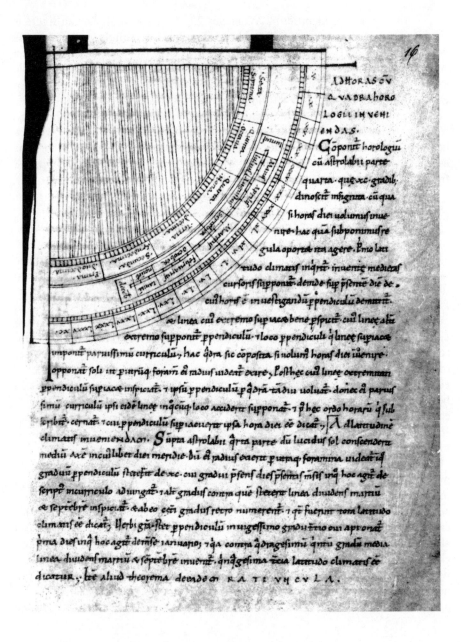

Figure 26: Ms. BnF lat. 7412 (A.27), f. 16r: universal horary quadrant. Image reproduced with the permission of the BnF from the microfilm stored in the Microfilm Archive of Medieval Scientific manuscripts (Munich University).

that a measure of altitude performed with the alidade of the astrolabe would have been rather unreliable, and that other devices should be used in practice. However, in written sources, the astrolabe received by far more attention than the horary quadrant. According to my thesis, this can be ascribed to the fact that the quadrant, just like the sinking bowl, although probably more efficient in practice, did not with its structure offer the possibility of understanding through geometrical imagination the rational patterns on which its function was based. The quadrant provided a means for estimating temporal hours and measuring the height of the sun, but did not make it possible to rationally connect experienced phenomena to the universal geometrical pattern of celestial motion: only the astrolabe could do this. Therefore, it was primarily the astrolabe, a philosophical instrument of the rational mind, that deserved to be discussed in writing, as appropriate for philosophical subjects.

5.5 THE EQUATORIAL SUNDIAL AND THE GEOMETRISATION OF THE FLOW OF TIME

5.5.1 The equatorial sundial (f. 19r)

The drawing on BnF lat. 7412 (A.27) f. 19r represents a sundial, as explicitly stated by the verse on its lower rim: "Sub radiis Phoebi sunt hec signacula plebi in quibus absque mora lucis dinoscitur hora" (fig. 27).[568] Very similar drawings appear in the manuscripts London BL Old Roy. 15.B.IX f. 77r (A.15) and BnF lat. 12117 f. 2v (A.30), and the drawing in the London manuscript is accompanied by the same verse as in BnF lat. 7412 (A.27).[569]

According to David Juste, the dial also appears in the manuscript Montpellier. Bibl. Interuniversitaire, Section de Mèdicine 48 (11th c., A.17) f. 3v.[570] On the dial in fig. 27, lines for temporal hours of prayers are clearly recognizable. However, this image does not represent a sundial of the well-known Greek-Roman type, where hyperbolic solsticial lines are joined by temporal hour-lines. Neither is this a dial of the kind which is most often found on the south walls of early and high medieval churches, where the lower half of a circle is divided into a number of equal parts, and provided with an horizontal gnomon in its centre.[571]

In our figure, instead, the daily path of the shadow for the different months of the year is indicated by seven concentric circles: the innermost and outermost cir-

---

568 My translation is: "Unter the rays of Phebus, these are the indicators for the common people in which the hour of daylight can be immediatlely recognized."

569 The drawings in BL Old Roy. 15.B.IX (A.15) are shown in Burnett (1997) p. 21, along with the Latin verse, which I translate as: An image of the drawing in BnF lat. 12117 (A.29) can be found in the online catalogue Mandragore. Its presence had already been noted by Marcel Destombes Destombes (1962) p. 45.

570 Juste (2007) p. 260. Having become aware of Juste's indication only a short time before comleting this book, I have had no opportunity to see a copy of the drawing.

571 On medieval sundials, see 5.1.4.

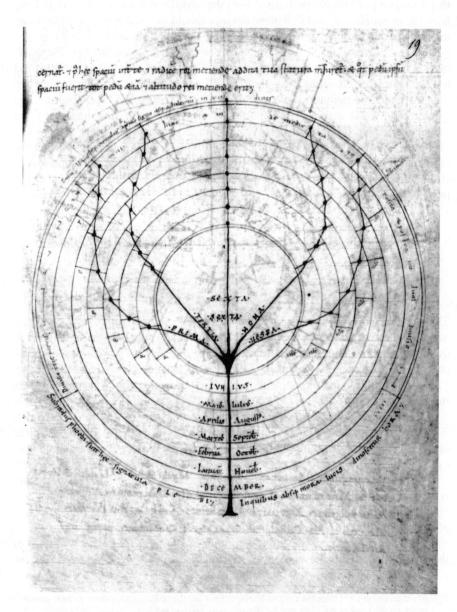

Figure 27: Ms. BnF lat. 7412 (A.27), f. 19r: lines of an equatorial sundial. Image reproduced with the permission of the BnF from the microfilm stored in the Microfilm Archive of Medieval Scientific manuscripts (Munich University).

cles correspond to the solsticial months of June and December, while the remaining months are represented in symmetrical pairs by each of the other circles. As it should be, months with longer days correspond to shorter shadows and therefore smaller circles. On each circle, sunrise and sunset are indicated by simple bars, while the canonical hours ('prima', 'tertia', 'sexta', 'nona' and 'messa' (i.e. vesper)) are noted also with points and joined into yearly hour-lines.

Thus, in this sundial, the shadow of the gnomon is expected to describe each day an arc of circle around its base, always maintaining the same length. This occurs only in 'equatorial sundials', i.e. sundials in which the gnomon points in the direction of the celestial north pole, while the surface on which its shadow falls is set perpendicular to it (and therefore parallel to the equator, whence the name).[572]

On an equatorial sundial, the shadow has a fixed length each day and revolves clockwise around the base of the gnomon with constant velocity, just like the hand of a mechanical clock (though, of course, only when the sun is shining). Lines for equal hours are easily drawn on the dial by dividing it into 24 equal parts. Drawing lines for the temporal hours is more difficult, but it can be done using a drawing of the fundamental lines of an astrolabe to determine the various arcs of circle corresponding to the varying period of daylight during the year, and then dividing each arc into twelve parts. The result is a drawing like that in fig. 27. The construction needed to do this is none other than the one discussed by Ptolemy in chapter 10 of his 'Planisphaerium' and drawn on f. 14v of our manuscript.[573] In my opinion, the dial on f. 19r was drawn using astrolabe lines like those on the astrolabe plate on f. 20r, i.e. the one corresponding to the seventh climate (latitude 48°, Northern Europe).[574]

If the dial is to be used as a sundial, though, there is one unrealistic feature in the drawing: since the sun remains six months above and six months below the equatorial plane, in a real artefact the hour-lines corresponding to the two periods would actually have to be drawn on opposite faces of the dial.[575] Of course, it is also possible to draw all lines on both faces of the device.

As discussed in 5.4.3, the construction from chapter 10 of the 'Planisphaerium' also allows the computation of the rising times of the zodiac signs for any latitude, and therefore the astrological ascendent. The dial drawn on f. 2v of the manuscript BnF lat 12117 clearly belongs in such an astrological context: above the drawing was copied from chapter 17 of the 'Liber Alchandraei', which explains how to connect the temporal hours of the day two by two with six zodiac signs, as seen in section 5.4.3. On the dial on f. 2v, each arc of circle corresponding to the period of daylight for a specific month was divided into twelve temporal hours, which were then labelled two by two with the names of the six zodiac signs

---

572 For a description of equatorial sundials and of how they work, see: Waugh (1973) p. 29–34; on their relation to the astrolabe and their use in constructing other sundials: Wählin (1931b).

573 See above, par. 5.4.3.

574 The dial drawn on Lon. BL Old Roy 15.B.IX f. 77v (A.15) seems to me to be drawn for an even larger latitude, which would fit the English provenance of the manuscript.

575 Waugh (1973) p. 30.

that would rise during daytime in that month. Thus, the dial became a table to compute the ascendent with the formula of the 'Liber Alchandraei'.

An equatorial sundial can also be used for telling the time of prayers, just as other types of sundials, but, thanks to the constant circular movement of its shadow, it embodies better than all other sundials the homogeneous flow of celestial time measurable in degrees of circle. In fact, an equatorial sundial, just like the astrolabe, is a device linking celestial and terrestrial time flow, because it uses the movement of sunlight to make visible and numerable the circular pattern of the experienced flow of time, as Abbo had proposed doing with the zodiac signs and a water-clock.[576] In this sense, the equatorial sundial and the astrolabe do not simply measure time, but rather make time measurable. I believe the presence of this drawing supports my thesis that Latin scholars were interested in constructions making evident to the senses and to geometrical imagination the necessary, rational patterns of creation. Once again, though, it is important to note that the medieval measure cannot be equated to the modern one. Associating specific portions of the day to the zodiac sign rising at the horizon provided a further connection between heaven and earth – a connection which was both quantitative and qualitative, as in the case of the seven climates.

Unity and order could be recognized in the plurality of phenomena not through passive contemplation, but thanks to an act of construction performed according to mathematical rules. The order of the 'superna machina' was accessible to man as an 'architectonica seu mechanica ratio'. Other than with the astrolabe, the equatorial sundial could be realized with a minimal construction effort.

5.5.2 Material evidence for high medieval equatorial sundials

Even if the image on f. 19r of BnF lat. 7412 (A.27) may be interpreted as an equatorial sundial, no indication is given in the drawing of the essential fact that the dial only works if aligned to the equatorial plane, with the gnomon perpendicular to it. What other evidence is there that equatorial sundials were actually known in the High Middle Ages?

In his survey of medieval sundials, Ernst Zinner placed the earliest evidence for sundials with a gnomon parallel to the north-south axis in the early fifteenth century, but the subject awaits further investigation.[577] Zinner also reported about a number of high medieval stone plates from England which present a hole in the middle, are engraved with a circular dial divided into equal parts (usually eight) around the hole and are sometimes even provided with a number of further concentric circles.[578] A stone dial of this type in Aldbrough bears an inscription dated 1066. These plates are often fixed to the south walls of churches, but Zinner suggests that they might have originally been used as free-standing sundials with a

576 See 5.3.1.
577 Zinner (1967) p. 54–56, Rau/Schaldach (1994) p. 283–284.
578 The following discussion is based on: Zinner (1964) p. 7–8.

gnomon placed in the hole. In fact, the Royal Scottish Museum in Edinburgh possesses such a stone with six circles around the central hole. Even more interesting is the dial in the church of Darlington: it has a 10-cm-hole in the middle of each face, surrounded by six circles on one side and eight circles on the other: these are very significant features, because, as already mentioned, equatorial sundials have to be engraved on both faces to take into account the fact that the sun is six months above and six months below the equator.

Although Zinner suggested that the stones might have been horizontal sundials, they could just as well have been equatorial ones. Constructing them was very easy and the only information needed to use them was the value of local latitude. In astrolabe texts, much attention was devoted to the problem of determining local latitude, and various more or less viable solutions were found, involving the astrolabe, the universal horary quadrant (in BnF lat. 7412 (A.27), three methods of finding latitude with a quadrant are explained) or the polar sighting tube. In fact, in both manuscripts in which it occurs, the drawing of the sundial is preceded by images of the polar sighting tube and of the sphere with sighting tubes.[579]

There is also material evidence that the interest in determining local latitude was not only literary: an eleventh-century sculpture from the cloister of St. Emmeram in Regensburg.[580] It is a round plate of stone standing on its rim, on which a cross-section of the heavenly sphere is engraved, with lines representing the celestial north-south axis, tropics, equator and zodiac. The north-south axis has an inclination of 48° (seventh climate) which also fits for Regensburg (49° latitude). Already in the Middle Ages, the sculpture, which has been dated to the period 1052–70, was associated to William, an astronomer who was in that period a monk in St. Emmeram, and later became abbot of Hirsau. Ever since the time in which he was a young monk in St. Emmeram, William had had an interest in astronomy and he is also credited with having built a 'naturale horologium ad exemplum celestis hemisperii', i.e. a natural 'horologium' (sundial?) after the example of the celestial hemisphere.[581] Possibly, William's 'horologium' was some kind of equatorial sundial.

5.5.3 Written evidence of high medieval equatorial sundials

Further evidence of the existence of equatorial sundials in the High Middle Ages can be found in the Latin text **eq** occurring on f. 94r–97r of the astrolabe manuscript Ripoll 225 (A.2).[582] As noted by Josep Casulleras, this text describes the

---

579 In the ms. BL Old Roy 15.B.IX (A.15), the polar sighting tube is on f. 76v and the sphere on f. 77r, together with the sundial.

580 The following discussion is based on: Zinner (1923), Morsbach (1989) and Wiesenbach (1991) p. 135–142.

581 Wiesenbach (1991) p. 110.

582 The text **eq** is edited in Millás Vallicrosa (1931) p. 318–320 (see also app. C). This part of the manuscript is possibly younger than the first part. Millás Vallicrosa discussed this text in Millás Vallicrosa (1931) p. 204–206, yet he did not interpret the dial as an equatorial de-

construction of an equatorial sundial.[583] The text **eq** tells the reader to take a flat, round stone and draw on it six circles, each of which shall correspond to two months. On each circle, the equal hours corresponding to the length of the day in those two months are represented symmetrically with respect to the southern line. Later on, a variant is suggested in which seven circles are drawn: the innermost for June, the outermost for December, and the others for two months each: this is exactly what is shown in BnF lat. 7412 (A.27) f. 19r.

The most important point, though, is that the text **eq** explicitly instructs the reader to set the stone so, that the gnomon is directed towards the celestial north pole. This should be done either by using an 'oroscopum' (probably a quadrant or an astrolabe) or by directly sighting the appropriate star:

> Ut vero directim sedere ipsa petra queat, die quo volueris eam facere directim sedere, horam in oroscopo pernoctabis VIam, id est quantum sol usque ad medium ascenderit diem. Quo invento facies tunc petram ipsam ita sedere, ut gnominis umbra per meridianam sursum veniat lineam. Si autem oroscopum minime habueris, facies ita sedere, ut per medium capitis gnomonis poli stellam videri possis. Sed multo melius est cum oroscopum.[584]

The first procedure described in this passage is not clear, but the second one indicates without any doubt that the gnomon has to be pointed at the star near the celestial north pole, which might have been found using a polar sighting tube. In fact, the expression "ut per medium capitis gnomonis poli stellam videri possit" would almost suggest that the gnomon itself was a sighting tube.[585]

Fragments from an eleventh-century Arabic treatise on sundials are preserved which describe a circular sundial divided into equal sections, although in this text no indications for positioning the shadow-surface parallel to the equator are given.[586] Casulleras interprets this fact as evidence that early medieval, pre-Arabic, knowledge of the equatorial sundial was transmitted both to the Arabic

---

vice. More recently Josep Casulleras has instead argued in this direction (Casulleras (1996)). On the dating of the various parts of Ripoll 225 (A.2): Beaujouan (1985) p. 658.

583 Casulleras (1996), especially p. 629–645.

584 Millás Vallicrosa (1931) p. 319. The first part of the passage is rather unclear. My translation is: "To lay said stone according to the right direction, on the day in which you want to lay it in the right direction, note in the astrolabe ('in oroscopo') the sixth hour, that is when the sun rises up to midday. Having found this, accomodate the stone so that the shadow of the gnomon comes up the southern line. In case you do not have an astrolabe, lay the stone so, that through the middle of the gnomon you can see the Pole Star. However, it will be much better to do it with an astrolabe". The first set of instructions might be interpreted in the sense that one should use the astrolabe to compute how long the equatorial shadow shoud be at midday on a particular day, and adjust the stone accordingly.

585 In his interpretation of this key passage, Millás Vallicrosa reads it as if it simply gave instructions to align the southern line of the dial to the north-south direction (Millás Vallicrosa (1931) p. 206), yet this was only one part of the positioning procedure. Casulleras interprets the passage in the same way I have done (Casulleras (1996) p. 641).

586 Samsó (1992) p. 101–104, Casulleras (1996) p. 613–629, King (1978) p. 367–368 and p. 387–388.

and to the Latin medieval cultures of Spain, and then with time decayed. Because of this fact, the dial of the equatorial device was, at first, erroneously arranged-horizontally and, later on, somehow adapted as to give reasonable readings in this new position.[587]

However, equatorial sundials might also have been introduced from the Arabic culture alongside the astrolabe and the universal horary quadrant. The drawings on BnF lat. 7412 (A.27) and London BL Old Roy. 15.B.IX (A.15), and the text **eq** suggest that knowledge on this subject did spread in Latin Europe parallel to astrolabe knowledge. This is hardly surprising since, thanks to the astrolabe, temporal hour lines for an equatorial sundial could be drawn for any latitude without the help of tables of daylight length.

5.6 THE ASTROLABE DRAWINGS IN MS. BnF LAT. 7412 (#4024)

5.6.1 Overview on the astrolabe drawings #4024

On f. 19v–23v of the manuscript BnF lat. 7412 (A.27), drawings of the various parts of an astrolabe appear, complete with Arabic inscriptions in kufi characters. These drawings are usually regarded as the manuscript copy (or possibly copy of a copy) of the parts of an Arabic astrolabe artefact from tenth-century al-Andalus. As such they have been assigned an entry in the Frankfurt Catalogue of Medieval Astronomical Instruments: **#4024**. Similar drawings are found in the manuscript Bern BB 196 (A.4) on f. 1r, 2v, 3v and 7v and also occurred on f. 30r of the manuscript Chartres 214 (A.6), which was destroyed in 1944.[588]

The Arabic inscriptions on the drawings have been discussed in detail by Paul Kunitzsch and, as far as the names of the months are concerned, by Kurt Maier.[589] The star positions on the rete have been analysed by Burkhard Stautz.[590] I shall shortly summarize some of their results and add a few comments on the possible relationship of these drawings to a Latin tradition of astrolabe studies. The parts of the astrolabe represented in BnF lat. 7412 (A.27) (**#4024**) are:

f. 19v: a rete, on whose star pointers are written Arabic star names in Latin transliteration. These names are repeated, in different spelling, at the bottom of the page (fig. 28);

f. 20r–23r: on each folio, a drawing of a plate for one of the seven climates occurs, starting from the seventh one (fig. 29). On each plate, one Arabic sentence is inscribed giving the latitude and the maximum length of daylight. These are the same sentences as in the Latin-Arabic climate table in **J 18** (f.

---

587 Casulleras (1996) p. 644–645.
588 On Bern BB 196, see 4.5.8 and appendix A, item A.4. The image from Chartres 214 (A.6) is reproduced in Van de Vyver (1931) pl. 2.
589 Kunitzsch (1998a), K. Maier (1996) p. 259–260.
590 Stautz (1997) p. 53–54.

8r–8v), but here they are written in kufi characters. On some plates, other inscriptions are present, too, e.g. the Arabic names of the twelve temporal hours;

f. 23v: back of an astrolabe, showing a calendric scale relating the zodiac signs to the months of the Latin calendar, whose names are written both in Latin and in Arabic language.[591] A shadow square is present and, in its middle, Kunitzsch recognized the signature of the astrolabe-maker: "The work of Khalaf ibn al-Mu'ādh", which appears to be a plausible high medieval Andalusi name.[592]

Although the inscriptions suggest that this drawing was based on an Arabic astrolabe from al-Andalus, its form is quite different from that of extant eleventh-century Andalusi artefacts and instead resembles the form of early Eastern Arabic ones.[593] On the other hand, the calendric scale with Latin months is a feature appearing on the back of eleventh-century Andalusi astrolabes, but not in the early Eastern Arabic artefacts.[594] Finally, neither Western nor Eastern Arabic-Islamic astrolabe artefacts usually had plates for the seven climates, although the oldest extant late ancient astrolabe treatises in Greek and old Syriac language describe devices with plates made for the seven climates.[595]

All extant Andalusi astrolabes which can be dated were constructed after the period of the civil war (ca. 1009–1031) that followed the fall of the Caliphate of Cordoba and, with it, of the Cordoba school of astronomy. Because of this, it is assumed that the drawings **#4042** are evidence of a tenth-century tradition of Andalusi astrolabe-making of which no artefacts are preserved. This older tradition would have been more closely linked to the early Eastern Arabic one, but was also influenced by the Latin culture, as the calendric scale with Latin months shows.[596] According to this interpretation, the late ancient Greek tradition of making plates for the seven climates influenced early Eastern Arabic astrolabes and, through them, also early Western Arabic ones. However, only one early Eastern Arabic astrolabe artefact with climates plates exists (**#3702**), and the only other astrolabe artefact that may be considered as produced in tenth-century al-Andalus (**#110 = #135**) does not have plates for the seven climates.[597]

While the drawings **#4024** clearly had an Arabic model, I would like to suggest that they also testify to Latin artists' capability of drawing astrolabe lines. I believe that the drawings were not executed by simply copying an original artefact

---

591 Image reproduced in Van de Vyver (1931) pl. 3.
592 Kunitzsch (1998a) p. 117–118.
593 Stautz (1997) p. 54. On early Andalusi astronomy and astrolabe-making, see 3.1.4 and 3.2.6.
594 King (1995) p. 377 and p. 384–385.
595 King (1995) p. 372–375. On the oldest treatises, see 3.2.2.
596 King (1995) p. 384–385. In a private communication, Prof. Kunitzsch has instead expressed the opinion that the Latin features were added by a Latin scribe to an orignally Arabic drawing.
597 Stautz (1994) (**#3702**), Stautz (1997) p. 54–55 (**#110=#135**), King (1999a) p. 7.

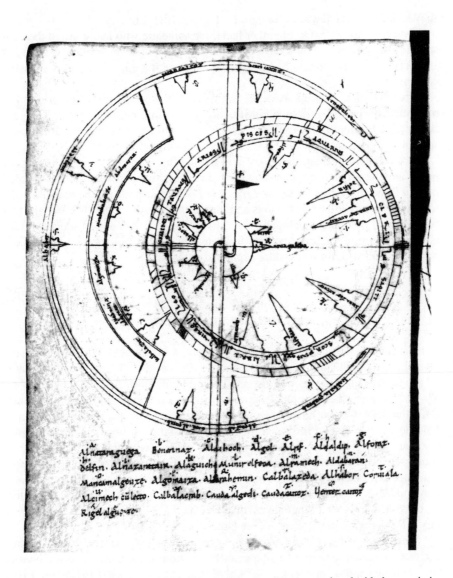

Figure 28: Ms. BnF lat. 7412 (A.27), f. 19v: astrolabe rete. Image reproduced with the permission of the BnF from from the microfilm stored in the Microfilm Archive of Medieval Scientific manuscripts (Munich University).

(or drawing), taking measures of all circles and lines: they were rather made by performing the geometrical construction described, for example, in **h2**.

For someone who had no understanding of how astrolabe lines were drawn, it would have been very difficult to copy so many of them so accurately (especially from an astrolabe artefact), because he would have been unable to notice when he was making small, but mathematically significant errors. As Bruce Eastwood has shown, for high medieval Latin scribes it was difficult to copy diagrams without understanding them.[598] On the other hand, for someone who had grasped the construction principles, it would have been much easier to use the original drawing or artefact as reference for performing his own construction, rather than slavishly copying it.

I believe there is some evidence in favour of this thesis in the drawings. The most difficult part of astrolabe construction is drawing altitude circles. They are constructed by finding for each circle the two points where it intersects the north-south axis of the astrolabe, and then drawing the circle from this diameter.

According to the instruction given in **h2**, the first step is finding all couples of points and the second one is joining them in circles two by two. The last, isolated point will be the zenith of the horizon. On f. 23r, it can clearly be seen how the artist erroneously joined two points and then had to erase a circle (fig. 29): the two points joined by mistake belonged respectively to the fourth and fifth circles from the zenith. In my opinion, this may indicate that he was not copying an artefact, but rather following the procedure explained in the text **h2**.

5.6.2 The astrolabe drawings #4024 and other early European astrolabe artefacts

If the artist who executed the astrolabe drawings **#4024** was able to construct latitude plates without simply copying them, maybe the idea of drawing seven plates for the seven climates was not just passive imitation of an Arabic artefact. Possibly, that choice was an original step taken in the context of transmission and assimilation of astrolabe knowledge in the Latin culture. In the tenth and eleventh centuries, thanks to the mediation of Arabic-speaking astronomers, Latin mathematicians had the possibility of coming into contact with the ample treasure of Greek-Arabic astronomy, isolating and assimilating those elements that most resonated with the new interests and methods that were being developed in Latin astronomy. As we have seen, the seven climates were a subject of particular interest to Latin scholars and the manuscript BnF lat. 7412 (A.27) contained all necessary information to draw the astrolabe plates found on f. 20r–23r.

Moreover, while extant Eastern and Western Arabic artefacts feature almost exclusively plates for sets of latitudes different from those of the seven climates, climate plates appear in a number of early European astrolabes (**#161**, **#166**, **#167**, **#202**, **#303**, **#420**, **#589**), again suggesting that it was medieval Latin scholars who revived the interest in the ancient Greek-Roman system of the climates.[599]

The oldest extant Latin astrolabe (**#3042**) only possesses five latitude plates which are labelled for latitudes 36°, 39°, 41° 30', 45° and 47° 30', although the

---

598 See 4.3.2.
599 King (1995) p. 372–375, King (1999a) p. 6–8.

values actually represented on the plates might be different.[600] These values, too, may be interpreted as relating to the four or five climates of highest latitudes.

The astrolabe drawings **#4024** also share a second characteristic with early European artefacts which is not found in Arabic ones: the form of the equatorial bar, i.e. the bar on the rete that joins the two equinoctial points cutting the zodiac into two unequal parts. On f. 19v, this bar has a single counter-change in the middle, i.e. a form like this:

This form is not found on tenth- or eleventh-century Arabic astrolabes, but, once again, it appears on a number of early European ones (**#161, #167, #420, #550**).[601] It also appears on another Latin drawing of a rete: the one in Vat. Reg. lat 598 f. 120r (**#4553**) which was discussed in 4.6.8. On the oldest Latin astrolabe **#3042**, on the other hand, the equinoctial bar has a more complex structure: it is counter-changed not only in the middle, but also at the intersections with the zodiac circle.[602]

The bar with a single counter-change was a very elegant solution when an artist was drawing a rete on parchment or paper, because one of the two diameters that had to be drawn in the beginning to divide the circle into four quarters could be re-used as central line for the counter-change. This was also done on the drawing on f. 19v, as seen by the fact that the line drawn in the beginning cuts the zodiac circle, while the other sides of the bar, drawn later, do not. The use of the counter-changed bar in both Latin drawings of a rete suggest that it might have been a common solution for drawing astrolabe lines on parchment or wood and was later also applied to metal artefacts.

In conclusion, the astrolabe parts drawn on BnF lat. 7412 (A.27) f. 19v–23v (**#4024**) were not copied from any Arabic artefact identical to the extant ones, but share two features with some of the earliest European astrolabes: plates for the seven climates and equinoctial bar counter-changed in the middle. The European astrolabes in question are:

#161: Astrolabe with Gothic numerals and letters. Its equinoctial bar has a single counter-change in the centre. The instrument is provided with three plates which, together with the mater, represent horizons for the latitudes: 15°, 23°, 30°, 36°, 41°, 45° and 48°. These values in all probability represent the seven climates (London, British Museum).[603]

---

600 The latitude values on the plates of **#3042** and their possible relationships to the seven climates are discussed in: d'Hollander (1995) p. 407–409, King (1995) p. 372–373, G. L'E. Turner (1995) p. 422.

601 King (1995) p. 366–369.

602 The peculiar structure of the equinoctial bar of **#3042** and its possible relationship with that of other astrolabe artefacts is discussed in King (1995) p. 367–369.

603 Description by S. Ackermann in online-catalogue Epact, Gunther (1931) p. 306, Ward (1981) p. 110 and 112. An image of this astrolabe is in King (1995) fig. 11.

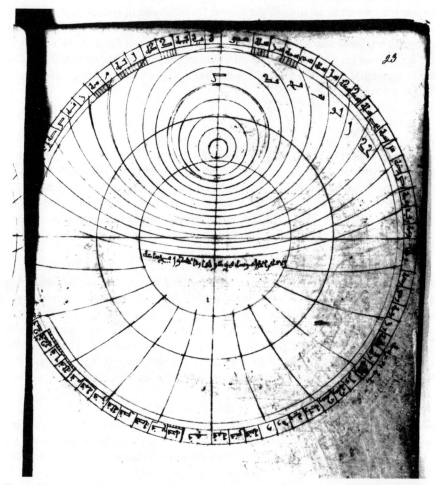

Figure 29: Ms. BnF lat. 7412 (A.27), f. 23r: horizon plate for the first climate. One of the altitude circles that has been wrongly drawn and erased. Counting from the smallest one, it is between the fourth and the fifth circles. Image reproduced with the permission of the BnF from the microfilm stored in the Microfilm Archive of Medieval Scientific manuscripts (Munich University).

#166: Astrolabe with script of Lombardic type. The instrument has five plates of which four are engraved on both sides with horizon lines explicitly marked as corresponding to the seven climates with values for the latitudes: 12°, 24°, 30°, 36°, 41°, 45° and 48° (?). All markings have the form: "CLIA 2 LAT 24". The back of one of the discs is engraved with an unmarked horizon corresponding to 51° according to Gunther's estimate, while the engravings on the last disc are unfinished (Oxford, Museum of the History of Science).[604]

#167: Astrolabe engraved in Gothic script and numerals and with some words of Italian origin. The equinoctial bar is counter-changed in the middle. The instrument is extant with only a single plate, which is inscribed on both sides with horizon lines marked respectively: '6, 45°' and '7, 48°'. The mater is inscribed with a horizon marked as '1, 16°'. This inscriptions indicate that the three latitudes represented the first, sixth and seventh climates and strongly suggest that two plates with the remaining four climates are missing (London, British Museum).[605]

#202: Astrolabe with inscription in Picard dialect of Old French, ca. fourteenth century. The instrument has four plates inscribed on both sides with horizons for latitudes 24°, 30°, 36°, 41°, 45°, 48°, 50° and 51°. The mater is only incompletely engraved, apparently for latitude 51°. The eight values on the plates serve six of the seven climates (from the second to the seventh one) plus two specific latitudes, 50° and 51°, that correspond well to Picardy, the region where the astrolabe was made (private collection).[606]

#303: An English astrolabe whose rete is missing. It is inscribed with Gothic and Lombardic letters and is possibly a re-worked oriental instrument. It has three plates engraved on both sides which, together with the mater, serve six of the seven climates, as explicitly stated by the engravings: "CLIMA I XVI" (mater), "CLIMA III XXX", "CLIMA IIII LATITUDO XXXVI", "CLIMA V LATITUDO XLI", "CLIMA VI LATITUDO XLV" and "CLIMA VII LATITUDO XLVIII". The seventh plate, instead of the second climate, serves the latitude of London: "LATITUDO LONDONIAE LII". The back of the instrument is equipped not only, as usual, with a calendric scale, but also with a lunar dial, now incomplete, which would have enabled the computation of the phase of the moon (Oxford, Merton College).[607]

#420: Astrolabe of uncertain provenance with equinoctial bar with a single counter-change in the centre. Its star pointers also have a shape similar to that of **#4042**. The instrument has two plates and an inscribed mater serving the upper climates, as explicitly indicated by the engravings "tersium" (!), "quartum", "quintum", "sextum clima" and "septimum clima" (Greenwich, National Maritime Museum).[608]

---

604 Description by I. Meliconi in online-catalogue Epact, Gunther (1931) p. 316–317, Stautz (1997) p. 84–85. An image of the astrolabe is in King (1995) fig. 10.
605 Description by S. Ackermann in online-catalogue Epact, Gunther (1931) p. 317, Ward (1981) p. 114.
606 King (2000), King (2001), especially p. 406–419.
607 Gunther (1923) p. 210–213, Gunther (1931) p. 482.
608 This astrolabe is described in Van Cleempoel (2005a), p 126–131 and in: Stautz (1997) p. 94. An image of the astrolabe is in King (1995) fig. 9.

#558: Astrolabe of uncertain provenance, with three plates engraved on both sides. Together with the mater, they serve the latitudes of the seven climates: 16°, 30°, 36°, 40°, 45°, 48° and 52° (Nürnberg, Germanisches Nationalmuseum).[609]

#589: Undated astrolabe, bought by Nicolaus Cusanus (1401–1464). The instrument has three plates engraved on both sides. Together with the mater, the seven horizons serve the seven climates, as unequivocally indicated by the inscriptions: CLI I L.XV (mater), II L.XXIII, III L. XXX, IIII L.XXXVI, V L.XLI, VI L.XLV, VIII L.XLVIII (six faces of the plates) (Bernkastel-Kues, Kusanus-Stift).[610]

#4556: unsigned, undated astrolabe. It has four plates engraved on both sides which, together with the mater, serve latitudes corresponding to the seven climates, two of which are represented with two latitude values. The values are: first climate 15°, second climate 24°, third climate 30°, fourth climate 36°, fifth climate 39° and 41°, sixth climate 43° and 45°, seventh climate 48° (private collection).[611]

I believe it is possible that the drawings **#4024** testify not only to an early Western Arabic tradition of astrolabe making, but also to an early Latin tradition of astrolabe-making or -drawing. A further indication of such a tradition is an image found in another eleventh-century manuscript, which I shall discuss in the next paragraph.

5.6.3 Abraham and his astrolabe in ms. BnF lat. 12117 f. 106r

The illuminated manuscript BnF lat. 12117 (A.30) was written and illustrated in the scriptorium of the Abbey of St.-Germain-des-Prés in Paris around the middle of the eleventh century (1031–1060).[612] BnF lat. 12117 (A.30) belongs to a group of manuscripts illustrated by an artist named Ingelard. As discussed in 5.5.1, on f. 2v of this manuscript the chapter from the 'Liber Alchandrei' on how to compute the ascendent appears, while an equatorial sundial is represented and used to demonstrate this method.[613]

The manuscript contains material of various kind, part of it of mathematical, astronomical and astrological interest:

---

609 Description by D. A. King in: Bott (1992) p. 574–576.
610 Hartmann (1919) p. 40–42, King (1999a) p. 27.
611 My description is based on Stautz (1997) p. 93. The instrument is described in Christie's (1994) p. 34–39, which I was unable to consult.
612 My discussion of this manuscript is primarily based on Deslandes (1955) as well as on copies of some folios acquired from the Bibliothèque Nationale de France. Images of illuminations from the manuscripts and a collation can be found online in Mandragore. Partial collations can be found in Leonardi (1959–60) p. 441 (with further bibliography) and in Jordanus.
613 Mandragore, Destombes (1962) p. 45, Juste (2007) p. 260.

Figure 30: Ms. BnF lat. 12117 (A.30), f. 106r: Abraham with astrolabe and compass in his hands. Image reproduced here with permission of the BnF.

- Rhemnius Fannius' poem on weight and measures;
- Ps.-Clemens, Recognitiones;
- Annals of St.-Germain-des-Prés, with many illuminations;
- Computus of Helperic of Auxerre;[614]
- Excerpt from the 'De astrologia', an astronomical and astrological manual from the late Roman Empire attributed to Hyginus (fl. 28 B.C. –10 A.D. This work is illustrated in the manuscript with images of the constellations.[615]

On f. 106r, illustrating part of a biblical genealogy, a figure holding a compass in his right hand and an astrolabe in his left one occurs (fig. 30). Since the first person mentioned on the page, right below the image, is Abraham, there can be no doubt that he is the person holding compass and astrolabe, as befits a master in the science of the stars.[616]

The astrolabe can be identified without doubt and is also described as such in the description in the on-line catalogue [Mandragore]: the teeth of the rete are clearly recognizable on the zodiac ring and in the center (fig. 31). The form and the pattern of the star pointers are very similar to those in early Latin astrolabe drawings (#4024, #4553), suggesting that the artist had at least seen such a drawing, if not a solid device made, for example, of copper or of wood and parchment. The artist has indicated that the rete is perforated by filling the space between the teeth with four crosses with dots. It would even seem that the equinoctial bar has a single counter-change in the middle.

Abraham holds the astrolabe in his hand and does not use it to look in the sky as is the case in most twelfth-century representations. The artefact does not even have a suspension ring: it is simply a revolving star chart, the function that it might have best performed in tenth- and eleventh-century Europe. This might be taken as a hint that eleventh-century Latin mathematicians did not actually attempt to use astrolabes to measure the altitude of celestial bodies. It is also important to note that, in his other hand, the patriarch is holding a compass, i.e. the mathematical instrument that was to become the standard attribute of 'deus geometra'.

Had Ingelard actually seen an astrolabe, either made out of metal or wood, or simply drawn on parchment? Or did he copy his Abraham from some other model? In her discussion of Ingelard's images of the constellations illustrating Hygin's 'De astrologia' in this same manuscript, Yvonne Deslandes says that the artist dressed most of the figures in contemporary clothing, instead of strictly following classic models, and that he represented best those animals that he had actually seen.[617] I believe that Ingelard had probably seen at least astrolabe drawings, if not an artefact. Moreover, whoever decided to have Abraham represented with an astrolabe also had some knowledge of what an astrolabe could be used

614 On astronomical observations in this treatise, see 5.1.5.
615 On Hygin, see: Brunhölzl (1999).
616 See 5.2.2.
617 Deslandes (1955) p. 10.

for, for example, computing the ascendent as on f. 2v. It may be noted that the following illuminations show a Nativity sequence in which the three Magi and the star have a very prominent role.

Figure 31: Ms. BnF lat. 12117 (A.30), f. 106r: Detail of the astrolabe in Abraham's hand. Image reproduced here with permission of the BnF.

6 SUMMARY AND OUTLOOK: THE ASPECTS OF THE ASTROLABE IN TWELFTH-CENTURY EUROPE

6.1 THE ASPECTS OF THE ASTROLABE AND THEIR INTERPLAY

In this study, I have approached the Latin medieval astrolabe by distinguishing in principle between its mathematical, practical and philosophical aspects. These three aspects of the astrolabe appear today as distinct and one may be tempted to classify them as 'theory', 'practice' and 'philosophical interpretation'. However, in the context of high medieval learning, these aspects interacted with each other in a much different way than mathematical theory with practice or with its possible philosophical interpretation. In fact, the aspects of the astrolabe can only be distinguished from each other as hermeneutic tools to analyse the sources: in their actual historical interplay they are connected into a unity that escapes modern definitions.

The mathematical aspects of the astrolabe are the structures I indicated as 'shadow triangle' and 'flat sphere'. They were discussed in chapter 2 in such a way as to show how easily mathematical knowledge of these subjects could be conveyed and stored by exercising and memorizing specific drawing procedures, eventually with the help of labelled drawings, solid models, mnemonic verses and short notes. In chapters 3 and 4, I have offered evidence that the composition of written texts describing those procedures in a clear way, as for example the treatise **h** by Herman of Reichenau, was not a premise, but rather a consequence of the spread of knowledge of the mathematical aspects of the astrolabe in tenth- and eleventh-century Europe. The modes of communication involved in the diffusion process were mainly non-written and non-verbal ones.

What I referred to as practical aspects of the astrolabe were instead its possible uses, e.g. as time-keeping device or as a computational tool for astronomy and astrology. Here, a further distinction is necessary: some of those possible uses may in fact have been practically impossible, for example, using the astrolabe to measure a time interval of a few minutes while standing in a boat.[618] Apart from limit cases like this one, though, it is usually not possible for us to tell exactly to which virtually possible uses of astrolabe artefacts could be realised in practice. However, as I have argued in chapter 5, virtual uses of the astrolabe could be as significant for Latin mathematicians as the real ones. For example, the fact that the astrolabe could virtually relate celestial and terrestrial time flow made it the ideal time-measuring device, even if it could not be used to tell time in practice. At the same time, though, it was important that there should be some feasible procedure to realize in practice those virtual aims: bodily, sensory experience had to come into play. I have stressed the possible relevance in this sense of other

618 See 5.3.7.

mathematical and astronomical devices, such as the universal horary quadrant, the polar sighing tube or equatorial sundials.

Finally, there are those aspects of the astrolabe which I have loosely referred to as philosophical. In analogy to other instruments of the mathematical arts (e.g. compass, abacus, monochord), the astrolabe could be perceived as an instrument of reason whose construction and use might allow rational creatures to gain knowledge of elements of cosmic order. In the case of the astrolabe, these were for example, the measuring and numbering of experienced time in equal hours, the division of the inhabited world in seven (or eight) climates or the recurring seasonal patterns of celestial and terrestrial phenomena.

The philosophical aspects of the astrolabe were partly based on – but not necessarily dependent from – the mathematical and practical ones. Thanks to its mathematical aspects, its virtual uses and its connection to experimental practise, the astrolabe could affirm itself as a philosophical instrument. However, its unifying potential greatly exceeded that of its purely mathematical and practical sides: in the context of medieval learning, the astrolabe as a philosophical instrument could prove itself effective not only thanks to experiment, but also to the authority of written tradition. An example is the scheme of the seven climates, whose quantitative and qualitative universality rested both on geometrical structures and on authoritative literary descriptions. By drawing astrolabe plates for the seven climates, the instrument could be made into an ordering measure of the variety of lands and peoples described by ancient authors. Vitruvius' statements on the 'architectonica ratio' of the world could be read as a proof that the astrolabe could deliver knowledge not only on how to build an astronomical machine, but also on how to understand the workings of the 'machina mundi'.

It was thanks to its philosophical aspects that the astrolabe found entrance into the literary culture of eleventh-century Europe: a large part of the earliest texts devoted to it show a greater preoccupation with underscoring its philosophical significance than with expounding its mathematical and practical aspects. At the same time, tools that might have been more useful in practice, such as the universal horary quadrant, the geometrical square or the sinking bowl, received much less attention in written sources. This situation reminds of the origin of mechanical clocks, where the first written sources informing us on weight-driven mechanisms are descriptions of extremely complex – and practically not very useful – astronomical clocks.[619]

As I argued in chapters 3 and 4, the earliest Latin astrolabe texts witness to a tension between the written word and other strategies of knowledge transfer. This tension was in my opinion an essential feature of tenth- and eleventh-century mathematics and had a great influence on the development of the mathematical and mechanical arts in the following centuries. In the concluding paragraphs of this study, I shall offer some tentative reflections on this subject, arguing that the astrolabe, which was the first mechanical device to be regarded as a possible source of philosophical knowledge, for some became the symbol of forbidden

619 For an overview and further bibliography Dohrn-van Rossum (1992) p. 49–55.

knowledge, while in other cultural circles it was a model of 'working rationality' to be imitated. At the same time, not all philosophers who considered the astrolabe a symbol of rational investigation had an interest to explore nature with mechanical constructions and experiments.

6.2 Twelfth-century astrolabe treatises and their fortunes

Around the middle of the twelfth century, a number of Latin treatises on astrolabe construction and/or use were composed or translated from the Arabic, among them the Latin version of Ptolemy's 'Planisphaerium'.[620] The authors and translators of the new material have in large part been identified, showing that their interest in the astrolabe was coupled to a passion not only for mathematics and astronomy, but also for astrology. Astrology had played a role in Latin culture since Carolingian times, but, as shown by the diffusion of the 'Alchandeana', the contact with Greek-Arabic works had prompted a new flourish of interest in this subject, whose influence on the Latin West can hardly be underestimated.[621] Twelfth-century astrology offered not only rational explanations of natural phenomena, but also the possibility of reconstructing, predicting and to some extent controlling the events of human history. In particular, the art of 'astrological elections' allowed to choose the appropriate time for specific activities, such as medical cures or military campaigns: the art of astrology soon became a constitutive part of medicine and politics.[622]

Raymond of Marseilles, author of an astrolabe treatise (written before 1141), was a practitioner and defender of astrology. Both Herman of Carinthia (fl. 1138–1143), the translator of the 'Planisphaerium', and John of Sevilla (fl. 1133–1153), who made a Latin version of ibn aṣ-Ṣaffār's (d. 1035) treatise on the use of the astrolabe, translated a masterwork of Arabic astrology that later had great influence on the Latin West: Abū Ma'shar's (786–866) 'Introductorium maius'.[623] Adelard of Bath (fl. 1116–1142) not only composed a treatise on astrolabe usage and made a Latin version of Euclid's 'Elements', but also translated from the Arabic astrological works, for example, parts of the 'Centiloquium', a collection of astrological advice which became very popular in the Latin Middle Ages.[624]

Finally, astrology had a central place in the works of two Jewish scholars from Spain who were among the main mediators between the Arabic and the Latin culture: Abraham ibn Ezra (d. 1167) and Abraham bar Hiyya/Savasorda (d. ca. 1136).[625] Abraham ibn Ezra composed or translated from the Arabic into He-

---

620 See app. D.
621 d'Alverny (1967) p. 36–39, Juste (2007) p. 219-223.
622 Gregory (1992a) especially p. 13–19.
623 Burnett (1978) p. 118–121, North (1986) p. 52–62, Sesiano (1999), Brentjes (2000) p. 277–279.
624 Clagett (1970), Burnett (1987).
625 Goldstein (1999).

brew a large number of works on astronomy and astrology.[626] He also wrote an astrolabe treatise of which a Latin version was made.[627] Abraham bar Hiyya collaborated with Plato of Tivoli (fl. 1133–1145) in translating both his own works and other astrological and mathematical texts.[628] The Jewish scholar probably also helped Plato in the translation of the astrolabe treatise of ibn aṣ-Ṣaffār.

Judging by modern criteria, twelfth-century texts were in general more readable then the earlier ones, but a comparison of their manuscript traditions with that of eleventh-century texts shows that the new material enjoyed a more limited diffusion than the old one. As we can seen in table 4, in twelfth- and early thirteenth-century manuscripts, all twelfth-century astrolabe treatises taken together still remained less popular than Herman of Reichenau's 'De mensura astrolabii' (**h**) alone. With time, the new treatises did not even tend to replace the old ones: in the thirteenth century, the older and the newer tradition were joined together in the same manuscripts. Significantly, this development exactly parallels the one noted by David Juste for the 'Alchandreana': in the twelfth century, instead of being supplanted by the new translations of Arabic classics of astrology, these older Latin astrological texts were copied with increasing frequency.[629]

Table 4 clearly shows how, by the twelfth century, the text **h** had become the most popular astrolabe treatise. That text, usually joined to **J** and **hv**, remained the main

Table 4: Diffusion of eleventh-, twelfth- and thirteenth-century astrolabe texts

	11^h-c. texts, **h** not among them	11th-c. texts, **h** among them	11th- and 12th-c. texts	12th-c. texts, or later ones	13th-c. texts, or later ones
11th c. to early 12th c.	16 mss	1 ms	---	---	---
12th c. to early 13th c.	4 mss	16 mss		4 ms	
13th c.		9 mss	3 mss	9 mss	more than 200 mss
14th c. to 15th c.	1 ms	4 mss	2 mss	20 mss, of which 11 with **PP**	

626 Langermann (1993), Goldstein (1996).
627 The treatise (AE) is edited in Millás Vallicrosa (1940). See also app. D.
628 Wigoder (1971), Minio-Paluello (1975), Brentjes (2000) p. 279–280.
629 Juste (2007) p. 271.

written reference for astrolabe studies until, around the middle of the thirteenth century, a new, anonymous treatise appeared, which was attributed to the Jewish-Arabic astronomer and astrologer Māshā'allāh (d. ca 815), known to the Latins as Messahalla.[630] For a long time historians regarded this text as a translation from the Arabic, possibly as 'the' translation which had been at the origin of the spread of astrolabe knowledge in Latin Europe. Once again, it was Paul Kunitzsch who corrected this long-standing error, showing beyond doubt that the Pseudo-Messahalla treatise was no translation, but an independent Latin composition from the thirteenth century, which partly owed a debt to the oldest Latin astrolabe texts.[631] Unfortunately, this text has up to now only been printed in a rather unsatisfactory way, using only a small part of its huge manuscripts tradition.[632]

6.3 The astronomer with astrolabe in Oxf. Bodl. 614, f.35r

One image from the English manuscript Oxf. Bodl. 614 (ca. 1100–1150, A.24) illustrates well how the astrolabe, as a rational instrument, could play a role in the rational natural philosophy which was characteristic of the early twelfth century.[633]

The image shows a man letting an astrolabe hang from his hand according to the pattern found in the old astrolabe texts, the only difference being that the astrolabe is hanging from the left hand and not from the right one. In the sky, one can recognize the Big Dipper (or possibly the Little Dipper, appearing so often in images of the polar sighting tube). Seen like this, the image would appear as a standard representation of an astronomer, but it acquires a special significance when reading the text written above it. It is a passage from the work 'De philosophia mundi' of William of Conches (d. after 1150), an author belonging to the trend in rational natural philosophy and theology which is often referred to as 'School of Chartres'.[634] These philosophers showed great interest in the power of reason, in the explanation of phenomena 'secundum physicam' and in the (re)discovery of Arabic and Greek knowledge.

The passage explains that falling stars are due to an optical illusion: some bright light in the sky makes a star invisible for a short time and, when the light is gone, the star seems to have disappeared. Yet it has not fallen down, as commonly believed, but has only moved a little away following celestial rotation. Thus, fall-

---

630 On Māshā'allāh and his works see Sezgin (1978) p. 127–129, where the Ps.-Messahalla treatise is still regarded as probably original.

631 Kunitzsch (1981b).

632 Ps.-Messahalla (1929). The part of the text on the use of the astrolabe has also been printed in Chaucer (1872) p. 88–105, which has since then been reprinted many times.

633 For bibliogaphy on this manuscript, see app. A, item A.24. My remanrks are based on a microfilm copy of the manuscript.

634 The work is edited in: William of Conches (1980). The passage is a shortened version of ch. 3, 12. On twelfth-century rational philosophy and theology see, for example: Stiefel (1985), Gregory (1992a), Gregory (1992b) especially p. 88–99 and Flasch (2000) p. 252–270.

ing stars can be explained as simple optical illusions if one is able to reconstruct the regular motion of the heavenly sphere. The text does not say how this can be done, but the image does: thanks to the astrolabe. The act represented in the picture is very implausible from a practical point of view: measuring a small change of position of a star with an astrolabe. However, the message of the picture is another one: it is thanks to the mathematical structures embodied in the astrolabe (i.e. the shadow triangle and the flat sphere) that it is in principle possible to grasp the ratio behind the phenomena and thus avoid being deluded by superficial sensual impressions.

6.4 The astrolabe as a symbol

While the scribe of Oxf. Bodl. 614 (A.24) appreciated the utility of the astrolabe, William of Conches did not mention it in his work. Twelfth-century rational philosophers did not show great interest in the astrolabe: it is true that Abelard and Heloise named their son 'Astralabius', but in Abaelard's work the science of the stars received little – and mostly negative – attention and the astrolabe was apparently not mentioned.[635] Since it was Heloise who chose the name Astralabius for her son, it might be that she was the one of the two particularly interested in astrology and astronomy.

In general, rational philosophers of the twelfth century seem to have taken less interest in building and using instruments than their colleagues of the previous century. Herman of Carinthia had a great faith in human reason as a means of investigating the work of the 'divine craftsman' and, in his main philosophical work 'De essentiis', he also used mathematical demonstrations taken (directly or indirectly) from the 'Almagest', developing them according to his own computations.[636] The subjects he discussed, though, were of purely mathematical character, as for example, computing the distances between the planets, and made no reference to observational or experimental practise with the astrolabe or other instruments.[637]

Adelard of Bath (fl. 1100–1150), who went to Southern Italy to learn from the Arabs the rational methods of natural philosophy, was more interested in finding and translating authoritative texts, among them Euclid's 'Elements', than in making experience with instruments.[638] Adelard's astrolabe treatise was probably written rather as a gift to the future king Henry II of England than as a natural phi-

635 d'Alverny (1975). According to Marie-Thérèse d'Alverny, "Ce n'est que dans l'un des derniers ouvrages d'Abélard, l'Expositio in Hexaemeron, écrit à la demande d'Héloise, que l'on trouve des allusions à la science qui était considerée par ses contemporains naturalistes comme le couronnement du quadrivium: l'astrologie" (d'Alverny (1975) p. 612).
636 Burnett (1988), esp. p. 393 on the 'divine craftsman'.
637 Burnett (1988) p. 395–396.
638 On Adelard of Bath, see the various contributions in: Burnett (1987). In particular, on his astrological interests: Adelard of Bath (1998) p. xi–xix. On his treatise on the astrolabe Dickey (1982), Burnett (1997) p. 31–46, Poulle (1987).

losophical work for scholars.[639] However, Adelard was probably the first one to make the astrolabe into the official instrument of Astronomy. In his work 'De eodem et diverso', the seven Liberal Arts appear one after the other and Astronomy is described so:

> Hec igitur quam vides splendore lucidam totumque corpus quodammodo oculeam, dextra radium, leva vero astrolabium gestantem, quicquid mobilis celestisque quantitatis infra applanon [i.e. the celestial sphere] continetur intelligentibus explicat.[640]

In the twelfth century, the astrolabe also started enjoying a certain popularity beyond scholarly circles, becoming an exotic but not unusual feature of literature, poetry, sculpture and painting, as an instrument both of astronomy and of astrology. As we already saw, in the Royal Portal in Chartres (ca. 1150) angels hold an astrolabe in their hands, more or less like Abraham did in the eleventh-century representation in BnF lat 12117 (A.30).[641] However, this kind of representation had no success: from the twelfth century onward, the standard way to represent the use of an astrolabe was the gesture of raising it as if to measure the altitude of a celestial body.

The place to look for astrolabes in twelfth-century artworks are cycles representing the seven Liberal Arts with their instruments. Such cycles became increasingly popular from ca. 1150 onward and they often occurred on church portals, as for example in Sens (ca. 1200).[642] In the 'Roman de Thebes' (ca.1150), too, the astrolabe is the tool of astronomy:

> Unne verge ot Geometrie,
> un astreleibe, Astronomie;
> l'une en terre me sa mesure,
> l'authre es estoiles met sa cure.[643]

The astrolabe was also represented as the tool of astronomers and astrologers, again according to the same iconography as seen in Oxf. Bodl. 614 (A.24), for example at the beginning of the (so-called) Psalter of Bianca de Castilla.[644]

---

639 Burnett (1997) p. 31–34.
640 Adelard of Bath (1998) p. 68, translated by Charles Burnett as: "This maiden whom you see shining with spendor, with her whole body in some way covered with eyes, and wielding a rod in her right hand, an astrolabe in her left, explains to those who are intelligent whatever mobile and celestial quantity is contained beneath the *aplanos*" (Adelard of Bath (1998) p. 69).
641 See 5.6.3.
642 On the representations of the Liberal arts see: Katzenellenbogen (1966), Verdier (1969), Tezmen–Siegel (1985). On representations of astrolabes: Van Cleempoel (2005b). Maybe the earliest extant sculptures of this kind are those on the Royal portal of Chartres, where astronomy uses some instrument to look in the sky. Unfortunately, her representation is damaged and it is not possible to determine which instrument it was (Houvet (1925) p. 46).
643 Roman de Thébes (1966) p. 156–157, vv. 4997–5000. On the 'Roman de Thebes': Schöning (1991).
644 Psautier of Bianca of Castilla (1910). For other examples of astronomers/astrologers with astrolabe see: Saxl/Meier (1953) p. 287–288 and p. 344, Tachau (1998).

The appearance of the astrolabe in the same role as the traditional instruments of the arts is, in my opinion, a sign that, already at the beginning of the twelfth century, this device was no longer regarded as a novelty to be studied by scholars, but rather as a well-understood tool of the mathematical arts. The astrolabe had acquired the status of a symbol, just like the monochord or the abacus, and was represented in its most symbolic function: providing a measure of celestial and terrestrial time flow. Possibly, the astrolabe had also started fascinating laymen and court-members thanks to the diffusion of Arabic artefacts and to the growing popularity of astrology.

6.5 Working knowledge in astrolabe lore

From the twelfth century onward, legends started surrounding the introduction of astrolabe knowledge in Latin Europe and in particular the figures of Gerbert of Aurillac and Herman of Reichenau. Gerbert of Aurillac had had many enemies during his lifetime, and he had become a devilish figure already in the second half of the eleventh century, in the context of the polemics centered on Pope Gregor VII: his successful career as bishop of Reims and then pope was attributed to a pact with the devil by cardinal Benno.[645] However, the astrolabe was not yet mentioned in this connection.

In the twelfth century, astrolabe texts that had up till then been circulating anonymously were ascribed to Gerbert, and he came to be associated with the introduction of the astrolabe into Latin Europe, this act being represented as part of his alleged pact with the devil.[646] The most interesting account on this subject is probably that of Michael Scotus (d. 1234): Gerbert was the best necromancer of France and he once borrowed an astrolabe from a giant-magician, but was unable to understand it.[647] He therefore conjured two demons and had them explain to him how to use the device. After that, he wrote a book on the subject. Following Gerbert's example, Michael Scot explained, others came in possession of the astrolabe, and wrote books based on their experiments.

While Gerbert was a disputed figure already during his lifetime, the same cannot be said for Herman of Reichenau.[648] Herman came from a noble family and had become a monk in the cloister Reichenau probably because of a bodily malformation. He was a scholar and writer of great capability, known for his works on history and music. By the twelfth century, not only the treatise **h**, but also a whole set of works on the astrolabe were ascribed to him. By the thirteenth century, the image of Herman with an astrolabe in his hand appeared alongside

---

645 Benno (1611).
646 On the legends surrounding the figure of Gerbert of Aurillac: Schulteß (1893) and Oldoni (1985). Some sources on this subject are edited and commented in: Guyotjeannin/Poulle (1996) p. 336–366. For further bibliographical references on Gerbert see 3.1.4.
647 The legend is summarized in Thorndike (1965) p. 93–94.
648 On Herman of Reichenau: Borst (1989) p. 77–84, on the legend: Borst (1989) p.90–91.

that of Euclid holding an armillary sphere and a sighting tube.[649] This picture can
be found in a number of manuscripts at the beginning of a treatise on astrological
predictions. Meanwhile, legends spread that Herman's bodily handicap was a
price paid in exchange for knowledge. In Herman's case, it was at least a price
paid to God and not to the devil:

> Dicitur quod iste [i.e. Herman] vir erat bonus et Deo carus. Quadam die angelus venit ad
> eum et ei duo proposuit, si vel corporis salutem sine magna sapientia, vel maximam
> scientiam cum corporis inbecillitate mallet; hanc igitur Hermannus elegit ideoque parali-
> ticus vel podager postmodum jacuit.[650]

6.6 Working knowledge, forbidden knowledge and the written word

Both in Gerbert's and in Herman's legend the astrolabe plays the role of a means
to attain knowledge of a somehow disreputable kind. In both cases it is clear that
the device is not a (magic) object to be possessed, but a (magic) tool to be used
and which, when used, can lead to a knowledge that works to change its user.
This knowledge was regarded as different from that found in books and appears to
have had a doubtful moral status: why otherwise should Herman, if he was so dear
to God, be made into a paralytic to compensate for it?

Despite the bad reputation of the astrolabe in some texts, though, Arabic astrolabe artefacts circulated in Latin Europe, and European craftsmen were learning
how to produce their own. The Pseudo-Messahalla treatise on the astrolabe became much more popular than all previous astrolabe treatises and, as we saw, astrolabes were even being represented in prayer books.

I believe the legends surrounding Gerbert, Herman and the astrolabe might
express growing tensions between concurring images of knowledge in Latin medieval culture. One image assigned epistemological and moral priority to the written word as a means of storing and communicating knowledge. This was the image of knowledge that would dominate in late medieval universities. The other
image is more difficult to grasp, as it was not well represented in written sources:
it may be characterized as that of late medieval hermetic philosophy, astrology
and also medicine – and probably also of late medieval art and mechanics. It was
an image of knowledge in which knowing was never a passive assimilation, but
rather a process in which both the subject knowing and the object known worked,
producing effects and eventually changing each other and the outer world. The
proximity of 'magical' and 'technical' disciplines in medieval and early modern

---

649 Saxl/Meier (1953) p. 287–288 and p. 344–345.
650 The text is edited from the manuscript Oxford Bodl. Lib. Digby 174, f. 210v (A.25) in Poulle
(1996b) p. 343 (with French translation and commentary). On the legend, see Borst (1989) p.
90–91 (text edited on p. 91, note 106). My translation of the quotation is: "They say that he
[i.e. Herman] was a good man and dear to God. One day, an angel came to him and asked him
which one of two alternatives he would prefer: either the health of the body without great
wisdom, or the greatest science with bodily impairment. Herman chose the second option and
thus, from that moment, was afflicted with paralysis or gout."

times has been often noted.[651] Richard Kieckhefer, writing about the 'specific rationality of medieval magic' underscores the importance of the effect in establishing rationality:

> To conceive magic as rational was to believe, first of all, that it could actually work (that its efficacy was shown by evidence recognized within the culture as authentic) and, secondly, that its workings were governed by principles (of theology or of physics) that could be coherently articulated. These principles need not always have been fully articulated or always articulated in the same way: conceptions of magic varied in their degree of specificity and in the specific types of principles they invoked.[652]

In technical as in magical disciplines, 'experimenta' and constructions of various kinds (e.g. ritual, alchemical, mechanical or mathematical) were a necessary component of the process of attaining knowledge.[653] In her study on the "body of the artisan" in the Renaissance, Pamela H. Smith has argued for the existence of a specific "artisanal epistemology", according to which "certainty was to be extracted from nature through bodily experience".[654] In the tradition giving priority to the written word, i.e. that of the universities, the liberal arts of the trivium became highly developed and the four mathematical arts stagnated, while outside of the universities the practise of mathematics flourished, maintaining its double speculative and practical character.[655]

6.7 The astrolabe as a model to be imitated and the development of astronomical clocks

The medieval experimental philosopher par excellence was Roger Bacon (ca. 1214/20–1292): for him, the mathematical arts were central to natural philosophy, but they were at the same time closely linked to material experience.[656] Roger Bacon explained that, while the armillary sphere could provide knowledge about heavenly motion, an expanded version of it would be desirable: a device capable of simulating all natural phenomena, including sea, flowing streams and the cycles of plants as they open and close in accordance with the motion of the sun.[657] In his opinion, such an invention would have been better than any astronomical instrument or clock and would have been worth the treasure of a king.

---

651 To quote only one title: Eamon (1983).
652 Kieckhefer (1994) p. 814.
653 Alessio (1965). The contribution of craftsmen to the Scientific Revolution is studied in this sense in P. H. Smith (2004).
654 P. H. Smith (2004) p. 155.
655 Molland (1993) p. 77–78. On Jordanus Nemorarius' failed attempt to incorporate the mathematical arts in the literate tradition of his time following the example of Greek mathematics, see Høyrup (1988).
656 On Roger Bacon see: Alessio (1985), in particular on mathematics as experience: p. 87–107, and also Gregory (1992a) p. 29–32.
657 Bacon (1897) vol. 1, p. 225.

Roger Bacon is well known for his philosophico-technological utopias, but, while he was writing about universal machines, European craftsmen were actually trying to realize them, as reported in the year 1271 by Robertus Anglicus in his Commentary to the 'Sphere' of John of Holywood:

> Conantur tamen artifices horologiorum facere circulum unum qui omnino moveatur secundum motum circuli equinoctialis, sed non possunt omnino complere opus eorum, quod, si possent facere, esset horologium verax valde et valeret plus quam astrolabium quantum ad horas capiendas vel aliud instrumentum astronomiae, si quis hoc sciret facere secundum modum antedictum.[658]

This enthusiasm for a possible machine-to-come can be compared to the ardour with which other scholars of the same age searched cloister libraries for the lost treasures of classical literature. It was in this spirit that European scholars and instrument-makers took interest in the magnetic compass and experimented in developing new kinds of universal astrolabes which, according to modern criteria, may sometimes appear as speculative as contemporary natural philosophy.[659] Following this path, medical astrologers built equatoria and astronomical clocks capable of computing and representing celestial movements.[660] The story of this enterprise and in particular of one of its most significant actors, Richard of Wallingford (1292–1336), has been told by John North in 'God's clockmaker. Richard of Wallingford and the invention of time' (2005).[661] Parallel to the development of mechanical clocks, a new type of sundial with equal hours appeared and rapidly spread across Europe.[662]

The early mechanical representations of celestial motion were no visual simulations: early astronomical clocks were not movable armillary spheres reproducing the movement of the celestial system on a small scale, but were rather like weight-driven astrolabes.[663] The close connection between astrolabes and mechanical clocks is still echoed in a nineteenth-century bas-relief in Aurillac, where Gerbert is represented along with a foliot-clockwork, which he had supposedly invented.[664] The construction of astronomical clockworks was 'working knowledge' in the sense that, by working, the machines proved that the constructors had grasped the 'architectonica ratio' of celestial movements. Medical astrologers also took an interest in more practically useful devices, such as military machines, hydraulic systems or musical instruments. Moreover, machines could also be de-

---

658 Thorndike (1949) p. 180. My translation is: "Those whose art is to make time-tellers, are trying to make a single circle which would move in all respects according to the motion of the Equator. Yet they are not able to bring their project to full completion. Should they be able to do it, though, that would be a really truthful time-teller and would be worth more than an astrolabe or other astronomical instrument for taking time, if someone really knew how to make it as I said above."

659 King (2003a).

660 White (1978b).

661 North (2005).

662 Zinner (1967) p. 52–69.

663 Wählin (1931a), H. C. King (1978) p. 42–61.

664 Poulle (1996a).

scribed in words or, as was often the case, simply sketched. Just as it is the case for astrolabes and sundials in the eleventh century, it is often difficult to tell whether a sketch could be – or was supposed to be – realizable. Still, each machine had its effect, virtual or not.

Among those who took an interest both in astrological medicine and in astronomical instruments was Jean Fusoris (ca. 1365–1436), who learned his father's metal-working trade and studied medicine at the university of Paris. He was both a practising physician and an instrument-maker, and sold or presented clocks and astrolabes to contemporary potentates, such as Henry V of England (r. 1413–1422) and pope John XXIII (r. 1410–1415).[665] Fusoris' work was highly influential and a large number of extant fifteenth-century astrolabe artefacts are described as 'Fusoris-type'.[666]

---

665 White (1978b) p. 308–309.
666 Frankfurt catalogue part 6.8.

7 APPENDIX A: ELEVENTH- AND TWELFTH-CENTURY MANUSCRIPTS CONTAINING ASTROLABE TEXTS AND/OR DRAWINGS

In compiling the following list, particular efforts were devoted to establishing where texts on astrolabe construction and relevant drawings occur. The sources of the information given are mostly published collations, but in some cases also direct inspection of the microfilm copies contained in the Microfilm Archive of Medieval Scientific Manuscripts at the Institute for the History of Science (LMU Munich). For the meaning of the abbreviations, see appendices C and D.

Some indications on further contents of the manuscripts and on recent bibliography are offered, with no attempt at completeness. When different datings for the same manuscript were found in the literature, the date is indicated with a question mark.

A.1 Avranches, Bibliothèque municipale 235
   date: 12th c.
   construction texts: **h2** (38v–39r), **h** (54r–58r), **pt** (68r–69v), **h'** (98r–99v), **a** (71v–72v)
   drawings: polar sighting tube (32v), **d1** (38v), **d2** (39v) (repr. in Bergmann (1985) p. 111–112)
   other: parts of **J'a, J'**, **hv**, star table type III (9v–10r)
   lit: Zinner (1925) p. 139, Van de Vyver (1931), Kunitzsch (1966) p. 25, Bergmann (1985) p. 111–112, Callebat/Desbordes (2000), Holtz (2000), Kunitzsch (2000a) p. 392, 395, 397, Kunitzsch (2000c) p. 183
   discussed on p. 123 (n. 339), 127, 138 (n. 380), 148, 195, 234, 239

A.2 Barcelona, Archivo de la Corona de Aragón Ripoll 225
   date: f. 1–65r: 11th c. (Borst (1989) p. 43, Burnett (1998) p. 332, Beaujouan (1985) p. 656–658); f. 65v–102v: possibly later than the 11th. c. A former, earlier dating (10th. c.) is still found in: Puigvert I Planagumà (2000) p. 175.
   construction texts: **h'** (1r–9r), **h3** (23r–23v)
   other: **eq** (94r–97), parts of **J, J'** and **J'a**
   lit: Van de Vyver (1931), Millás Vallicrosa (1931) esp. p. 150–211, Beaujouan (1985), Bergmann (1985) p. 243–245, Kunitzsch (1997) p. 197, Martinez-Gasquez (2000), Puigvert I Planagumà (2000), Kunitzsch (2000b) p. 248, Kunitzsch (2000c) p. 182
   discussed on p. 85–86, 119 (n. 330), 127, 201

A.3 Berlin, Staatsbibliothek zu Berlin Preussischer Kulturbesitz lat. fol. 307
   date: (?) 1150–1200 (Toneatto (1995) p. 1052 and p. 1056); 13th c. (Bubnov (1899) p. XX)
   construction texts: parts of **h** (11r–11v)
   other: parts of **J, GIA**
   lit: Bubnov (1899) p. XX–XXII, Zinner (1925) p. 139, Toneatto (1995) p. 1052–1057, Kunitzsch (2003) p. 11

A.4 Bern, Burgerbibliothek (Stadtbibliothek, Bibliotheca Bongarsiana)196, f. 1–8
   date: ca. 1000. I am indebted to Prof. Arno Borst for confirming to me this dating, which is found in (Sezgin/ Neubauer (2003) p. 92); a dating to the 9th–10th century (Hagen

(1874) p. 246, Mostert (1989) p. 62) is incorrect as far as this initial part of the manuscript is concerned.

drawings: astrolabe plate (1r), astrolabe rete (2v), back of astrolabe (3v), astrolabe mater (7r) (colour reproductions in Sezgin/ Neubauer (2003) p. 93)

other: parts of **J'**, **J 19**

lit: Hagen (1874) p. 246–247, Zinner (1925) p. 44, Van de Vyver (1931), Thorndike (1948) p. 53–54 (n. 18), Kunitzsch (1982) p. 24–25, Kunitzsch (1987), Mostert (1989) p. 62, Kunitzsch (1997) p. 197, Sezgin/Neubauer (2003) p. 92–93

discussed on p. 138, 149, 203

A.5 Cambridge, University Library Ii.6.5
date: early 12th c.
construction texts: **a** (121r–122r)
other: parts of **J**
lit: Burnett (1998) p. 343–345

A.6 [Chartres, Bibliothèque municipale 214] (destroyed in 1944)
date: 12th c.
drawings: astrolabe mater with arabic characters (30r) (repr. in Van de Vyver (1931) pl. 2) polar sighting tube (repr. in Wiesenbach (1994) p. 380)
lit.: Bubnov (1899) p. XXV–XXVI, Zinner (1925) p. 139, Van de Vyver (1931), Burnett (1984) esp. p.140–142, Kunitzsch (1997) p. 197
discussed on p. 193, 203

A.7 [Cheltenham (Thirlestaine House), Library of Sir Thomas Philipps 4437] (now in private collection)
date: 11th–12th c. (Toneatto (1995) p. 1039)
construction texts: **h3** (53v), **h2** (54r–55r)
other: **dpp**, parts of **GIA**
lit.: Bubnov (1899) p. XXVIII–XXX, Toneatto (1995) p. 1037–1040, Jacquemard (2000) esp. p. 86. A photocopy of the ms. has been deposited at the British Library (Toneatto (1995) p. 1040)
discussed on p. 127

A.8 Darmstadt, Hessische Landes- und Hochschulbibliothek 947
date: 12th c.
construction texts: **h** (166r–171v)
other: parts of **hv**, **J** , star table type III (169v–170r)
lit.: Zinner (1925) p. 108, 139, 140, Kunitzsch (1966) p. 26, Bergmann (1985) p. 226–228

A.9 Göttingen, Niedersächsische Staats- und Universitätsbibliothek phil. 42
date: 12th c.
construction texts: **h** (1r–4r), **h2** (13v–14r)
other: parts of **J**, **J'**, **J'a**
lit.: Bubnov (1899) p. XXXII, Zinner (1925) p. 108, 139, Bergmann (1985) p. 226–228, Jordanus
discussed on p. 127

A.10 Karlsruhe, Badische Landesbibliothek 504
date: 11th–12th c.
other: parts of **hv**
lit.: Zinner (1925) p. 140, Bergmann (1985) p. 226

A.11 Karlsruhe, Badische Landesbibliothek Augusta 146
date: 11th. c.
other: star table type III (113r)

lit.: Holder (1888), Zinner (1925) p. 140, Kunitzsch (1966) p. 26, Bergmann (1980) p. 90

A.12 Konstanz, Stadtarchiv Fragmentensammlung Mappe 2 Umschlag 8 Stück 7
   date: ca. 1000
   other: parts of **J, J', J'a**
   lit.: Borst (1989) esp. p. 112–119 and p. 120–127, Kunitzsch (1997) p. 197, Kunitzsch (2000b) p. 246, 248
   discussed on p. 89 (n. 217)

A.13 Leiden, Bibliotheek der Rijksuniversiteit Scalig. 38
   date: 11th c.
   construction texts: **h3** (46r), **h'** (41r–43r)
   other: parts of **J, J', J'a,** star table type III (41r-43r)
   lit.: Bubnov (1899) p. XXXIV–XXXV, Zinner (1925) p. 108, Van de Vyver (1931), Kunitzsch (1966) p. 25, Bergmann (1985) p. 241–242, Kunitzsch (2000b) p. 249, Jordanus
      discussed on p. 119 (n. 330), 124, 127, 239

A.14 London, British Library Add. 17808
   date: 11th c.
   construction texts: **h', a, pt**
   other: parts of **J, J'a,** Alchandreana
   lit.: Burnett (1997), Juste (2007) p. 320–322

A. 15 London, British Library Old Royal 15.B.IX
   date: (?) 11th. (Bubnov (1899) p. XXXVI); 11th/12th c. (Bergmann (1985) p. 231); first half of 13th c. (Toneatto (1995) p. 533 and 539)
   construction texts: **h3** (70v), **h''** = **h2+dz1** (75r–76r),
   drawings: astrolabe mater and quadrans (71r) (repr. Burnett (1997) p. 20, Bergmann (1985) p. 233), **d1** (75r) (repr. Bergmann (1985) p. 109), **d2** (75v) (repr. Bergmann (1985) p. 110), polar sighting tube (76v), equatorial sundial and sphere with sighting tubes (77r) (repr. Burnett (1997) p. 21, Bergmann (1985) p. 234)
   other: parts of **J,** climate table (71r)
   lit.: Bubnov (1899) p. XXXVI–XXXVIII, Zinner (1925) p. 139, Millás Vallicrosa (1931) p. 303, Bergmann (1985) p. 231–235, Toneatto (1995) p. 533–540, Burnett (1997) p. 17–21, Kunitzsch (1997) p 197, Jacquemard (2000) p. 87, Jordanus
      discussed on p. 124 (n. 341), 125, 127, 192, 197, 199 (n. 574), 203

A.16 London, British Library Cotton Vesp. A II, f. 27–40
   date: 12th–13th c.
   construction texts: **RB** (3rB–37vA)
   drawings: construction drawings to **RB**, discussed in Lorch (1999) p. 75–77
   other: **AE** (37vA–40vB)
   lit.: Millás Vallicrosa (1940), Lorch (1999) p. 58–59, Jordanus
   discussed on p. 240

A.17 Montpellier. Bibl. Interuniversitaire, Section de Mèdicine 48
   date: 11th c.
   drawings: probably drawing of equinoctial sundial (f. 3v)
   other: chapter 17 of the 'Liber Alchandrei'
   lit: Juste (2007) p. 260, 339
   discussed on p. 197

A.18 Munich, Bayerische Staatsbibliothek Clm 560, f. 1–88
   date: 11th c.
   other: star table type III, **pa,** Alchandreana, parts of **J, J', J'a**

lit.: Bubnov (1899) p. XLI, Zinner (1925) p. 108, Van de Vyver (1931), Bergmann (1985) p. 236–237, Toneatto (1995) p. 877–881, Kunitzsch (2000b) p. 246, 248, Juste (2007) p. 340–342, Jordanus

A.19 Munich, Bayerische Staatsbibliothek Clm 13021
date: 12th c.
construction texts: **h** (69r–72r)
other: parts of **J, hv**
lit.: Bubnov (1899) p. XLI–XLII, Zinner (1925) p. 108, 139, 140, Kunitzsch (1991–1992) p. 12, Toneatto (1995) p. 1123–1128

A.20 Munich, Bayerische Staatsbibliothek Clm 14689
date: 12th c.
construction texts: **h** (81v–84v)
drawings: sketch of the 'disc' of William of Hirsau (1v) (repr. in Wiesenbach (1991) p. 132)
other: parts of **J, pa, J'a**
lit.: Bubnov (1899) p. XLIII–XLV, Zinner (1925) p. 108, Van de Vyver (1931), Kunitzsch (2000a) p. 395

A.21 Munich, Bayerische Staatsbibliothek Clm 14763
date: 12th. c.
construction texts: **h** (189r–202r), **n** (204r–206r), **h2** (212r–213r, without introductory lines), **dz2** (213r)
other: parts of **J**, star table type III, **hv**
lit.: Bubnov (1899) p. XLV–XLVI, Zinner (1925) p. 108, 140, Van de Vyver (1931), Bergmann (1985) p. 239–240, Kunitzsch (2000b) p. 248
discussed on p. 127, 239

A.22 Munich, Bayerische Staatsbibliothek Clm 14836
date: 11th–12th c.
construction texts: **h** (16v–24r), **h'** (incl. **h1, dz3** and **dz4**) (156v–159r)
drawings: Plinian latitude diagrams with stereographic projection (124) (Eastwood (1995))
other: **hv**, parts of **GIA**, star table type III (16v)
lit.: Bubnov (1899) p. XLVI–XLVIII, Zinner (1925) p. 108, 139, 140, Van de Vyver (1931), Kunitzsch (1966) p. 26, Bergmann (1985) p. 240–241, Eastwood (1995), Toneatto (1995) p. 975–984 and p. 1041–1051, Kunitzsch (1999–2000) p. 59, Kunitzsch (2000b) p. 245, Kunitzsch (2000c) p. 182, Jordanus

A.23 Neaples, Biblioteca Nazionale Vittorio Emanuele II VIII.C.50
date: 12th c.
construction texts: **RB** (80r–83v)
drawings: construction drawings to **RB**, discussed in Lorch (1999) p. 75–77
lit.: Burnett (1988), Lorch (1999)
discussed on p. 240

A.24 Oxford, Bodleian Library Bodl. 614
date: 1100–1150
drawings: illuminations on almost every page. On f. 35v: astronomer with astrolabe with text from William of Conches, Philosophia mundi 3,12 (i.e. III, VII §22 in William of Conches (1980))
lit.: Saxl/Meier (1953) p. 313–316. My discussion of this manuscript is based on microfilm copies purchased from the Bodleian Library.
discussed on p. 218–220

A.25 Oxford, Bodleian Library Digby 174
date: 12th c.

construction texts: **h** (196r–199r)
other: parts of **J**, **hv**, attribution of **J** to Gerbert (210v)
lit.: Bubnov (1899) p. LI and p. 113–114, Zinner (1925) p. 139, Kunitzsch (1991/1992) p. 10, Kunitzsch (1993b) p. 206, 209, Poulle (1996b)
discussed on p. 222 (n. 65)

A.26 Oxford, Corpus Christi College 283, f. 66–113
   date: 11th c. (Burnett (2000b) p. 63)
   construction texts: **h1** (without philosophical prologue, 81v–85r), **dz2**, **dz3**, **h3** (89v), **a** (95v–97r), **h2** (107v–108v)
   drawings: back of astrolabe similar to that in Vat. Reg. lat. 598, but with no text (82r), **d1** (108r), **d2** (unfinished, 108v)
   other: parts of **J'**, **J'a**, **J**, **GIA**
   lit.: Van de Vyver (1931), Burnett (2000b); Jacquemard (2000), Jordanus
   discussed on p. 124 (n. 341), 125, 127, 144 (n. 395)

A.27 Paris, Bibliothèque Nationale de France lat. 7412, f. 1–23
   date: 11th c.
   this ms. is discussed in detail in chapter 5 of this work, its contents are listed in appendix E
   construction texts: fragments of Ptolemy's 'Planisphaerium' (11r–11v), **h2** (f. 11v–12r)
   lit.: Van de Vyver (1931), Destombes (1962) p. 43–45, Poulle (1964), Kunitzsch (1966) p. 26 Kunitzsch (1993a), Toneatto (1995) p. 1262–1263, Kunitzsch (1998a), Jacquemard (2000), Kunitzsch (2004) 1993a p. 97–99, 1997 p. 197, 1998 p. 114–120, 2000a p. 395–396, Kunitzsch/Dekker (1996) p. 656–657, 664–667, 669, Kunitzsch (1999–2000) p. 58, 62, 68–69, Kunitzsch (2000b) p. 244–245, 247–250, Kunitzsch (2000c) p. 181–182, 185, Jordanus

A.28 Paris, Bibliothèque Nationale de France lat. 11248, f. 1–33
   date: middle of the 11th c.
   construction texts: **h2** (22v–24v)
   drawings: **d1** (26r), **d2** (26v) (repr. in Millás Vallicrosa (1931) pl. 14 and 15 and from there in Burnett (1998) p. 355)
   other: climate table (18v), parts of **J'**, **J'a**
   lit.: Bubnov (1899) p. LIX–LXI, Millás Vallicrosa (1931) pl. 14 and 15, Van de Vyver (1931), Bergmann (1985) p. 235–236, Burnett (1998) p. 355, Kunitzsch (1998a) p. 114–115, Jacquemard (2000), Jordanus
   discussed on p. 119, 123, 125, 127, 134–135

A.29 Paris, Bibliothèque Nationale de France lat. 11248, f. 33ff.
   date: 12th c.
   construction texts: **h**
   other: parts of **J**, **hv** (21r–21v)
   lit.: Bubnov (1899) p. LIX–LXI, Zinner (1925) p. 139, Van de Vyver (1931), Kunitzsch (1966) p. 4 (note 6), Bergmann (1985) p. 235–236, Jordanus

A.30 Paris, Bibliothèque Nationale de France lat. 12117
   date: 11th c.
   other: Alchandreana
   drawings: many illuminations by Ingelard, among them: Abraham with astrolabe (106r) (Deslandes (1955) p. 9–11). On f. 2v the drawing of an equatorial sundial
   lit.: Van de Vyver (1936) p. 672, Leonardi (1959–60) p. 441, Destombes (1962) p. 45, Juste (2004) p. 193 n. 35, Juste (2007) p. 361–362, Mandragore. On the illuminations: Deslandes (1955). My discussion of this manuscript is also based on phtographic copies of some folios acquired from the Bibliothèque nationale de France
   discussed on p. 79, 197, 210–211, 213, 220

A.31 Paris, Bibliothèque Nationale de France lat. 14056, f. 47r–52v
date: 11th–12th c.
other: star tables, Alchandreana, parts of **J**
lit.: Bubnov (1899) p. LXIII–LXIV, Bergmann (1985) p. 227, Gautier Dalché (1996), Juste (2007) p. 362–364

A.32 Paris, Bibliothèque Nationale de France lat. 15078
date: 12th c.
construction texts: **h** (33r–39r)
other: parts of **J**
lit.: Bubnov (1899) p. LXIV, Bergmann (1985) p. 207

A.33 Paris, Bibliothèque Nationale de France lat. 16201
date: 12th c.
construction texts: **h** (1r–2v)
other: parts of **J, hv**
lit.: Bubnov (1899) p. LXV, Zinner (1925) p. 139, Bergmann (1985) p. 227, Kunitzsch (1993b) p. 206–208

A.34 Paris, Bibliothèque Nationale de France lat. 16208
date: (?) 12th c. Bubnov (1899) p. LXV; first half of 13th c. Toneatto (1995) p. 1086; late 12th–early 13th c. Juste (2007) p. 363
construction texts: **h** (83v–85r)
other: star table type III, Alchandreana, parts of **J, hv**
lit.: Bubnov (1899) p. LXV–LXVI, Zinner (1925) p. 139, Bergmann (1980) p. 87, Toneatto (1995) p. 1083–1087, Juste (2007) p. 363–364

A.35 Paris, Bibliothèque Nationale de France nouv. acq. lat. 229
date: 12th c.
construction texts: **h** (19r–25v)
other: parts of **J, hv**
lit.: Bubnov (1899) p. LXX–LXXI, Zinner (1925) p. 108, 139, 140

A.36 Pommersfelden, Gräflich Schönbornsche Bibliothek 60 nr. 2
date: 12th c.
construction texts: **h**
lit.: Zinner (1925) p. 139

A.37 Pommersfelden, Gräflich Schönbornsche Bibliothek 2640
date: 12th c.
other: **SJ**
lit.: Zinner (1925) p. 22, Kunitzsch (1982) p. 486–487
discussed on p. 240

A.38 Rostock, Universitätsbibliothek philol. 18
date: 12th c.
construction texts: **h3** (65r), parts of **h'** (65r–66v)
other: parts of **GIA**
lit.: Folkerts (1970) p.31–32, Toneatto (1995) p. 502–511, Jacquemard (2000), Jordanus
discussed on p. 127

A.39 Salzburg , Stiftsbibliothek St. Peter (Erzabtei) a. V. 2, f. 82–101
date: late 12th c.
other: **AB** (82r–101r)
lit.: Dickey (1982) p. 118–123, Kunitzsch (1993d) p. 200
discussed on p. 240

A.40 Salzburg , Stiftsbibliothek St. Peter (Erzabtei) a. V. 7
    date: 12th c.
    construction texts: **h3** (107r), **h2** (107r–109r)
    drawings: **d1** (109r), **d2** (109v) (rotated)
    other: star table type III, parts of **J**, **hv**, **GIA**
    lit.: Bubnov (1899) p. LXXXVII–LXXXVIII, Zinner (1925) p. 108, 139, 140, Kunitzsch
        (1966) p. 26, Toneatto (1995) p. 1067–1070, Jacquemard (2000), Jordanus
    discussed on p. 124 (n. 341), 125, 127, 129, 238

A.41 Salzburg , Stiftsbibliothek Sankt Peter (Erzabtei) a. V. 32
    date: 12th c.
    construction texts: **h** (94r–96v)
    lit.: Bubnov (1899) p. LXXXVII–LXXXVIII, Zinner (1925) p. 139

A.42 Vatican, Biblioteca Apostolica Vaticana Ottob. lat. 1631
    date: 12th c.
    construction texts: 'opusculum de mensuris astrolabio agendis' (16r–17r)
    other: parts of **J**, **GIA**
    lit.: Toneatto (1995) p. 1071–1075

A.43 Vatican, Biblioteca Apostolica Vaticana Reg. lat. 598
    date: 11th c.
    construction texts: **h'** (incl. **h1**, **dz3** and **dz4**) (117r–118v), **h3** (119v)
    drawings: back of astrolabe (119v), rete of astrolabe (120r) (i.e. **#4553**, repr. in Millás Val-
        licrosa (1931) pl. 10 and 11 and from there in Burnett (1998) p. 357 and 358)
    other: parts of **J**, **J'**, **J'a**, star table type III (117r–118r)
    lit.: Bubnov (1899) p. LXXXI, Millás Vallicrosa (1931), Van de Vyver (1931), Kunitsch
        (1966) p. 25, Bergmann (1980), Bergmann (1985) p. 245–246, Jacquemard (2000), Borst
        (1986) p. 293–294, Jordanus
    discussed on p. 119 (n. 330), 124 (n. 341), 127, 138, 149 (n. 408 and 414), 245

A.44 Vatican, Biblioteca Apostolica Vaticana Reg. lat. 1661
    date: 11th–12th c.
    construction texts: **h3** (64r), **h'** (incl. **h1**, **dz3** and **dz4**) (73v–75), **h2** (79r–79v), fragment of
        a Burnett (2000a)
    drawings: sphere with fistulae (60r, 77v)
    other: Gerbert's letter 'de spera', **J'**, **J'a**, **pa**; short text on sphere with sighting tubes, star ta-
        ble type III (73v-75v)
    lit.: Bubnov (1899) p. LXXXII–LXXXIII, Van de Vyver (1931), Kunitzsch (1966) p. 25,
        Bergmann (1985) p. 246–248, Jacquemard (2000), Jordanus
    discussed on p. 124 (n. 341), 127, 192, 239

A.45 Vienna, Österreichische Nationalbibliothek 12600, f. 7–41
    date: 12th c.
    construction texts: **h** (19r–21r)
    other: star table type III (20r, 21r)
    lit.: Toneatto (1995) p. 1076–1081, Zinner (1925) p. 139, Kunitzsch (1966) p. 26

A.46 Zurich, Zentralbibliothek Car C 172
    date: (?)11th–12th c. (Bergmann (1985), Jordanus; 11th c. (Van de Vyver (1931))
    construction texts: **h2** (without beginning, 65v–66v), **dz2** (66v), **n** (71r–74r)
    other: star table type III, climate table
    lit.: Zinner (1925) p. 108 (? as C 170), Van de Vyver (1931), Bergmann (1985) p. 248–250,
        Jordanus
    discussed on p. 127, 239

8 APPENDIX B: THIRTEENTH-, FOURTEENTH-, AND FIFTEENTH-CENTURY MANUSCRIPTS CONTAINING ELEVENTH- AND TWELFTH-CENTURY ASTROLABE TEXTS

The present list does not contain manuscripts in which only late medieval astrolabe texts occur. For the meaning of the abbreviaions, see appendices C and D.

B.1 Cambridge, Library of the Fitzwilliam Museum McClean 165 (12th–13th c.)
   drawings: diagrams and tables of star coordinates
   other: **AB,** star table type III
   lit.: Kunitzsch (1966) p. 27, Dickey (1982) p. 113–118

B.2 Cambridge, Library of Jesus College Q.G.29
   construction texts: **AR**
   lit.: Lorch (1999) p. 93

B.3 Cambridge, University Library Kk.1.1 (13th c.)
   construction texts: Ps.-Messahalla on astrolabe construction
   other: **SJ**
   lit.: Kunitzsch (1982) p. 486–487, Jordanus

B.4 Cambridge, Library of Trinity Collge 1144 (15th c.)
   construction texts: **RB** (126r–128v, only construction)
   drawings: construction drawings to **RB**, discussed in Lorch (1999) p. 75–77
   lit.: Lorch (1999) p. 59–60

B.5 Cracow, Biblioteka Jagiellońska 1924
   construction texts: **PP** (165–189)
   lit.: Jordanus de Nemore (1978) p. 48 n. 23

B.6 Dresden, Sächsische Landesbibliothek Db 86 (14th c.)
   construction texts: **PP** (214r–219v)
   lit.: Jordanus de Nemore (1978) p. 48 n. 23, Burnett (1978) p. 108, Jordanus

B.7 Erfurt, Wissenschaftliche Allgemein-Bibliothek Amplon. 4° 351
   date: from end of 12th c. to 2nd half of 14th c.
   construction texts: **h** (24r–26r)
   other: **J, hv**
   lit.: Schum (1887) p. 587–590, Zinner (1925) p. 109, 139, 140

B.8 Erfurt, Wissenschaftliche Allgemein-Bibliothek Amplon. 4° 357 (late 13th-early 14th c.)
   construction texts: **h** (101r–113v)
   lit.: Schum (1887) p. 597–600, Zinner (1925) p. 139, Kunitzsch (1966) p. 77

B.9 Erfurt, Wissenschaftliche Allgemein-Bibliothek Amplon. 4° 363 (13th c.)
   construction texts: **h** (97r–98r)
   other: **J**
   lit.: Schum (1887) p. 607–609, Zinner (1925) p. 109, 140

B.10 Erfurt, Wissenschaftliche Allgemein-Bibliothek Amplon. 4° 369 (ca. 1325)
   construction texts: **h** (231r–234r)
   other: **J, hv**

lit.: Schum (1887) p. 617–621, Zinner (1925) p. 109, 140 (twice) , Kunitzsch (1966) p. 40, 53, 54, 88

B.11 Fermo, Biblioteca Comunale 85 (late 13th c.)
construction texts: **JH** (99rb–106v)
lit.: Lorch (1999) p. 93, possibly the same as 'Fermo 3' in Bubnov (1899) p. XXXI–XXXII, Kunitzsch (1993c) p. 197, 201–207

B.12 Hanover, G.W. Leibniz-Bibliothek/Niedersächsische Landesbibliothek [formerly Öffentliche Bibliothek] 194 (13th c.)
other: **J**
lit.: Zinner (1925) p. 109

B.13 Kassel, Gesamthochschul-Bibliothek [formerly Landesbibliothek und Murhardsche Bibliothek der Stadt Kassel] ast. 8° 4 (14th c.)
construction texts: **h**
lit.: Zinner (1925) p. 140

B.14 Copenhagen, Kongelike Bibliotek Gl. kgl. S. 277 fol. (13th c.)
construction texts: **h** (68r–69v)
other: parts of **J**, **GIA**
lit.: Toneatto (1995) p. 933–940, Kunitzsch (1991–1992) p. 12

B.15 Leiden, Bibliotheek der Rijksuniversiteit Bibl. publica Leidensis XVIII 191 E (13th c.)
construction texts: **h** (157r–162r)
other: parts of **J, hv**
lit.: Bubnov (1899) p. XXXV, Zinner (1925) p. 109, 140

B.16 Leipzig, Universitätsbibliothek 1473 (15th c.)
other: **J**
lit.: Zinner (1925) p. 109

B.17 London, British Library Arundel 339
date: (?) first half of 13th c. (Toneatto (1995) p. 1144); late 12th–early 13th c. (Juste (2007) p. 323); on other datings: Toneatto (1995) p. 1148, Juste (2007) p. 323 n. 3
other: Alchandreana, parts of **GIA**, **AW**, **AWG**
lit.: Kunitzsch (1982) p. 453–454, Toneatto (1995) p. 1144–1149, Juste (2007) p. 323–324

B.18 London, British Library Arundel 377 (13th c.)
construction texts: **h**
other: **AB**, **AE**
lit.: Bubnov (1899) p. XXXVIII, Zinner (1925) p. 140, Dickey (1982) p. 124–130

B.19 London, Royal College of Physicians of London 383 (13th c.)
construction texts: **RB** (only construction, 160r–161v)
drawings: construction drawings to **RB**, discussed in Lorch (1999) p. 75–77
lit.: Lorch (1999) p. 59

B.20 Lyon, Bibliothèque de la ville 328 (14th c.)
construction texts: **PP** (47r–58r)
lit.: Jordanus de Nemore (1978) p. 48 n. 23, Burnett (1978) p. 108, Jordanus

B.21 Madrid, Biblioteca Nacional 10009 (13th c.)
other: **TQ**, **NN,** star table type III
lit.: Kunitzsch (1966) p. 26, Kunitzsch (1982) p. 495–497, Jordanus

B.22 Madrid, Biblioteca Nacional 10053 (13th c.)
construction texts: Jordanus Nemorarius' 'Planisphaerium' (13th c.)
other: **SJ**
lit.: Kunitzsch (1966) p. 40, 77, Millás Vallicrosa (1942) p. 180–202, Kunitzsch (1982) p. 486–487

B.23 Milan, Biblioteca Ambrosiana A.183.inf (16th c.)
construction texts: **PP** (14r–19v)
lit.: Jordanus de Nemore (1978) p. 48 n. 23, Jordanus

B.24 Milan, Biblioteca Ambrosiana H 109.sup (15th. c.)
other: **RO** (10v–17v)
lit.: Haskins (1924) p. 122–123, Kunitzsch (1982) p. 492, Jordanus

B.25 Munich, Bayerische Staatsbibliothek Clm 15957
date: 15th c., but with letters in much earlier style (Jordanus)
construction texts: **h, AW**
drawings: with **h**
lit.: Kunitzsch (1982) p. 493–494, Jordanus

B.26 Oxford, Bodleian Library Ash. 304 (13th c.)
drawings: Euclid with armilla-sphere and sighing tube, Hermannus [of Reichenau] with astolabe (2v)
lit.: Murdoch (1984) p. 187, fig. 161; Saxl/Meier (1953) p. 287–288

B.27 Oxford, Bodleian Library Auct. F.5.28 (13th c.)
construction texts: **PP** (88r–95r)
lit.: Jordanus de Nemore (1978) p. 48 note 23, Burnett (1978) p. 108, Kunitzsch (1993b) p. 207, Jordanus

B.28 Oxford, Bodleian Library Canonici misc. 61 (15th c.)
other: **RO** (12r–22v)
lit.: Kunitzsch (1982) p. 492

B.29 Oxford, Bodleian Library Canonici misc. 340 (14th c.)
construction texts: **JH** (49–53v)
other: star table type III
lit.: Kunitzsch (1966) p. 27, 31, Lorch (1999) p. 93

B.30 Oxford, Bodleian Library Digby 51 (13th c.)
construction texts: **h** (18r–21r), **RB** (26ra–28ra, only construction)
other: **SP, J, hv**
lit.: Bubnov (1899) p. L–LI, Zinner (1925) p. 139, Lorch/Brey/Kirschner/Schöner (1994) p. 134–136, Lorch (1999) p. 57–58, Jordanus

B.31 Oxford, Merton College 259 (13th c.)
construction texts: **h** (without prologue)
other: **SJ, NN**
lit.: Bubnov (1899) p. LIII, Zinner (1925) p. 139, Poulle (1954) p 102, Kunitzsch (1982) p. 486–487, Jordanus

B.32 Paris, Bibliothèque National de France lat. 7214 (14th c.)
construction texts: **PP** (211r–217v)
lit.: Jordanus de Nemore (1978) p. 48 n. 23, Burnett (1978) p. 108, Jordanus

B.33 Paris, Bibliothèque National lat. 7292 (15th c.)
other: **SJ**, Ps-Messahalla on astrolabe usage
lit.: Poulle (1954) p. 102–103, Jordanus

B.34 Paris, Bibliothèque National de France lat. 7293A (13th–14th. c.)
construction texts: **JH**
lit.: Poulle (1954) p. 102–103, Kunitzsch (1966) p. 19, 31, Kunitzsch (1982) p. 488, Jordanus

B.35 Paris, Bibliothèque National de France lat. 7377B (14th–15th. c.)
construction texts: **PP** (73r–80v)
lit.: Jordanus de Nemore (1978) p. 48 n. 23, Burnett (1978) p. 108, Jordanus

B.36 Paris, Bibliothèque National de France lat. 7399 (14th–15th c.)
construction texts: **PP** (1r–12v)
lit.: Jordanus de Nemore (1978) p. 48 n. 23, Burnett (1978) p. 108, Jordanus

B.37 Paris, Bibliothèque National de France lat. 10266
date: 1486 but, according to Poulle (1954), copy of much older material
construction texts: **h**, **RM** (Raymond of Marseilles on construction and use)
other: anonymous on the astrolabe, star table type III
lit.: Poulle (1954) p. 85–86 and p 102–103, Poulle (1964) p. 872–873, Kunitzsch (1966) p.27,
    76, 89

B.38 Paris, Bibliothèque National de France lat. 16652 (first half of 13th c.)
construction texts: **JH** (2r–6v), **h** (11r–14v), **RB** (24r–28r), **AR** (28r–37v),
other: **J**, **hv**, star table type III
lit.: Bubnov (1899) p. LXVI, Zinner (1925) p. 139, Poulle (1954) p. 100–101, Kunitzsch
    (1966) p. 27, 77, Lorch (1999) p. 59 and 93

B.39 Paris, Bibliothèque Mazarine 3642 (13th c.)
construction texts: **h** (55r–69v)
other: Alchandreana, **J**, **hv**
lit.: Molinier (1890) p. 151–152, Zinner (1925) p. 139, Juste (2007) p. 350–351

B.40 Parma, Biblioteca Palatina 984
construction texts: **PP** (106r–115r)
lit.: Jordanus de Nemore (1978) p. 48 n. 23

B.41 Stuttgart, Württembergische Landesbibliothek [formerly Öffentliche Bibliothek] mat. 4° 33
    (13th c.)
construction texts: **h** (73v–78r)
other: **J**
lit.: Zinner (1925) p. 109, 140, Kunitzsch (1966) p. 76

B.42 Toledo, Archivo y Biblioteca Capitular [formerly Biblioteca del Cabiedo] Cat. 98–27
    (15th c.)
construction texts: **JH**
other: **SJ**
lit.: Millás Vallicrosa (1942), Kunitzsch (1982) p. 486–488

B.43 Vatican, Biblioteca Apostolica Vaticana lat. 3096
construction texts: **PP** (3r–14r)
lit.: Jordanus de Nemore (1978) p. 48 n. 23, Burnett (1978) p. 108

B.44 Vatican, Biblioteca Apostolica Vaticana lat. 4087
construction texts: **JH** (86v–93v)
lit.: Lorch (1999) p. 93

B.45 Vatican, Biblioteca Apostolica Vaticana Ottob. lat. 309 (early 14th c.)
construction texts: **h**, Ps.-Messahalla
other: **SP**

lit.: Lorch/Brey/Kirschner/Schöner (1994) p. 131–134

B.46 Vatican, Biblioteca Apostolica Vaticana Reg. lat. 1285
    construction texts: **PP** (153r–162r)
    lit.: Jordanus de Nemore (1978) p. 48 n. 23, Burnett (1978) p. 108

B.47 Vatican, Biblioteca Apostolica Vaticana Reg. lat. 4539 (15th–16th. c.)
    construction texts: **h2** (95v–96v), **h3** (111r)
    other: parts of **J'**, **J GIA**, **J'a**
    lit.: Toneatto (1995) p. 1098–1110, Jacquemard (2000)

B.48 Vienna, Österreichische Nationalbibliothek 5496
    construction texts: **PP** (1r–11v)
    lit.: Jordanus de Nemore (1978) p. 48 n. 23

B.49 Wolfenbüttel, Herzog August Bibliothek 3549 (13th c.)
    construction texts: **h** (91v–96r)
    other: **J**
    lit.: Zinner (1925) p. 108, 140

9 APPENDIX C: ABBREVIATIONS USED IN THIS WORK TO INDICATE ELEVENTH-CENTURY ASTROLABE-RELATED TEXTS AND DRAWINGS

C.1 Abbreviations introduced by Bubnov, Millás Vallicrosa and Bergmann

a: treatise on astrolabe construction by Ascelin of Augsburg: ed.: Burnett (1998) p. 345–349 (followed by English transl.), Bergmann (1985) p. 223–225, from ms. Avranches 235 (A.1)

Arabic/Latin climate table: ed.: Millás Vallicrosa (1931) p 290–292 (l. 28), analysed in Kunitzsch (1998a), Kunitzsch (2000a).

dpp ('de probanda profunditate'), text on measuring water depths with a sinker and an astrolabe, ed.: Jacquemard (2000) p. 105 (long version) and 106 (short version). The short version is ed. also in Millás Vallicrosa (1931) p. 303, l. 15–34 (VI, second passage)

GIA ("Geometria incertis auctoris'): a group of passages on geometrical subjects edited by Bubnov among Gerbert of Aurillac's 'opera incerta' as book III and IV of a geometrical work whose other parts are not extant (Bubnov (1899) p. 317–365). For further references on the subject, see: Jacquemard (2000)

GIA III 20–25: passages describing surveying procedures based on the properties of similar triangles, edited by Bubnov in the 'Geometria incerti auctoris' (**GIA**), book III, chapters 20–25 (Bubnov (1899) p. 330, l. 6 – 334 l. 11)

h ('de mensura astrolabii'): treatise on astrolabe construction by Herman of Reichenau (1013–1054), ed. and comm.: [Herman of Reichenau 1931] p. 203–212, from ms. Munich Clm 14836. Earlier edition as part of **h+J+hv** in: Herman of Reichenau (1853), from ms. Salz. S. P. a. V. 7 (A.40)

h' (containing **h1**, **dz2** (as variant), **dz3** and **dz4**): text on astrolabe construction, ed: Millás Vallicrosa (1931) p. 296–305 (V)

h" (= **h2** + **dz1**): text on astrolabe construction, ed.: Millás Vallicrosa (1931) p. 293–295 (III)

J ('de utilitatibus astrolabii'): collection of passages on the use of the astrolabe, ed.: Bubnov (1899) p. 114–147

J' ('Sententiae astrolabii'): collection of passages on the use of the astrolabe, ed.: Millás Vallicrosa (1931) p. 275–288 (II A–C)

J'a: passages on the use of a 'spera rotunda', i.e. a celestial globe or possibly a spherical astrolabe, ed. Millás Vallicrosa (1931) p. 288–290 (II D)

k = hv ('de horologio viatorum'): collection of passages describing, among other things, a portable vertical sundial, ed.: Ps.-Herman of Reichenau (1853b), among the works of Herman of Reichenau as 'de utilitatibus astrolabii liber secundus'

Ps.-Beda, Libellus de astrolabio: collection of texts on astrolabe construction (the first one is **h2** without the initial sentence) edited in Ps.-Beda (1904a). According to Bergmann, some parts of this text are not found in the manuscripts known today (Begmann (1985) p. 96–99)

Star table type III: list of 27 star names along with coordinates expressing their position in a way particularly useful to locate them on the rete of an astrolabe, ed. and comm: Kunitzsch (1966) p. 23-30.

C.2 Abbreviations introduced in this work:

cq ('cuiuslibet quantitatis'): passage on construction and use of a particular kind of geometrical square, ed.: Millás Vallicrosa (1931) p. 302, l. 1– 303 l. 14 (VI, first passage) and in Bubnov (1899) p. 365 note 5a

d1 and d2: construction drawings for astrolabe lines: tropics and zodiac (**d1**) and horizon (**d2**). They are associated to **h2** in a number of manuscripts (table. 1 on p. 127). Various versions of **d1** are reproduced in figs. 19a–e

dz1 (printed by Millás Vallicrosa as the last part of **h''**): passage on the division of the zodiac on the astrolabe, ed.: Millás Vallicrosa (1931) p. 295, l. 45–55

dz2 (printed by Millás Vallicrosa as part of **h'**): passage on the division of the zodiac on the astrolabe, ed.: Millás Vallicrosa (1931) p. 300, note to lines 137–138, from Vat. Reg. lat. 1661 (A.44) f. 75r. Also ed. in Herman of Reichenau (1931) p. 214 and Bergmann (1985) p. 249 n. 8 (here from Zür. Car. C 172 (A.46) f. 66v). The text also appears in Mün. Clm 14763 (A.21) f. 213r (Bergmann (1985) p. 240)

dz3 (printed by Millás Vallicrosa as part of **h'**): passage on the division of the zodiac on the astrolabe, ed.: Millás Vallicrosa (1931) p. 302 lines 168–169band Herman of Reichenau (1931) p. 214

dz4 (printed by Millás Vallicrosa as part of **h'**): passage on the division of the zodiac on the astrolabe, ed.: Millás Vallicrosa (1931) p. 302 lines 169–171 (up to: "virgo a LXIIIes.") and Herman of Reichenau (1931) p. 214

eq text on how to make an equatorial sundial, ed.: Millás Vallicrosa (1931) p. 318–320 (X)

h1 (initial part of **h'**): Millás Vallicrosa (1931) p. 296–300 l. 138

h2 (initial part of **h''**): Millás Vallicrosa (1931) p. 293–295 (l. 45). The manuscripts in which **h2** occurs are listed in table. 1,p. 127

h3 short, sketchy text ('fragment') on astrolabe construction, ed: Millás Vallicrosa (1931) p. 294, note. In 4.4.4, I give the version found in Leid. Scal 38 (A.13) on f. 46r. The manuscripts in which **h3** occurs are listed in table. 1, p. 127

n (Inc. "In compositione astrolabii tres circuli sunt necessarii", Exc. "deinde unumquemqe quadrantem in XVIII intervalla"): anonymous text on astrolabe construction, not yet edited, occurring in two manuscripts: Zurich Car C 172 (71r–74r) (11th/12th c., A.46) and Munich Clm 14763 (204r–206r) (12th c., A.21)

pa (Prologue 'Ad intimas'): anonymous text on the philosophical significance and practical utility of the astrolabe, usually regarded as a prologue to a (lost) astrolabe treatise, ed. Millás Vallicrosa (1931) p. 271–275 (I) and Bubnov (1899) p. 370–375

pt ('Iubet rex Ptolomaeus'): anonymous text on astrolabe construction, ed: Millás Vallicrosa (1931) p. 322–324 (XII), from Avranches 235 (A.1)

uhq collection of passages on the use of a universal horary quadrant (quadrant with cursor), ed.: Millás Vallicrosa (1931) p. 304–308 (VII A–C)

10 APPENDIX D: MOST DIFFUSED TWELFTH-CENTURY ASTROLABE TREATISES

Listed below are the twelfth-century astrolabe texts identified by Paul Kunitzsch, indicated using the letters introduced by him in Kunitzsch (1982) p. 475–476. The manuscripts in which each text appears are indicated by the numbers I have assigned to them in appendix A and appendix B. The most recent editions or commentaries of each text are also listed.

AB: Adelard of Bath (fl. 1100–1150), treatise on astrolabe use
3 mss.: A.39, B.1, B.18
lit.: Haskins (1924) p. 28–29, Poulle (1964), Dickey (1982) (ed. and comm.), Kunitzsch (1982) p. 488–489, Burnett (1997) p. 31–34

AE: Abraham ibn Ezra, treatise on astrolabe use
2 mss.: A.16, B.18
lit.: Millás Vallicrosa (1940) (ed. and comm.), Levy (1942), Millás Vallicrosa (1950), Kunitzsch (1982) p. 497

AR: 'Arialdus', treatise on astrolabe construction and use.
2 ms.: B.2, B.38
lit.: Poulle (1954) 87–88, Kunitzsch (1982) p. 492–493, Lorch (1999) p. 93, not edited

AW:'Liber de wazalkora' on astrolabe usage (possibly 13th c.)
2 mss.: B.17, B.25
lit.: Kunitzsch (1982) p. 493–494, not edited

AWG: List of 18 Arabic astrolabe terms (possibly 13th. c.)
1 ms.: B.17
lit.: Kunitzsch (1982) p. 494–495, not edited

JH: Johannes Hispalensis/of Sevilla (fl. 1120–1153), treatise on astrolabe construction
at least 8 mss. (Lorch (1999) p. 93), among them: B.11, B.29, B.34, B.38, B.42, B.44
lit.: Millás Vallicrosa (1942) p. 322–327 (partial ed.), Kunitzsch (1982) p. 488, Lorch (1999) p. 9

NN: 'Astrologiae speculationis exercitium', treatise on astrolabe construction
2 ms.:B.21, B.31
lit.: Millás Vallicrosa (1942) p. 313–321 (ed. together with **TQ**), Kunitzsch (1982) p. 496–497

PP: Herman of Carinthia (fl. 1138–1143), translation form the Arabic of Ptolemy's 'Planisphaerium' and Maslama's notes to it (1143)
12 mss.: a B.5, B.6, B.20, B.23, B.27, B.32, B.35, B.36, B.40, B.43, B.46, B.48 (Jordanus de Nemore (1978) p. 48, in note)
lit.: Ptolemy (1907) (ed.), Drecker (1927–28) (German transl. and comm.), Burnett (1978) p. 108–112 (comm., ed. and Engl. tranls. of Herman's preface), Sinisgalli/Vastoia (1992) (repr. of 16th–c. eds. with Italian transl.), Kunitzsch/Lorch (1994)

RB: Rudolf of Bruges, treatise on construction and use of the astrolabe (wr. 1144).
6 mss.: A.16, A.23, B.4, B.19, B.30, B.38

lit.: Kunitzsch (1982) p. 485–486, Lorch (1999) (ed. and comm.)

RC: treatise on the universal astrolabe, possibly by 'Robertus Cestrensis'
   On the many uncertainties sorrounding this text and its manuscript tradition see: Kunitzsch
   (1982) p. 489–491, which contains further bibliographical references

RM: Raymond of Marseilles (fl. 1140), treatise on construction and use of the astrolabe.
   1 ms.: B.37
   lit.: Poulle (1964) (ed. and comm.), Kunitzsch (1982) p. 483–484

RO: 'Robertus Cestrensis', treatise on astrolabe use (ca. 1150)
   2 ms.: B.24, B.28
   lit.: Haskins (1924) p. 122–123 (comm.), Kunitzsch (1982) p. 492 (comm.), not edited

SJ: Johannes Hispalensis/of Sevilla (fl. 1133–1153), translation of ibn aṣ-Ṣaffār's (d. 1035) treatise on astrolabe use
   14 mss. or more (Kunitzsch (1982) p. 487, Poulle (1954) p. 102), among them: A.37, B.3,
   B.22, B.31, B.33, B.42
   lit.: Millás Vallicrosa (1942) p. 261–284 (ed.), Poulle (1954) p. 102, Kunitzsch (1982) p. 487

SP: Plato of Tivoli (fl. 1133–1145), translation of ibn aṣ-Ṣaffār's (d. 1035) treatise on astrolabe
   use
   2 mss.: B.30, B.45
   lit.: Kunitzsch (1982) p. 487, Lorch/Brey/Kirschner/Schöner (1994) (ed. and comm.)

TQ: 'De recta imaginatione spere', treatise on astrolabe construction, possibly translation of an
   Arabic original of Thābit ibn Qurra
   1 ms.: B.21
   lit.: Millás Vallicrosa (1942) p. 313–321 (ed. together with **NN**), Kunitzsch (1982) p. 495–
   496

11 APPENDIX E: CONTENTS OF THE MANUSCRIPT PARIS, BIBLIOTHÈQUE NATIONALE DE FRANCE LAT. 7412, F. 1r–23v (11th C.)

For the meaning of the abbreviations used her, see appendix C

1r–4v	**J 1** to **J 8, 3** (up to "da unicuique undeviginti"). In Bubnov's edition, **J 3** contains a table of solar longitudes which in this ms. instead appears isolated on f. 6r ed.: Bubnov (1899) p. 114 – 132, l. 13
5r	empty
5v	star table type III ed.: Kunitzsch (1966) p. 28; Bergmann (1980) p. 84. The latter edition is criticized as 'unsatisfactory' by Paul Kunitzsch (Kunitzsch (2000b) p. 244, n. 6) lit.: On the star table and its unique features as evidence of a connection between early Latin astrolabe texts and the Andalusi astronomer Maslama (d. 1007), see: Samsó (2000), esp. p. 512
	J', chapter 'ut scias quota pars hore transacta sit, vel quota remaneat' ed.: Millás Vallicrosa (1931) p. 287
6r	table of solar longitudes which in Bubnov's edition is part of **J 3** ed.: Bubnov (1899) p. 125
6v	empty
6v–8v	**J 8, 3** (continued from f. 4v) to **J 18** (with climate table both in Arabic and Latin and not only in Latis as in Bubnov's edition) ed.: Bubnov (1899) p. 132, l. 13 – 142, with climate table as in Millás Vallicrosa (1931) p. 290–292, l. 28.
8v–9r	**J 19, 7** and **8** (i.e. only 7th climate), **J 21** ed.: Bubnov (1899) p. 145, l. 19 –146, l. 9 and p. 147
9r–10r	parts of **J'a** and Arabic/Latin climate table ed.: Millás Vallicrosa (1931) p. 288 l. 384 (from 'Ianus et Apollo') – p. 292 l. 28
10r–10v	"octo circulos terrae... noctesque per vices" and "Umbilici quem... obtinente": two extracts from: Byrrtferth of Ramsey, Commentary to Beda's 'De temporum ratione' ed. in Migne PL: Byrtferth of Ramsey (1904) c. 445–446 and c. 431–432, among 'Glossae et scholia' under: Brid. Rames. Glossae.
11r	"Haec distributiones climatum et latitudines secundum arabicos fiunt auctores et compendiose dicatur taliter latini disponunt auctores"
	Martianus Capella, De nuptiis VIII, 876–877
	"Si quis autem studio flagrantior praetaxata climata lucesque illae diligentius discriminare voluerit, excitetur a somno et orbivagum (?) secet (?) item. singula peragrat (?) climata phebeos cursus recursusque oculetur. his ad integritatis iure regresso probetur locus confessionis."
11r–11v	Latin version of some sentences from Ptolemy's 'Planisphaerium': beginning of chapter 2 and of chapter 3 ed.: Kunitzsch (1993a) p. 98–99
11v	"Ceteros almucantarat... si V V": final part of **h2**

	ed.: Millás Vallicrosa (1931) p. 295 l. 33–39
11v–12r	"Philosophi qui sua sapientia... a usque m.": initial part of **h2**
	ed: Millás Vallicrosa (1931) p. 293 – p. 295 l. 33
	on 12r also drawings **d1** and **d2** and notes on the margin (Kunitzsch (1993a) p. 101), repr. fig. 19
12v	**dpp**: 'de probanda profunditate', long version
	ed.: Jacquemard (2000) p. 105
12v–14v	**GIA III 20–23** with figures
	'construe quadratum': text on geometric square
	GIA III 24–25 with figure
	ed.: Bubnov (1899) p. 331–334, p. 365 and figs. 58–64 (GIA 20–25); Bubnov (1899) p. 365 l. 7–21 ('construe quadratum ')
14v	drawings: Euclid's 'Elements', prop. I.1–10, 'Planisphaerium' ch. 10 and alidade, repr. in fig. 24
15r	drawing of an hemisphere with sighting tubes and of a polar sighiting tube, repr. in fig. 25
15v	empty
16r	'componitur horologium': universal horary quadrant and drawing, repr. in fig. 26
	'ad latitudinem climatis inveniendam': finding latitude with the quadrant
	ed.: Millás Vallicrosa (1931) p. 304 l. 1 – p. 305 l. 14 and p. 307 l. 73 – p. 308 l. 82.
16v–18r	prologue 'Ad intimas' (**pa**), preceded by an introduction and with a note on the margin
	ed.: Millás Vallicrosa (1931) p. 271–275 (I) and Bubnov (1899) p. 370–375 (**pa**); Destombes (1962) p. 44 (introduction and note)
18r–18v	**J 19, 1–6** (i.e. first part only, with decriptions of climates one to six)
	ed. : Bubnov (1899) p. 142 l. 10 – p. 145 l. 11
18v–19r	"Sumpto astrolapsu... rei metiende erit": two methods for finding latitude with a quadrant.
	ed.: Millás Vallicrosa (1931) p. 308 l. 84–95
19r	drawing of an equatorial sundial, repr. fig. 27
19v	drawing of the rete of an astrolabe (repr. fig. 28), with Arabic names of zodiac signs and stars, along with their Latin translation (zodiac signs) or transliteration (stars, transliteration written below the drawing and indicated with letters corresponding to the pointers)
	lit.: on this and on the following images see Kunitzsch (1966) p. 26, Kunitzsch (1998a), Maier (1996) p. 259–260, Stautz (1997) p. 53–54
20r–23r	drawings of six horizon plates and the mater of an astrolabe, with horizons corresponding to the climates seven to one (plate for 1st climate repr. fig. 29)
	on each plate, the Arabic formula stating the latitude and maximum day length for the corresponding climate is written on the plate. It is the same formula as in the Arabic-Latin climate table on f. 10r, but here it is written in Arabic characters. On some plates, further Arabic inscriptions occur (e.g. names of unequal hours, values of the altitude circles etc.)
23v	back of an astrolabe with calendric scale and shadow square, inscribed in Arabic characters, repr. in Van de Vyver (1931) pl.3

12 APPENDIX F: ASTROLABE ARTEFACTS AND ASTROLABE DRAWINGS DISCUSSED IN THIS STUDY

Listed below are the astrolabe artefacts and drawings referred to in this study, with indications of the sections in which they are discussed. Artefacts and drawings are quoted using their number in the Frankfurt Catalogue of Medieval Astronomical Instruments (see 3.2.2.)

#2: Byzantine astrolabe, signed by Sergius and dated 1062, discussed in sec. 3.1.2, 3.2.1

#110 (=#135): Western Islamic astrolabe, unsigned and undated (London, British Museum), discussed in sec. 5.6.1

#116: Western Islamic astrolabe signed by Muḥammad ibn aṣ-Ṣaffār and dated 420 H (=1029/1030) (Berlin, Staatsbibliothek Preußicher Kulturbesitz, Orientabteilung), discussed in sec. 2.3.3, 3.1.4, 3.2.6 and fig. 16

#118: Western Islamic astrolabe dated 460 H (=1067/1068) (Oxford, Museum of the History of Science), discussed in sec. 3.2.6

#161: Astrolabe with Gothic numerals and letters (London, British Museum), discussed in sec. 3.2.4, 5.6.2

#166: Astrolabe with script of Lombardic type (Oxford, Museum of the History of Science), discussed in sec. 3.2.4, 5.6.2

#167: Astrolabe engraved in Gothic script and numerals and with some words of Italian origin (London, British Museum), discussed in sec. 3.2.4, 5.6.2

#202: Astrolabe with inscription in Picard dialect of Old French, ca. fourteenth century (private collection), discussed in sec. 3.2.4, 5.6.2

#303: English astrolabe with missing rete (Oxford, Merton College), discussed in sec. 3.2.4, 5.6.2

#420: European astrolabe of uncertain provenance with equinoctial bar with a single counterchange in the centre (Greenwich, National Maritime Museum), discussed in sec. 3.2.4, 5.6.2

#550: German astrolabic plate dated 1468, said to have been owned by Regiomontanus (Nürnberg, Germanisches Nationalmuseum), discussed in sec. 5.6.2

#558: Astrolabe of uncertain provenance, with three plates engraved on both sides (Nürnberg, Germanisches Nationalmuseum), discussed in sec. 3.2.4, 5.6.2

#589: Undated astrolabe, bought by Nicolaus Cusanus (1401–1464). (Bernkastel-Kues, Kusanus-Stift), discussed in sec. 3.2.4, 5.6.2

#1090: Western Islamic astrolabe signed by Abū-Bakr ibn Yūsuf, Marrakesch and dated 613 H (=1216/1217) (Toulouse, Musée Paul Dupuy), discussed in sec. 4.6.11

#1099: Western Islamic astrolabe dated 472 H (=1079/1080) (Nürnberg, Germanisches Nationalmuseum), discussed in sec. 3.2.6

#3042: Oldest European astrolabe ('Destombe's astrolabe', Paris, Institut du Monde Arabe), discussed in sec. 3.2.1, 3.2.2, 3.3.3, 3.2.4, 3.2.5, 4.1.6, 4.6, 4.6.8, 4.6.10, 4.6.11, 5.6.2 and app. A and E

#3622: Western Islamic astrolabe dated 446 H (=1054/1055) with later inscriptions in Catalan (Cracow Jagellonian University Museum), discussed in sec. 3.2.6

#3650: Western Islamic astrolabe signed by Muḥammad ibn aṣ-Ṣaffār and dated 417 H (=1026/1027) (Edinburgh, Royal Scottish Museum), discussed in sec. 3.1.4

#3702: Late astrolabe copied from an early Eastern Islamic one (Baghdad, Archeological Museum) , discussed in sec. 5.6.1

#4024: Drawings of astrolabe parts in ms. BnF lat. 7412 (A.27), discussed in sec. 4.6, 4.6.8, 5.6, 5.6.1, 5.6.2, 5.6.3

#4025: Plate signed by Muḥammad ibn aṣ-Ṣaffār (Palermo, Museo Nazionale), discussed in sec. 3.1.4

#4553: Drawings of astrolabe parts in ms. Vat. Reg. lat. 598 (A.43), discussed in sec. 4.6, 4.6.8, 5.6.2, 5.6.3 and app. A

#4556: European astrolabe, unsigned and undated, with four plates engraved on both sides (private collection), discussed in sec. 3.2.4, 5.6.2

#4560: Astrolabe with markings in Arabic, Latin and Hebrew ('Toledo astrolabe', private collection), discussed in sec. 3.2.6

13 BIBLIOGRAPHY

Abgrall (2000): Abgrall, Philippe, La géométrie de l'astrolabe au Xe siècle, *Arabic sciences and philosophy: a historical journal* 10 (2000) p. 7–78.

Abry (2000): Abry, Josèphe, Martianus Capella: la diffusion du livre 8 du 'de nuptiis' dans les florilèges astronomiques, in: L. Callebat and O. Desbordes, eds., *Science antique, science médiévale (autour d'Avranches 235)* (Hildesheim: Olms-Weidmann, 2000) p. 191–202.

Ackermann (2001): Ackermann, Silke, Dormant treasures: the Zinner-archive at Frankfurt University, *Nuncius* 16 (2001) p. 711–722.

Adelard of Bath (1998): Adelard of Bath, De eodem et diverso, in: Adelard of Bath, *Conversations with his nephew: On the same and the different, Questions on natural science and On birds*, ed. and transl. by C. Burnett (Cambridge: Cambridge University Press, 1998) p.1–79.

Alessio (1965): Alessio, Franco, La filosofia e le 'artes mechanicae' nel secolo XII, *Studi medievali* VI–I (1965) p. 71–161.

Alessio (1985): Alessio, F., *Introduzione a Ruggero Bacone* (Rome: Editori Laterza, 1985).

d'Alverny (1967): d'Alverny, Marie-Thérèse, Astrologues et théologiens au XIIe siècle, in: *Mélanges offerts à M.-D. Chenu, maître en théologie* (Paris: Vrin, 1967) p. 31–50, repr. in: d'Alverny, M.-T., *La transmission des textes philosophiques et scientifiques au moyen âge* (Aldershot: Variorum, 1994).

d'Alverny (1975): d'Alverny, M.-T., Abélard et l'astrologie, in: *Pierre Abélard, Pierre le Vénérable* (Paris: Éd. du Centre nationale de la recherche scientifique, 1975) p. 611–630, repr. in: d'Alverny, M.-T., *La transmission des textes philosophiques et scientifiques au moyen âge* (Aldershot: Variorum, 1994).

d'Alverny (1991): d'Alverny, M.-T., Translations and translators, in: R. L. Benson, G. Constable and C. D. Lanham, eds., *Renaissance and renewal in the twelfth century* (Toronto: Toronto University Press, 1991) p. 421–462.

Anagnostakis (1984): Anagnostakis, Christopher, *The Arabic version of Ptolemy's 'Planisphaerium'*, Dissertation presented to the faculty of the graduate school of Yale University (1984) (Ann Arbor: University Microfilms International, 1986).

Antoine (1992): Antoine, Jean-Philippe, L'arte della memoria e la trasformazione dello spazio pittorico in Italia nel Duecento e Trecento, in: L. Bolzoni and P. Corsi, eds., *La cultura della memoria* (Bologna: Società editrice Il Mulino 1992) p. 99–115.

Arnaldi/Schaldach (1997): Arnaldi, Mario and Karlheinz Schaldach, A Roman cylinder dial: witness to a forgotten tradition, *Journal for the history of astronomy* 28 (1997) p. 107–117.

Ascher (1991): Ascher, Marcia, *Ethnomathematics. A multicultural view of mathematical ideas* (Pacific Grove: Brooks/Cole Pub. co., 1991).

Bacon (1897): Bacon, Roger, *Opus maius*, vol. 1 edited with introduction and analytical table by J. H. Bridges (Oxford, 1897, repr. Frankfurt a. M.: Minerva-GmbH, 1964).

Barton (2004): Barton, Simon, Spain in the eleventh century, in: D. Luscombe and J. Riley-Smith, eds., *The new Cambridge medieval history. Vol IV: c. 1024–1198*, part 2 (Cambridge: Cambridge University Press, 2004) p. 154–190 and p. 804–808.

Beaujouan (1972): Beaujouan, Guy, L'enseignement du 'Quadrivium', in: *La Scuola nell'Occidente latino dell'Alto Medioevo*, vol. 2 (Spoleto: presso la sede del Centro di Studi, 1972) p. 719–723 (Settimane di studio 19,2).

Beaujouan (1974): Beaujouan, G., Réflexions sur les rapports entre théorie et pratique au moyen âge, in: J. E. Murdoch and E. D. Sylla, eds., *The cultural context of medieval learning* (Boston: D. Reidel Pub. Co., 1975) p. 437–484.

Beaujouan (1985): Beaujouan, G., Les apocryphes mathématiques de Gerbert, in: M. Tosi, ed., *Gerberto: scienza, storia e mito* (Bobbio: editrice degli A.S.B., 1985) p. 645–658.

Beaujouan (1991): Beaujouan, G., The transformation of the Quadrivium, in: R. L. Benson, G. Constable and C. D. Lanham, eds., *Renaissance and renewal in the twelfth century* (Toronto: Toronto University Press, 1991) p. 463–487.

Beaujouan (1995): Beaujouan, G., L'authenticité de l'astrolabe dit 'carolingien', *Physis* 32 (1995) p. 439–450.

Bäuml (1980): Bäuml, Franz H., Varieties and consequences of medieval literacy and illiteracy, *Speculum* 55 (1980) p. 237–265.

Benno (1611): Bennonis cardinalis rom. de vita et gestis Hildebrandi papae libri II, in: M. Goldast, ed., Apologia pro imp. Heinrico IV franco (Hanover: Villerionus, 1611) p. 1–25.

Berger (2002): Berger, Karol, The Guidonian hand, in: M. Carruthers and J. M. Ziolowski, eds., *The medieval craft of memory. An anthology of texts and pictures* (Philadelphia: University of Pennsylvania Press, 2002) p. 71–82.

Bergmann (1980): Bergmann, Werner, Der Traktat 'De mensura astrolabii' von Hermann von Reichenau, *Francia* 8 (1980) p. 65–103.

Bergmann (1985): Bergmann, W., Innovationen des 10. und 11. Jahrhunderts: Studien zur Einführung von Astrolab und Abakus im lateinischen Mittelalter (Stuttgart: Steiner Verlag, 1985).

Beumann (1997): Beumann, Helmut *Die Ottonen*. Fourth edition (Stuttgart: Verlag W. Kohlhammer, 1997).

Biagioli (2006): Biagioli, Mario, From print to patents: living on instruments in early modern Europe, *History of Science* 45 (2006) p. 139–186.

Bischoff (1971): Bischoff, Bernhard, Die Überlieferung der technischen Literatur, in: *Artigianato e tecnica nella società dell'alto medioevo occidentale*, vol. 1 (Spoleto: presso la sede del Centro di Studi, 1971) p. 267–296 (Settimane di studio 18,1).

Blaine (1976): Blaine, Bradford B., The enigmatic water-mill, in: B. S. Hall and D. C. West, eds., *On premodern technology and science. A volume of studies in honor of Lynn White, jr.* (Malibu: Undena Publ., 1976) p. 163–176.

Block (1981a): Block, Ned, Introduction, in: N. Block, ed., *Imagery* (Cambridge MA: MIT Press, 1981) p. 1–18.

Block (1981b): Block, N., ed., *Imagery* (Cambridge MA: MIT Press, 1981).

Block Frieman (1974): Block Friedman, John, The architect's compass in creation miniatures of the Later Middle Ages, *Traditio* 30 (1974) p. 419–429.

Bloor (1991): Bloor, David, *Knowledge and social imagery*. Second edition (Chicago: University of Chicago Press, 1991).

Boase (1992): Boase, Roger, Arab influences on European love-poetry, in: S. K. Jayyusi, ed., T*he legacy of muslim Spain*, vol. 2 (Leiden: Brill, 1992) p. 457–482.

Boehm (1993): Boehm, Laetitia, Artes mechanicae und artes liberales im Mittelalter. Die praktischen Künste zwischen illiteraler Bildungstradition und schriftlicher Wissenschaftskultur, in: K. R. Schnith and R. Pauler, eds., *Festschrift für E. Hlawitschka zum 65. Geburtstag* (Kallmünz, Opf.: Lassleben, 1993) p. 419–444.

Boethius (1867): Boethius, Ancius Manlius Torquatus Severinus, *De institutione arithmetica libri duo*, ed. by G. Friedlein (Leipzig: Teubner, 1867).

Bolzoni (1992): Bolzoni, Lina, Costruire immagini. L'arte della memoria tra letteratura e arti figurative, in: L. Bolzoni and P. Corsi, eds., *La cultura della memoria* (Bologna: Società editrice Il Mulino, 1992) p. 53–97.

Bolzoni/Corsi (1992): Bolzoni, L. and Pietro Corsi, eds., *La cultura della memoria* (Bologna: Società editrice Il Mulino, 1992).

Borrelli (2006): Borrelli, Arianna, The flat sphere, in: S. Zielinski and D. Link, eds., *Variantology 2. On deep time relations of arts, sciences and technologies* (Cologne: Velag der Buchhandlung Walther König, 2006) p. 145–166.

Borst (1986): Borst, Arno, *Das mittelalterliche Zahlenkampfspiel* (Heidelberg: Winter 1986).

Borst (1989): Borst, A., *Astrolab und Klosterreform an der Jahrtausendwende* (Heidelberg: Winter, 1989).

Borst (1999): Borst, A., *Computus: Zeit und Zahl in der Geschichte Europas*. Second edition (Munich: Deutscher Taschenbuch Verlag, 1999).

Bott (1992): Bott, Gerhard, ed., *Focus Behaim Globus. Ausstellungskatalog*, 2 vols. (Nürnberg: Verlag des Germanischen Nationalmuseum, 1992).

Bottazzini/Dahan Dalmedico (2001): Bottazzini, Umberto and Amy Dahan Dalmedico, Introduction, in: U. Bottazzini and A. Dahan Dalmedico, eds., *Changing images in mathematics. From the French Revolution to the new millennium* (London: Routledge, 2001) p. 1–14.

Bourgain (1994): Bourgain, Pascale, Le tourrnant littéraire du milieu du XIIe siècle, in: *Mutations et renouveau en France dans la première moitié du XIIe siècle* (Paris: Le Léopard d'Or, 1994) p. 303–323.

Bourin (1992): Bourin, Monique, Quel jour, en quelle année? A l'origine de la 'revolution calendaire' dans le Midi de la France, in: B. Ribémont, ed., *Le temps, sa mesure et sa perception au moyen âge* (Caen: Paradigme, 1992) p. 37–56.

Brentjes (2000): Brentjes, Sonja, Reflexionen zur Bedeutung der im 12. Jh. angefertigten lateinischen Übersetzungen wissenschaftlicher Texte für die europäische Wissenschaftsgeschichte, in: J. Cobet, C. F. Gethmann and D. Lau, eds., *Europa. Die Gegenwärtigkeit der antiken Überlieferung* (Aachen: Shaker, 2000) p. 269–305.

Brieux/Maddison (unpublished): Brieux, Alan und Francis Maddison, *Répertoire des facteurs d'astrolabes et de leurs oeuvres Ier partie: Islam plus Byzance, Armenie, Géorgie et Inde*, unpublished

Brunhölzl (1999): Brunhölzl, Franz, Hyginus im Mittelalter, in: *Lexikon des Mittelalters* 5 (1991) c. 244.

Bubnov (1899): Bubnov, Nicolaus, ed., *Gerberti postea Silvestri papae opera mathematica* (Berlin 1899, repr. Hildesheim: Olms, 1963).

Burnett (1978): Burnett, Charles, Arabic into Latin in twelfth century Spain: the works of Hermann of Carinthia, *Mittellateinisches Jahrbuch* 13 (1978) p. 100–134.

Burnett (1984): Burnett, C., The contents and affiliation of the scientific manuscripts written at, or brought to, Chartres in the time of John of Salisbury, in: M. J. Wilks, ed., *The world of John of Salisbury* (Oxford: Blackwell, 1984) p. 127–160.

Burnett (1987): Burnett C., ed., Adelard of Bath: an English scientist and arabist of the twelfth century (London: The Warburg Institute, 1987).

Burnett (1988): Burnett, C., Hermann of Carinthia, in: P. Dronke, ed., *A history of twelfth-century western philosophy* (Cambridge: Cambridge University Press, 1988) p. 386–404.

Burnett (1995–96): Burnett, C., Give him the white cow: notes and note-taking in the universities in the twelfth and thirteenth centuries, *History of universities* 14 (1995–1996) p. 1–30.

Burnett (1997): Burnett, C., *The introduction of Arabic learning into England* (London: The Warburg Institute, 1997).

Burnett (1998): Burnett, C., King Ptolemy and Alchandreus the philosopher: the earliest texts on the astrolabe and Arabic astrology at Fleury, Micy and Chartres, *Annals of science* 55 (1998) p. 329–368.

Burnett (2000a): Burnett, C., Addendum to'King Ptolemy and Alchandreus the philosopher: the earliest texts on the astrolabe and Arabic astrology at Fleury, Micy and Chartres', *Annals of science* 57 (2000) p. 187.

Burnett (2000b): Burnett, C., Avranches, B. M. 235 et Oxford, Corpus Christi College, 283, in: L. Callebat and O. Desbordes, eds., *Science antique, science médiévale (autour d'Avranches 235)* (Hildesheim: Olms-Weidemann, 2000) p. 63–70.

Byrtferth of Ramsey (1904) Byrtferth of Ramsey, Commentary to Beda's 'De temporum ratione', in: J.-P. Migne, ed., *Patrologiae cursus completus. Series latina* 90 (Paris, 1862, repr. Paris: Migne, 1904) c. 297–518 and 685–702.

Callebat/Desbordes (2000): Callebat, Louis and Olivier Desbordes, eds., *Science antique, science médiévale (autour d'Avranches 235)* (Hildesheim: Olms-Weidmann, 2000).

Camerota (2004): Camerota, Filippo, Reinassance descriptive geometry: the codification of drawing methods, in: W. Lefèvre, ed., *Picturing machines 1400–1700* (Cambridge MA: MIT Press, 2004) p. 175–208.

Carmody (1956): Carmody, Francis J., *Arabic astronomical and astrological sciences in Latin translation. A critical bibliography* (Berkeley: University of California Press, 1956).

Carruthers (1990): Carruthers, Mary J., *The book of memory: a study of memory in medieval culture* (Cambridge: Cambridge University Press, 1990).

Carruthers (2000): Carruthers, M. J., *The craft of thought. Meditation, rhetoric and the making of images, 400–1200* (Cambridge: Cambridge University Press, 2000).

Carruthers/Ziolowski 2002): Carruthers, M. J. and Jan M. Ziolowski, eds., *The medieval craft of memory. An anthology of texts and pictures* (Philadelphia: University of Pennsylvania Press, 2002).

Casulleras (1996): Casulleras, Josep, El último capítulo del kitāb al-asrār fī natā'iŷ al-afkār, in: J. Casulleras and J. Samsó, eds., *From Baghdad to Barcelona: Studies in the Islamic Exact Sciences in Honour of Prof. Juan Vernet – De Bagdad a Barcelona: Estudios sobre Historia de las Ciencias Exactas en el Mundo Islámico en honor del Prof. Juan Vernet*, vol. 2 (Barcelona: Univ. de Barcelona, Fak. de Filologia, 1996) p. 612–651.

Charbonnel/ Iung (1997): Charbonnel, Nicole and Jean-Eric Iung, eds., *Gerbert l'européenn. Actes du colloque d'Aurillac 4. –7. Juin 1996* (Aurillac: Gebert, 1997).

Chaucer (1872): Chaucer, Geoffrey, *A treatise on the astrolabe addressed to his son Lowys A.D. 1391*, ed. dy W. Skeat (London: 1872, repr. London: Oxford University Press, 1968).

Chejne (1980): Chejne, Anwar, The role of al-Andalus in the movement of ideas between Islam and the West, in: K. I. Semaan, ed., *Islam and the medieval West, vol. 2: aspects of intercultural relations* (Albany: state University of New York Press, 1980) p. 110–133.

Christie's (1994): *Christie's London 29.9.94: Catalogue*.

Clagett (1970): Clagett, Mashall, Adelard of Bath, in: *Dictionary of scientific biography* 1 (1970) p. 61–64.

Clanchy (1999): Clanchy, Michael, Introduction, in: M. Mostert, ed., *New approaches to medieval communication* (Turnhout: Brepols, 1999) p. 3–13.

Conant (1971): Conant, Kenneth J., Observations on the practical talents and technology of the medieval Benedictines, in: N. Hunt, ed., *Cluniac monasticism in the central Middle Ages* (London: MacMillan, 1971) p. 77–84.

Conzelmann/Hess (1980–81): Conzelmann, Peter and Marianne Hess, Zur Bedeutung des Astrolabs in den Schriften Hermanns des Lahmen von Reichenau, *Archiv für Kulturgeschichte* 62–63 (1980–81) p. 49–63.

Comnena (2001): Comnena, Anna, *Annae Comnenae Alexias*, ed. by D. R. Reinsch and A. Kamblys (Berlin: de Gruyter, 2001) (Corpus fontium historiae Byzantinae 40).

Couloubaritsis (1998): Couloubaritsis, Lambros, *Histoire de la philosophe ancienne et médiévale, figure illustres* (Paris: Grasset, 1998).

Dalton (1926): Dalton, O. M., The byzantine astrolabe at Brescia, *Proceedings of the British academy* 12 (1926) p. 133–146.

D'Ambrosio (1999): D'Ambrosio, Ubiratan, Ethnomathematics. The art or technique of explaining and knowing, *Max-Planck-Institut für Wissenschaftsgeschichte Preprint* 116 (1999) p. 1–75.

D'Ambrosio (2000): D'Ambrosio, U., A historiographical proposal for non-Western mathematics, in: H. Selin, ed., *Mathematics across cultures. The history of non-western mathematics* (Dordrecht: Kluwer, 2000) p.79–92.

D'Ambrosio (2006): D'Ambrosio, U., *Ethnomathematics. Link between traditions and modernity* (Rotterdam: Sense Publishers, 2006).

Damerow (2007): Damerow, Peter, The material culture of calculation, in: U. Gellert and E. Jablonka, eds., *Mathematisation and dematematisation. Social, philosophical and educational ramifications* (Rotterdam: Sense Publishers, 2007) p. 19–56.

Dannanfeldt (1976): Dannanfeldt, Karl H., Synesius of Cyrene, *Dictionary of scientific biography* 13 (1976) p. 225–226.

Delambre (1817): Delambre, Jean Baptiste Joseph, *Histoire de l'astronomie ancienne*, vol. 2 (Paris, 1817, repr. New York: Johnson, 1965).

Deslandes (1955): Deslandres, Yvonne, Les manuscrits decorés au XIe siècle à Saint-Germain-des-Prés par Ingelard, *Scriptorium* 9 (1955) p. 3–16.

Destombes (1962): Destombes, Marcel, Un astrolabe carolingien et l'origine de nos chiffres arabes, *Archives internationales d'histoire de sciences* 5 (1962) p. 3–45.

Dickey (1982): Dickey, Bruce G., *Adelard of Bath. An examination based on heretofore unexamined manuscripts* (Toronto 1982) (unpublished dissertation).

Dohrn-van Rossum (1992): Dohrn-van Rossum, Gerhard, *Die Geschichte der Stunde. Uhren und moderne Zeitordnung* (Munich: Deutsche Taschenbuch Verlag, 1992).

Drachmann (1954): Drachmann, Aage G., The plane astrolabe and the anaphoric clock, *Centaurus* 3 (1954) p. 183–189.

Drecker (1927–28): Drecker, Jospeh, Das Planisphaerium des Claudius Ptolemaeus. Übersetzung von J. Drecker, *Isis* 9 (1927–28) p. 255–278.

Dreyfus/Eisenberg (1996): Dreyfus, Tommy and Theodore Eisenberg, On different facets of mathematical thinking, in: R. J. Sternberg and T. Ben-Zeev , eds., *The nature of mathematical thinking* (Mahwah: Erlbaum, 1996) p. 253–284.

Eamon (1983): Eamon, William, Technology as magic in the Late Middle Ages and the Renaissance, *Janus* 70 (1983) p. 171–212.

Eastwood (1995): Eastwood, Bruce Stansfield, Latin planetary studies in the IXth and Xth century, *Physis* 32 (1995) p. 217–226.

Eastwood (1997): Eastwood, B. S., Astronomy in Christian Latin Europe, *Journal for the history of astronomy* 28 (1997) p. 235–258.

Eastwood (1999): Eastwood, B. S., Calcidius's commentary on Plato's 'Timaeus' in Latin astronomy of the ninth to the eleventh centuries, in: L. Nauta und A. Vanderjagt, eds., *Between demonstration and imagination: essays in the history of science and philosophy presented to John D. North* (Leiden: Brill, 1999) p. 171–209.

Eisler (1949): Eisler, Robert, The polar sighing tube, *Archives internationales d'histoire des sciences* 6 (1949) p. 312–332.

Elkana (1981): Elkana, Yehuda, A programmatic attempt at an anthropology of knowledge, in: E. Mendelsohn and Y. Elkana, eds., *Science and cultures: anthropological and historical studies of the sciences* (Dordrecht: D. Reidel, 1981) p. 1–76 (Sociology of the sciences 5).

Epact: *Epact: scientific instruments of medieval and Renaissance Europe. Oxford, London, Florence, Leiden*, online resource: www.mhs.ox.ac.uk/epact.

Euclid (1883): Euclid, *Opera omnia*, ed. by J. L. Heiberg and H. Menge, vol. 1 (Leipzig: Teubner, 1883).

Evans (1976a): Evans, Gillian R., The Rithmomachia: a medieval mathematical teaching aid?, *Janus* 63 (1976) p. 257–273.

Evans (1976b): Evans, G. R., Duc oculum. Aids to understanding in some medieval treatises on the abacus, *Centaurus* 19 (1976) p. 252–263.

Evans (1977a): Evans, G. R., 'More geometrico': the place of the axiomatic method in the twelfth century commentaries on Boethius' opuscula sacra, *Archives internationales d'histoire des sciences* 27 (1977) p. 207–221.

Evans (1977b): Evans, G. R., 'Difficillima et ardua': theory and practice in treatises on the abacus, 950–1150, *Journals of medieval history* 3 (1977) p. 21–38.

Evans/Peden (1985): Evans, G. R. and A. M. Peden, Natural science and the liberal arts in Abbo of Fleury's commentary on the Calculus of Victorinus of Aquitaine, *Viator* 16 (1985) p. 109–127.

M. Evans (1980): Evans, Michael, The geometry of the mind, *Architectural association quarterly* 12 (1980) p. 32–55.

al-Farghānī (2005): al-Farghānī, *On the astrolabe*, edited by R. Lorch (Stuttgart: Steiner, 2005).

Fischbein (1987): Fischbein, Efraim, *Intuition in science and mathematics: an educational approach* (Berlin: Springer Netherland,1987).

Flasch (2000): Flasch, Kurt, *Das philosophische Denken im Mittelalter*. Second edition (Stuttgart: Reclam, 2000).

Flint (1991): Flint, Valerie I. J., *The rise of magic in early medieval Europe* (Oxford: Clarendon, 1991).

Folkerts (1970): Folkerts, Menso, *Boethius' Geometrie II: ein mathematisches Lehrbuch des Mittelalters* (Wiesbaden: Steiner, 1970).

Frank (1920): Frank, Josef, *Zur Geschichte des Astrolabs* (Erlangen 1920), repr. in: F. Sezgin, ed., *Abū l-Rayḥān al-Bīrūnī (d. 440/1048): Texts and studies*, vol. 2 (Frankfurt a. M.: Institute for the history of Arabic-Islamic science, 1998) (Islamic mathematics and astronomy 35).

Frankfurt catalogue: *Catalogue of medieval astronomical instruments to ca. 1500, Johann Wolfgang Goethe Universität, Frankfurt a. M*, online resource: www.web.uni-frankfurt.de /fb13/ign/instrument-catalogue.hml.

Fulbert (1976): Fulbert of Chartres, The letters and poems of Fulbert of Chartres, edited and translated by F. Behrends (Oxford: Clarendon Press, 1976).

Fumagalli Beonio Brocchieri/Parodi (1998): Fumagalli Beonio Brocchieri, Mariateresa e Massimo Parodi, *Storia della filosofia medievale. Da Boezio a Wyclif* (Bari: Editori Laterza, 1998).

Galloni (1998): Galloni, Paolo, *Il sacro artefice. Mitologie degli artigiani medievali* (Roma: Editori Laterza, 1998).

Gandz (1927): Gandz, Solomon, The astrolabe in Jewish literature, *Hebrew Union College annual* 4 (1927, repr. 1968) p. 469–486.

Garin (1976): Garin, Eugenio, *Lo Zodiaco della vita: La polemica sull'astrologia dal Trecento al Cinquecento* (Bari: Editori Laterza, 1976).

Gasparini (2001): Gasparini, Giovanni, *Tempo e vita quotidiana* (Bari: Editori Laterza, 2001).

Gautier Dalché (1996): Gautier Dalché, Patrick, L'espace cosmologique: la table des climats du 'De utilitatibus astrolabii' du Pseudo-Gerbert, in: O. Guyotjeannin and E. Poulle, eds., *Autour de Gerbert D'Aurillac, le pape de l'an mil: album de documents commentés* (Paris: Ecole de chartes, 1996) p. 330–334.

Georges (1992): Georges, Karl Ernst, *Ausführliches Lateinisch-deutsches Handwörterbuch*, 8th ed., 2 vols. (Darmstadt: Wissenschaftliche Buchgesellschaft, 1992).

Goldstein (1996): Goldstein, Bernerd R., Astronomy and astrology in the works of Abraham ibn Ezra, *Arabic science and philosophy* 6 (1996) p. 9–21.

Goldstein (1999): Goldstein, B. R., Astronomy in the medieval Spanish Jewish community, in: L. Nauta and A. Vanderjagt, eds., *Between demonstration and imagination: essays in the history of science and philosophy presented to John D. North* (Leiden: Brill, 1999) p. 225–241.

Grafton (2000): Grafton, Anthony, Starry messengers: Recent work in the history of western astrology, *Perspectives on science* 8 (2000) p. 70–83.

Grattan-Guinness (1997): Grattan-Guinness, Ivor, *The Fontana history of mathematical sciences. The rainbow of mathematics* (Glasgow: Fontana, 1997).

Grattan-Guinness (2003): Grattan-Guinness, I., The mathematics of the past: distinguishing its history from our heritage, *Historia mathematica* 31 (2003) p. 163–185.

Grautze/Barrandon (1995): Grautze, B. and Jean-Noël Barrandon, Nouvelles analyses de l'astrolabe latin AI. 66–31, *Physis* 32 (1995) p. 433–438.

Greek-English Lexicon (1996): *A Greek-English Lexicon*, compiled by H. G. Liddel and R. Scott. With a revised supplement 1996 (Oxford: Clarendon Press, 1996).

Green (1990): Green, Dennis H., Orality and reading: the state of research in medieval studies, *Speculum* 65 (1990) p. 267–280.

Green (2000): Green, D. H., Review of M. Carruthers, The craft of thought: meditation, rhetoric, and the making of images 400–1200, *Arbitrium* 18 (2000) p. 34–37.

Gregory (1992a): Gregory, Tullio, Forme di conoscenza e ideali di sapere nella cultura medievale, in: Tullio Gregory, *Mundana sapientia. Forme di conoscenza nella cultura medievale* (Roma: Ed. di storia e letteratura, 1992) p. 1–59.

Gregory (1992b): Gregory, T., L'idea di natura nella filosofia medievale prima dell'ingresso della fisica di Aristotele. Il secolo XII, in: Tullio Gregory, *Mundana sapientia. Forme di conoscenza nella cultura medievale* (Roma: Ed. di storia e letteratura, 1992) p. 77–114.

Gunther (1923): Gunther, R. T., *Early science in Oxford. Vol. 2: Astronomy* (London, 1923, repr. London: Holland, 1967).

Gunther (1931): Gunther, Robert T., *Astrolabes of the world* (London, 1931, repr. London: Holland, 1976).

Guyotjeannin/Poulle 1996): Guyotjeannin, Olivier and Emmanuel Poulle, eds., *Autour de Gerbert D'Aurillac, le pape de l'an mil: album de documents commentés* (Paris: Ecole des chartes, 1996).

Haage (1981): Haage, Bernhard Dietrich, "Astrolabium planum" deutsch, *Sudhoffs Archiv* 65 (1981) p. 117–143.

Hagen (1874): Hagen, Hermann, ed., *Catalogus codicum Bernensium (Bibliotheca Bongsariana)* (Bern, 1874, repr. Hildesheim: Olms, 1974).

Hageman (1999): Hageman, Mariélle, Between the imperial and the sacred: the gesture of coronation in Carolingian and Ottonian images, in: M. Mostert, ed., *New approaches to medieval communication* (Turnhout: Brepols, 1999) p. 127–163.

Haire/Jacquemard (2000a): Hairie, Alain and Catherine Jacquemard, Les sources orientales du 'de profunditate maris vel fluminis probanda', Avranches, BM 235, f. 36, in: L. Callebat and O. Desbordes, eds., *Science antique, science médiévale (autour d'Avranches 235)* (Hildesheim: Olms-Weidmann, 2000) p. 223–235.

Haire/Jacquemard (2000b): Hairie, A. and C. Jacquemard, Avant le sondeur de Puelher (1563): le 'de profunditate maris vel fluminis probanda' (XIe siècle), *Archives internationales d'histoire des sciences* 50 (2000) p. 244–255.

Haire/Jacquemard (2000c): Hairie, A. and C. Jacquemard, Études théorique et expérimentale du sondeur sans fil sécrit dans le manuscrit latin Avranches, bm 235, *Archives internationales d'histoire des sciences* 50 (2000) p. 256–263.

Hamburger (1999): Hamburger, Jeffrey, Review of M. Carruthers, The craft of thought: meditation, rhetoric, and the making of images 400–1200, *Medium aevum* 68 (1999) p. 307–309.

Hamel (2007): Hamel, Jürgen, *Inventar der historischen Sonnenuhren in Mecklenburg-Vorpommern* (Frankfurt a. M.: Verlag Harri Deutsch, 2007)

Harari (2003): Harari, Orna, The concept of existence and the role of constructions in Euclid's elements, *Archive for history of exact sciences* 57 (2003) p. 1–23.

Hart (2003): Hart, Cyril, *Learning and culture in Late Anglo-Saxon England and the influence of Ramsey Abbey on the major English monastic schools. A survey of the development of mathematical, medical, and scientific studies in England before the Norman conquest*, 2 vols. (Lewiston: Edwin Mellen, 2003).

Hartmann (1919): Hartmann, Johannes Franz, Die astronomischen Instrumente des Kardinals Nikolaus Cusanus, *Abhandlungen der Königlichen Gesellschaft der Wissenschaften zu Göttingen. Math.-Phys. Klasse N. F.* 10,6 (Berlin: Weidmann, 1919).

Hartner (1939): Hartner, Willy, The principle and use of the astrolabe, in: W. Hartner, *Oriens-occidens. Ausgewählte Schriften zur Wissenschafts- und Kulturgeschichte. Festschrift zum 60. Geburtstag*, vol. 1 (Hildesheim: Olms, 1968) p. 287–311, repr. from: A. Upham Pope, ed., *A survey of Persian art*, vol. 3 (London: Oxford University Press, 1939).

Hartner (1979): Hartner, W., Asturlab, *Encyclopaedia of Islam*. New edition 1 (1979) p. 722–728.

Haskins (1924): Haskins, Charles Homer, *Studies in the history of medieval science* (Cambridge MA: Harvard University Press, 1924).

Heilbron (1998): Heilbron, John L., *Geometry civilized: history, culture and technique* (Oxford: Clarendon Press, 1998).

Herman of Reichenau (1853): [Hermannus Contractus =Herman of Reichenau], De mensura astrolabii liber, in: J.-P. Migne, ed., *Patrologiae cursus completus. Series latina* 143 (Paris: Migne, 1853) c. 379–390.

Herman of Reichenau (1931): Drecker, Joseph, Hermannus Contractus. Über das Astrolab, *Isis* 16 (1931) p. 200–219.

Hiscock (2000): Hiscock, Nigel, *The wise master builder: platonic geometry in plans of medieval abbeys and cathedrals* (Aldershot: Ashgate, 2000).

Høyrup (1988): Høyrup, Jens, Jordanus de Nemore, 13th century mathematical innovator: an essay on intellectual context, achievement, and failure, *Archive for history of exact sciences* 38 (1988) p. 307–363.

Høyrup (1994): Høyrup, J., Varieties of mathematical discourse in premodern sociocultural contexts: Mesopotamia, Greece, and the Latin Middle Ages, in: J. Høyrup, *In measure, number and weight: studies in mathematics and culture* (Albany: State University of New York Press, 1994) p. 1–43.

Holder (1888): Holder, A., Ein Brief des Abts Bern von Reichenau, *Neues Archiv* 13 (1888) p. 630–632.

d'Hollander 1995): d'Hollander, Raymond, Étude comparative entre l'astrolabe dit 'carolingien' et l'astrolabe d'Abū-Bakr ibn Yusuf de Toulouse, *Physis* 32 (1995) p. 405–419.

Holtz (2000): Holtz, Louis, Ms. Avranches, B. M. 235: étude codocologique, in: L. Callebat and O. Desbordes, eds., *Science antique, science médiévale (autour d'Avranches 235)* (Hildesheim: Olms-Weidmann, 2000) p. 19-62, figures in appendix.

Honigmann (1929): Honigmann, Ernst, *Die sieben Klimata und die πόλεις ἐπίσημοι. Eine Untersuchung zur Geschichte der Geographie und Astrologie im Altertum und Mittelalter* (Heidelberg: Winter, 1929).

Houvet (1925): Houvet, Etienne, *Cathédrale de Chartres, Portail occidental ou royal* (Chartres: Acad. de Beaux-Arts, 1925).

L'Huillier (1994): L'Huillier, Ghislaine, Practical geometry in the Middle Ages and the Renaissance, in: I. Grattan-Guinness, ed., *Companion encyclopedia of the history and philosophy of the mathematical sciences* (London: Routledge, 1994) p. 185–191.

Illich (1993): Illich, Ivan, *In the vineyard of the text: a commentary to Hugh's 'Didascalion'* (Chicago: University of Chicago Press, 1993).

Isidor of Sevilla (1911): Isidor of Sevilla, *Etymologiarum sive originum libri XX*, 2 vols. edited by W. M. Lindsay (Oxford: Clarendon 1911, repr. 1957).

Jacquemard (2000): Jacquemard, Catherine, Recherches sur la composition et la transmission de la 'Geometria incerti auctoris'. A propos du 'De profunditate maris vel fluminis probanda', Avranches, BM 235, f. 36, in: L. Callebat and O. Desbordes, eds., *Science antique, science médiévale (autour d'Avranches 235)* (Hildesheim: Olms-Weidmann, 2000) p. 81–119.

Jordanus: *Jordanus: an international computer catalogue of medieval scientific manuscripts*, Institute for the History of Science (Munich) and Max Planck Institute for the History of Science (Berlin), online resource: www.jordanus.ign.uni-muenchen.de/cgi-bin/iccmsm .

Jordanus de Nemore (1978): Jordanus de Nemore, *Jordanus de Nemore and the mathematics of astrolabes: 'de plana spera'*, edited with introduction, translation and commentary by R. B. Thomson (Toronto: Pontifical institute of medieval studies, 1978).

Juste (2000): Juste, David, Les doctrines astrologiques du 'Liber Alchandrei', in: I. Draelants, A. Tihon, B. and van den Abeele, eds., *Occident et Proche Orient contacts scientifiques au temps des Croisades* (Turnhout: Brepols, 2000) p. 277–311.

Juste (2004): Juste, D., Neither observation nor astronomical tables: an alternative way of computing the planetary longitudes in the Early Western Middle Ages, in: C. Burnett et al., eds., *Studies in the history of the exact sciences in honour of David Pingree* (Leiden: Brill, 2004) p. 181–222.

Juste (2007): Juste, David, *Les 'Alchandreana' primitifs. Étude sur les plus anciens traités astrologiques latins d'origine arabe (Xe siècle)* (Leiden: Brill, 2007).

Kassis (1999): Kassis, Hanna, A glimpse of openness in medieval society: al-Hakam II of Córdoba and his non-muslim collaborators, in: B. Nagy and M. Sebők, eds., ... *the man of many devices, who wandered full many ways... Festschrift in honor of Jámes M. Bak* (Budapest: Central European University Press, 1999) p. 160–166.

Katzenellenbogen (1966): Katzenellenbogen, Adolf, The representation of the seven liberal arts, in: M. Clagett, G. Post und R. Reynolds, eds., *Twelfth-century Europe and the foundations of modern society* (Madison: University of Winsconsin Press, 1966) p. 39–55.

Kieckhefer (1994): Kieckhefer, Richard, The specific rationality of medieval magic, *American historical review* 99,1 (1994) p. 813–836.

King (1978): King, David A., Three sundials from Islamic Andalusia, *Journal for the history of Arabic science* 2 (1978) p. 358–392, repr. in: D. A. King, *Islamic astronomical instruments* (London: Variorum, 1987).

King (1981): King, D. A., The origin of the astrolabe according to the medieval islamic sources, *Journal for the history of Arabic science* 5 (1981) p. 43–83. (Nachdr. in: D. A. King, *Islamic astronomical instruments* (London: Variorum, 1987).

King (1991): King, D. A., Medieval astronomical instruments: a catalogue in preparation, *Bulletin of the scientific instruments society* 31 (1991) p. 3–7.

King (1993): King, D. A., Some medieval astronomical instruments and their secrets, in: R. G. Mazzolini, ed., *Non-verbal communication in science prior to 1900* (Firenze: Olschki,1993) p. 29–52.

King (1994): King, D. A., Astronomical instruments between East and West, in: H. Kühnel, ed., *Kommunikation zwischen Orient und Okzident: Alltag und Sachkultur* (Wien: Verlag der österreichischen Akademie der Wissenschaften, 1994) p. 143–198

King (1995): King, D. A., The earliest known european astrolabe in the light of other early astrolabes, *Physis* 32 (1995) p. 359–404.

King (1997): King, D. A., Der Frankfurter Katalog mittelalterlicher astronomischer Instrumente, in: G. Endress and R. Kruk, eds., *The ancient tradition in Christian and Islamic Hellenism. Studies on the transmission of Greek philosophy and sciences dedicated to H. J. Drossaart Lulofs on his ninetieth birthday* (Leiden: Research school CNWS, 1997) p. 145–164.

King (1999a): King, D. A., Bringing astronomical instruments back to earth: the geographical data on medieval astrolabes (to ca. 1100), in: L. Nauta and A. Vanderjagt, ed., *Between demonstration and imagination: essays in the history of science and philosophy presenzed to John D. North* (Leiden: Brill, 1999) p. 3–53.

King (1999b): King, D. A., *World-maps for finding the direction and distance to Mecca: innovation and tradition in islamic science* (Leiden: Brill, 1999).

King (2001): King, D. A., *The ciphers of the monks: a forgotten number notation of the Middle Ages* (Stuttgart: Steiner Verlag, 2001).

King (2002): King, D. A., A 'vetustissimus' arabic treatise on the 'quadrans vetus', *Journal for the history of astronomy* 33 (2002) p. 237–255.

King (2003a): King, D. A., A remarkable Italian astrolabe from ca. 1300 – witness to an ingenious tradition of non-standard astrolabes, aus: M. Beretta, P. Galluzzi and C. Triarico, eds., *Musa musaei. Studies on scientific instruments and collections in honour of Mara Miniati* (Firenze: Olschki, 2003) p. 29–52.

King (2003b): King, D. A., The negelcted astrolabe, in: K. Grubmüller and M. Stock, ed., *Automaten in Kunst und Literatur des Mittelalters und der Frühen Neuzeit* (Wiesbaden: Harrassowitz, 2003) p. 45–55.

King (2003c): King, D. A., An astrolabe from 14th-century Christian Spain with inscriptions in Latin, Hebrew and Arabic. A unique testimonial to an intercultural encounter, *Suhayl* 3 (2003) p. 9–156.

King (2004): King, D. A., Towards a history from Antiquity to the Renaissance of sundials and other instruments for reckoning time by the sun and stars, *Annals of science* 61 (2004) p. 375–388.

King (2005) King, D. A., *In synchrony with the heavens. Studies in astronomical timekeeping and instrumentation in medieval Islamic civilization. Vol. 2: Instrument of mass calculation (studies X–XVIII)* (Leiden: Brill, 2005).

King/Turner (1994): King, D. A. and Gerard L'E. Turner, The astrolabe presented by Regiomontanus to cardinal Bessarion in 1462, *Nuncius* 9 (1994) p. 165–206.

H. C. King (1978): King, Henry C., *Geared to the stars. The evolution of planetariums, orreries and astronomical clocks* (Toronto: University of Toronto Press, 1978).

Kitcher (1984): Kitcher, Philip, *The nature of mathemtical knowledge* (New York: Oxford University Press, 1984).

Klinkenberg (1986): Klinkenberg, Hans Martin, Zum Problem des Zahlbegriffs im früheren Mittelalter (9.–12. Jahrhundert), in: H. Witthöft et al., eds., *Die historische Metrologie in den Wissenschaften* (St. Katharinen: Scripta-mercaturae-Verlag, 1986) p. 31–38.

Köhler (1994): Köhler, Bärbel, *Die Wissenschaft unter den ägyptischen Fatimiden* (Hildesheim: Olms, 1994).

Kuchenbuch (1999): Kuchenbuch, Ludolf, Kerbhölzer in Alteuropa – zwischen Dorfschmiede und Schatzamt, in: B. Nagy and M Sebők, eds., *...the man of many devices, who wandered full many ways... Festschrift in honor of Janos M. Bak* (Budapest: Central European University Press, 1999) p. 303–325.

Kunitzsch (1966): Kunitzsch, Paul, *Typen von Sternenverzeichnissen in astronomischen Handschriften des zehnten bis vierzehnten Jahrhunderts* (Wiesbaden: Harrassowitz, 1966).

Kunitzsch (1981a): Kunitzsch, P., Observations on the arabic reception of the astrolabe, *Archives internationales d'histoire des sciences* 31 (1981) p. 245–252, repr. in: Kunitzsch (1989).

Kunitzsch (1981b): Kunitzsch, P., On the authenticity of the treatise on the composition and use of the astrolabe ascribed to Messahalla, *Archives internationales d'histoire des sciences* 31 (1981) p. 42–62.

Kunitzsch (1982): Kunitzsch, P., *Glossar der arabischen Fachausdrücke in der mittelalterlichen europäischen Astrolabliteratur* (Göttingen: Vandenhoeck & Ruprecht, 1982).

Kunitzsch (1987): Kunitzsch, P., al-Khwārizmī as a source for the Sententiae astrolabii, in: D. A. King and G. Saliba, eds., *From deferent to equant. A volume of studies in honor of E. S. Kennedy* (New York: New York academy of sciences, 1987) p. 227–236, repr. in: Kunitzsch (1989).

Kunitzsch (1989): Kunitzsch, P., *The Arabs and the stars: texts and traditions on the fixed stars, and their infuence in medieval Europe* (Northampton: Variorum, 1989).

Kunitzsch (1991/1992): Kunitzsch, P., Letters in geometrical diagrams Greek-Arabic-Latin, *Zeitschrift für Geschichte der arabischen-islamischen Wissesnchaften* 7 (1991/92) p. 1–20, repr. in: Kunitzsch (2004).

Kunitzsch (1993a): Kunitzsch, P., Fragments of Ptolemy's 'Planisphaerium' in an early Latin translation, *Centaurus* 36 (1993) p. 97–101, repr. in: Kunitzsch (2004).

Kunitzsch (1993b): Kunitzsch, P., 'The peacock's tail': on the names of some theorems of Euclid's 'Elements'. in: *Vestigia Mathematica: studies in medieval and early modern mathematics in honour of H.L.L. Busard* (Amsterdam: Rodopi, 1993) p. 205–214, repr. in: Kunitzsch (2004).

Kunitzsch (1993c): Kunitzsch, P., Zur Problematik der Astrolabsternen: eine weitere unbruchbare Sterntafel, *Archives internationales d'histoire des sciences* 43 (1993) p. 197–208, repr. in: Kunitzsch (2004).

Kunitzsch (1993d): Kunitzsch, P., On six kinds of astrolabes: a hitherto unknown Latin treatise, *Centaurus* 36 (1993) p. 200–208, repr. in: Kunitzsch (2004).

Kunitzsch (1994): Kunitzsch, P., The second Arabic manuscript of Ptolemy's 'Planisphaerium', *Zeitschrift für Geschichte der arabisch-islamischen Wissenschaften* 9 (1994) p. 83–89, repr. in: Kunitzsch (2004).

Kunitzsch (1995): Kunitzsch, P., The role of al-Andalus in the transmission of Ptolemy's 'Planisphaerium' and 'Almagest', *Zeitschrift für Geschichte der arabisch-islamischen Wissenschaften* 10 (1995) p. 147–157.

Kunitzsch (1996): Kunitzsch, P., Das Astrolab, in: U. Lindgren, ed., *Europäische Technik im Mittelalter, 800 bis 1200: Tradition und Innovation, ein Handbuch* (Berlin: Gebr. Mann, 1996) p. 399–404.

Kunitzsch (1997): Kunitzsch, P., Les relations scientifiques entre l'Occident et le monde arabe à l'époque de Gerbert, in: N. Charbonnel and J.-E. Iung, eds., *Gerbert l'européen. Actes du colloque d'Aurillac 4.–7. Juin 1996* (Aurillac: Gebert, 1997) p. 193–203, repr. in: Kunitzsch (2004).

Kunitzsch (1998a): Kunitzsch, P., Traces of a Tenth-Century Spanish-Arabic Astrolabe, *Zeitschrift für Geschichte der Arabisch-Islamischen Wissenschaften* 12 (1998) p. 113–120, repr. in: Kunitzsch (2004).

Kunitzsch (1998b): Kunitzsch, P., The astronomer al-Ṣūfī as a source for Uluǵ Beg's star catalogue (1437), in: Ž. Vesel, H. Beikbaghban and B. Thierry de Crussol des Epesse, eds., *La science dans le monde iranien à l'époque islamique* (Teheran: Institut français de récherche en Iran, 1998) p. 41–47, repr. in: Kunitzsch (2004).

Kunitzsch (1999–2000): Kunitzsch, P., Three dubious stars in the oldest European table of astrolabe stars, *Zeitschrift für Geschichte der Arabisch-Islamischen Wissenschaften* 13 (1999–2000) p. 57–69, repr. in: Kunitzsch (2004).

Kunitzsch (2000a): Kunitzsch, P., La table des climats dans le corpus des plus anciens textes latins sur l'astrolabe, in: L. Callebat and O. Desbordes, eds., *Science antique, science médiévale (autour d'Avranches 235)* (Hildesheim: Olms-Weidmann, 2000) p. 391–399, repr. in Kunitzsch (2004).

Kunitzsch (2000b): Kunitzsch, P., The chapter on the stars in an early European treatise on the use of the astrolabe (ca. AD 1000), *Suhayl* 1 (2000) p. 243–250, repr. in: Kunitzsch (2004).

Kunitzsch 2000c): Kunitzsch, P., A note on Ascelinus' table of astrolabe stars, *Annals of science* 57 (2000) p. 181–185, repr. in: Kunitzsch (2004).

Kunitzsch (2003): Kunitzsch, P., The transmission of Hindu-Arabic numerals reconsidered, in: J. P. Hogendijk and A. I. Sabra, eds., *The enterprise of science in Islam: new perspectives* (Cambridge MA: MIT Press, 2003) p. 3–21.

Kunitzsch (2004): Kunitzsch, P., *Stars and numbers: astronomy and mathematics in the medieval Arab and western world* (Aldershot: Ashgate, 2004).

Kunitzsch (2005): Kunitzsch, P., The stars on the astrolabe, in: K. Van Cleempoel, ed., *Astrolabes at Greenwich. A catalogue of the astrolabes in the Netional Maritime Museum* (Oxford: Oxford University Press, 2005) p. 41–46.

Kunitzsch/Dekker (1996): Kunitzsch, P., and Elly Dekker, The stars on the rete of the so-called 'carolingian astrolabe', in: J. Casulleras and J. Samsó, eds., *From Baghdad to Barcelona: Studies in the Islamic Exact Sciences in Honour of Prof. Juan Vernet – De Bagdad a Barcelona: Estudios sobre Historia de las Ciencias Exactas en el Mundo Islámico en honor del Prof. Juan Vernet*, vol. 2 (Barcelona: Univ. de Barcelona, Fak. de Filologia, 1996) p. 655–672, repr. in: Kunitzsch (2004).

Kunitzsch/Lorch (1994): Kunitzsch, P. and Richard P. Lorch, *Maslama's notes on Ptolemy's Planisphaerium and related texts* (Munich: Beck, 1994) (Sitzungsberichte der Bayerischen Akademie der Wissenschaften 1994,2).

Langermann (1993): Langermann, Y. Tzvi, Some astrological themes in the thought of Abraham ibn Ezra, in: I. Twersky and J. M. Harris, eds., *Rabbi Abraham ibn Ezra: Studies in the writings of a twelfth-century polymath* (Cambridge MA: Harvard University Press, 1993) p. 28–85.

Lecoq (1992): Lecoq, Danielle, Le temps et l'intemporel. Sur quelques représentations mediévales du monde au XIIe et au XIIIe siècle, in: B. Ribémont, ed., *Le temps, sa mesure et sa perception au moyen âge* (Caen: Paradigme, 1992) p. 113–149.

Lefèvre (2004): Lefèvre, Wolfgang, ed., *Picturing machines 1400–1700* (Cambridge MA: MIT Press, 2004).

Leonardi (1959–60): Leonardi, Claudio, *I codici di Marziano Capella* (Milano: Società Editrice Vita e Pensiero, 1959–60).

Levy (1942): Levy, R., The authorship of a Latin treatise on the astrolabe, *Speculum* 17 (1942) p. 566–569.

Levy (2002): Levy, Tony, De l'arabe à l'hébreu: la constitution de la littérature mathématique hébraïque (XIIe–XVIe siècle), in: Y. Dold-Samplonius et al., eds., *From China to Paris: 2000 years transmission of mathematical ideas* (Stuttgart: Steiner, 2002) p. 307–326.

Lexikon Gräzität (2001): *Lexikon zur byzantinischen Gräzität: besonders des 9. –12. Jahrhunderts* 1 (2001).

Lindgren (1976): Lindgren, Uta, *Gerbert von Aurillac und das Quadrivum. Untersuchungen zur Bildung im Zeitalter der Ottonen* (Wiesbaden: Steiner, 1976).

Lindgren (1985): Lindgren, U., Ptolémée chez Gerbert d'Aurillac, in: M. Tosi, ed., *Gerberto: scienza, storia e mito* (Bobbio: Editrice degli A.S.B., 1985) p. 619–644.

Long (2001): Long, Pamela O., *Openness, secrecy, authorship. Technical arts and the culture of knowledge from Antiquity to the Renaissance* (Baltimore: John Hopkins University Press, 2001).

Lorch (1995): Lorch, Richard P., Ptolemy and Maslama on the transformation of circles into circles in stereographic projection, *Archive for the history of exact sciences* 49 (1995) p. 271–284.

Lorch (1999): Lorch, R. P., The treatise on the astrolabe by Rudolf of Bruges, in: L. Nauta und A. Vanderjagt, eds., *Between demonstration and imagination: essays in the history of science and philosophy presented to John D. North* (Leiden: Brill, 1999) p. 55–100.

Lorch (2005): Lorch, R. P., The literature of the astrolabe to 1450, in: K. Van Cleempoel, ed., *Astrolabes at Greenwich. A catalogue of the astrolabes in the Netional Maritime Museum, Greenwich* (Oxford: Oxford University Press, 2005) p. 23–30.

Lorch/Brey/Kirschner/Schöner (1994): Lorch, R. P., Gerhard Brey, Stephan Kirschner und Christoph Schöner, Ibn-aṣ -Ṣaffārs Traktat über das Astrolab in der Übersezung von Plato von Tivoli, in: B. Fritscher and G. Brey, eds., *Cosmographica et geographica: Festschrift für Heribert M. Nobis zum 70. Geburtstag*, vol. 1 (Munich: Institut für Geschichte der Naturwissenschaften, 1994) p. 125–180.

Maddison/Savage-Smith (1997): Maddison, Francis and Emilie Savage-Smith, eds., *Science, tools and magic*, 2 vols. (London: The Nour Foundation, 1997).

Mahoney (1987): Mahoney, Michael S., Mathematics, in: *Dictionary of the Middle Ages* 8 (1987) p. 205–222.

Mahoney (2004): Mahoney, M. S., Drawing mechanics, in: W. Lefèvre, ed., *Picturing machines 1400–1700* (Cambridge MA: MIT Press 2004) p. 281–306.

K. Maier (1994): Maier, Kurt, Bemerkungen zu romanischen Monatsnamen auf mittelalterlichen Astrolabien, in: A. von Godstedter, ed., 'Ad radices': Festband zum fünfzigjährigen Bestehen des Instituts für Geschichte der Naturwissenschaften der Johann Wolfgang Goethe-Universität Frankfurt am Main (Stuttgart: Steiner, 1994) p. 237–254.

K. Maier (1996): Maier, K., Zeugen der Mehrsprachigkeit: mittelalterliche romanische Monatsnamen auf islamischen astronomischen Instrumenten, in: J. Lüdtke, ed., *Romania arabica. Festschrift für Reinhold Kontzi zum 70. Geburtstag* (Tübingen: Narr, 1996) p. 251–270.

K. Maier (1999): Maier, K., Ein arabisches Astrolab aus Córdoba mit späteren altkatalanischen Inschriften, in: P. Eisenhardt, ed., *Der Weg der Wahrheit – Aufsätze zur Einheit der Wissenschaften – Festschrift zu Walter G. Saltzers 60. Geburtstag*, vol. 2 (Hildesheim: Olms, 1999) p. 119–133.

P. H. Maier (1999): Maier, Peter Herbert, Raumgeometrie mit Raumvorstellung – Thesen zur Neustrukturierung des Geometrieunterrichts, *Der Mathematik Unterricht* 45 (1999) p. 4–18.

Mandragore: *Mandragore: base iconographique du département des manuscripts*, Bibliothèque nationale de France, online resource: mandragore.bnf.fr.

Martianus Capella (1987): Martianus Capella, *De nuptiis philologiae et Mercurii: liber IX* (Latin/Italian), ed. by L. Cristante (Padova: Editore Antenore, 1987).

Martinez-Gasquez (2000): Martinez-Gasquez, José, Le monastère de Ripoll dans le nord-est péninsulaire: point de rencontre des cultures arabe et chrétienne, in: L. Callebat and O. Desbordes, eds., *Science antique, science médiévale (autour d'Avranches 235)* (Hildesheim: Olms-Weidmann, 2000) p. 241–253.

Mauersberger (1994): Mauersberger, Klaus, Visuelles Denken und nichtverbales Wissen im Maschinenbau, *Beiträge zur Geschichte von Technik und technischer Bildung* 9 (1994) p. 3–28.

Mayer (1956): Mayer, Leo Ary, *Islamic astrolabists and their works* (Geneva: Kundig, 1956).

Mayer (1959): Mayer, L. A., Islamic astrolabists: some new material, in: E. Kuehnel, ed., *Aus der Welt der islamischen Kunst* (Berlin: MAnn, 1959) p. 293–296.

Mayer/Hegarty (1996): Mayer, Richard E. and Mary Hegarty, The process of understanding mathematical problems, in: R. J. Sternberg and T. Ben-Zeev, eds., *The nature of mathematical thinking* (Mahwah: Erlbaum, 1996) p. 29–53.

McCluskey (1998): McCluskey, Stephen C., *Astronomies and cultures in early medieval Europe* (Cambridge: Cambridge University Press, 1998).

McCluskey (2003): McCluskey, S. C., Changing contexts and criteria for the justification of computistical knowledge and practice, *Journal for the history of astronomy* 34 (2003) p. 201–217.

McKitterick (2000): McKitterick, Rosamond, Books and science before print, in: M. Frasca-Spada and N. Jardine, eds., *Books and the science in history* (Cambridge: Cambridge University Press, 2000) p. 13–34.

McVaugh/Behrends (1971) McVaugh, Michael and Frederick Behrends, Fulbert of Chartres' notes on arabic asctronomy, *Manuscripta* 15 (1971) p. 172–177.

Michel (1947): Michel, Henri, *Traité de l'astrolabe*, préface de Francis Maddison (Paris, 1947, repr. Paris: Brieux, 1976).

Millás Vallicrosa (1931) Millás Vallicrosa, José Maria, *Assaig d'història de les idèes fisiques i matemàtiques a la Catalunya medieval* (Barcelona: Inst.Patxot, 1931).

Millás Vallicrosa (1940): Millás Vallicrosa, J. M., Un nuevo tratado de astrolabio de R. Abraham ibn Ezra, *al-Andalus* 5 (1940) p. 1–29.

Millás Vallicrosa (1942): Millás Vallicrosa J. M., ed., *Las traducciones orientales en los manuscritos de la biblioteca catedral de Toledo* (Madrid: Inst. Arias Montanas, 1942).

Millás Vallicrosa (1950): Millas-Vallicrosa, J. M., Encore une note sur 'Abrahismus', *Archives internationales d'historie des sciences* 29 (1950) p. 856–858.

Miller/ Paredes (1996): Miller Kevin F., and David R. Paredes, On the shoulders of giants: cultural tools and mathematical development, in: R. J. Sternberg and T. Ben-Zeev, eds., *The nature of mathematical thinking* (Mahwah: Erlbaum, 1996) p. 83–117.

Minio-Paluello (1975): Minio-Paluello, Lorenzo, Plato of Tivoli, *Dictionary of scientific biography* 11 (1975) p. 31–33.

Molinier (1890): Molinier, A., ed., *Catalogue général des manuscrits des bibliothèques publiques de France. Paris: Bibliothèque Mazarine*, vol. 3 (Paris: Plon, 1890).

Molland (1983): Molland, George, Continuity and measure in medieval Natural Philosophy, in: A. Zimmermann, ed., *Mensura – Mass, Zahl, Zahlensymbolik im Mittelalter*, vol. 1 (Berlin: de Gruyter, 1983) p. 132–144.

Molland (1993): Molland, G., The 'quadrivium' in the universities: four questions, in: I. Craemer-Ruegenberg and A. Speer, eds., *'scientia' und 'ars' im Hoch- und Spätmittelalter*, vol. 1 (Berlin: de Gruyter, 1993) p. 66–78.

Molland (1996): Molland, G., Semiotic aspects of medieval mathematics, in: M. Folkerts, ed., *Mathematische Probleme im Mittelalter: der lateinische und arabische Sprachbereich* (Wiesbaden: Harrassowitz, 1996) p. 1–16.

Moore (2001): Moore, Patrick, *Stargazing. Astronomy without a telescope*. Second edition (Cambridge: Cambridge University Press, 2001).

Morelon (1996): Morelon, Régis, General survey of Arabic astronomy, in: R. Rashed et al., eds., *Encyclopedia of the history of arabic science*, vol. 1 (London: Routledge, 1996) p. 1–19.

Morsbach 1989): Morsbach, Peter, Wilhelm von Hirsau, in: *Ratisbona sacra: das Bistum Regensburg im Mittelalter* (Zurich: Schnell u. Steiner, 1989) p. 192–194.

Mostert (1989): Mostert, Marco, *The library of Fleury: a provisional list of manuscripts* (Hilversum: Verloren, 1989).

Mostert (1999a): Mostert, M., ed., *New approaches to medieval communication* (Turnhout: Brepols, 1999).

Mostert (1999b): Mostert, M., New approaches to medieval communication?, in: M. Mostert, ed., *New approaches to medieval communication* (Turnhout: Brepols, 1999) p. 15–37.

Mundó (1971): Mundó, Anscari M., Monastic movements in the east Pyrenees, in: N. Hunt, ed., *Cluniac monasticism in the central Middle Ages* (London: McMillan, 1971) p. 98–122.

Mundó (1995): Mundó, A. M., Analyse paléographique de l'astrolabe 'carolingien', *Physis* 32 (1995) p. 303–321.

Murdoch (1984): Murdoch, John E., *Album of science: Antiquity and the Middle Ages* (New York: Scribner, 1984).

Nallino (1930): Nallino, Carlo Alfonso, Astrolabio, in: C. A. Nallino, *Raccolta di scritti editi ed inediti, vol 5: Astrologia-Astronomia-Geografia* (Roma: Ist. per l'Oriente, 1944) p. 345–34, repr. from: *Enciclopedia italiana* 5 (1930) p. 96–97.

Nallino (1944): Nallino, C. A., Astrologia e astronomia presso i Musulmani, in: C. A. Nallino, *Raccolta di scritti editi ed inediti, vol 5: Astrologia-Astronomia-Geografia* (Roma: Ist. per l'Oriente, 1944) p. 1–87.

Netz (2003): Netz, Reviel, Introduction: the history of early mathematics – ways of re-writing, *Science in context* 16 (2003) p. 275–286.

Neugebauer (1949): Neugebauer, Otto, The early history of the astrolabe. Studies in ancient astronomy IX, *Isis* 40 (1949) p. 240–256, repr. in: O. Neugebauer, *Astronomy and history* (New York: Springer, 1983).

Neugebauer (1975): O. Neugebauer, *A history of ancient mathematical astronomy: in 3 parts* (Berlin: Springer, 1975).

Nobis (1969): Nobis, Heribert M., Die Umwandlung der mittelalterlichen Naturvorstellung. Ihre Ursachen und ihre wissenschaftsgeschichtlichen Folgen, *Archiv für Begriffsgeschichte* 13 (1969) p. 34–57.

North (1974): North, John D., The astrolabe, *Scientific american* 230 (January 1974) p. 96–106.

North (1975): North, J. D., Monasticism and the first mechanical clocks , in: J. T. Fraser and N. Lawrence, eds., *The study of time* (Berlin: Springer, 1975) p. 381–393, repr. in: J. D. North, *Stars, minds and fate: essays in ancient and medieval cosmology* (London: Hambledon, 1989).

North (1987): North, J. D., Medieval concepts of celestial influence: a survey, in: P. Curry, ed., *Astrology, science and society: historical essays* (Woodbridge: The Beydell Press, 1987) p. 1–4.

North (2005): North, J. D., *God's clockmaker. Richard of Wallingford and the invention of time* (London: Hambledon and London, 2005).

Noss (1996): Noss, Richard, *Windows on mathematical meaning: learning and computers* (Dordrecht: Kluwer, 1996).

Oldoni (1985): Oldoni, Massimo, 'Imago' e 'fantasma': l'incantesimo storiografico di Gerberto, in: M. Tosi, ed., *Gerberto: scienza, storia e mito* (Bobbio: Editrice degli A.S.B., 1985) p. 747–768.

Ong (1982): Ong, Walter J., *Orality and literacy: the technologizing of the world* (London: Methuen, 1982).

Oxford Companion (1995): *The Oxford companion to philosophy* (Oxford: Oxford University Press, 1995).

Oxford Latin Dictionary (1982): *Oxford Latin Dictionary* (Oxford: Clarendon Press, 1982).

Parodi (1998) Parodi, Massimo, La retorica del mondo. Ordine, linguaggio e comunicazione nell'alto medievo, in: M. Galuzzi, G. Micheli e M. T. Monti, eds., *Le forme della comunicazione scientifica* (Milan: Franco Angeli, 1998) p. 69–83.

Pedersen (1974): Pedersen, Olaf , *A survey of the 'Almagest'* (Odense: Odense University Press, 1974).

Peiffer (2004): Peiffer, Jeanne, Projections embodied in technical drawings: Dürer and his followers, in: W. Lefèvre, ed., *Picturing machines 1400–1700* (Cambridge MA: MIT Press, 2004) p. 245–275.

Philoponos (1981): Philoponos, John, *Traité de l'astrolabe*, ed. by H. Hase, comm. and transl. by A.–P. Segonds (Paris: Societè International de l'Astrolabe, 1981).

Pier Damiani (1943): Pier Damiani, De sancta simplicitate scientiae inflanti anteponenda, in: P. Damiani, *De divina omnipotentia e altri opuscoli. Testo critico con introduzioni e note*, ed.by P. Brezzi and transl. by B. Nardi (Firenze: Vallecchi, 1943) p. 164–201.

Pingree (1987): Pingree, David, Astorlab, *Encyclopaedia iranica* 2 (1987) p. 854–857.

Pingree (1990): Pingree, D., The 'Preceptum canonis Ptolomei', in: J. Hamesse and M. Fattori, eds., *Rencontres de cultures dans la philosophie médiévale. Traductions et traducteurs de l'antiquité tardive au XVIe siècle* (Louvain-la neuve: Inst. d'Etudes Médiévales de l'Univ. Catholique de Louvaine, 1990) p. 355–375.

Pingree/Kazhan (1991): Pingree, D. und A. Kazhan, Astrology, *The Oxford dictionary of Byzantium* 1 (1991) p. 214–216.

Polanyi (1958): Polanyi, Michael, *Personal knowledge: towards a post-critical philosophy* (Chicago: University of Chicago Press, 1958).

Polanyi (1967): Polanyi, M., *The tacit dimension* (New York: Doubleday &co., 1967).

Popplow (1993): Popplow, Marcus, Die Verwendung von lat. *machina* im Mittelalter und in der Frühen Neuzeit – vom Baugerüst zu Zoncas mechanischem Bratenwender, *Technikgeschichte* 60 (1993) p. 7–26.

Popplow (2002): Popplow, M., Models of machines: A 'missing link' between early modern engineering and mechanics?, *Max-Planck-Institut für Wissenschaftsgeschichte Preprint* 225 (2002), to appear in: D. Paven und W. Krohn, *Reappraisal of the Zilsel thesis* (in preparation).

Poulle (1954): Poulle, Emmanuel, L'astrolabe médiéval d'après les manuscripts de la Bibliothèque nationale, *Bibliotheque de l'ecole des chartes* 112 (1954) p. 81–103.

Poulle (1955): Poulle, E., La fabrication des astrolabes au moyen âge, *Techniques et civilisations* 4 (1955) p. 117–128.

Poulle (1964): Poulle, E., Le traité d'astrolabe de Raymonde de Marseille, *Studi Medievali* s. III 5 1964) p. 866–900.

Poulle (1981): Poulle, E., *Les sources astronomiques (textes, tables, instruments)* (Turnhout: Brepols, 1981).

Poulle (1985): Poulle, E., L'astronomie de Gerbert, in: M. Tosi, ed., *Gerberto: scienza, storia e mito* (Bobbio: Editrice degli A.S.B., 1985) p. 597–617.

Poulle (1987): Poulle, E., Le traité de l'astrolabe d'Adélard de Bath, in: C. Burnett, ed., *Adelard of Bath: an english scientist and arabist of the twelfth century* (London: Warburg Inst. Univ. of London, 1987) p. 119–132.

Poulle (1991): Poulle, E., L'instrumentation astronomique médiévale, in: B. Ribémont, ed., *Observer, lire, écrire le ciel au moyen âge* (Paris: Klincksieck, 1991) p. 253–281.

Poulle (1994): Poulle, E., L'astrolabe sphérique dans l'Occident latin, in: W. D. Hackmann and A. J. Turner, eds., *Learning, language and invention. Essays presented to Francis Maddison* (Aldershot: Variorum, 1994) p. 223–237.

Poulle (1995): Poulle, E., La littérature astrolabique latine jusqu'au XIIIe siècle, *Physis* 32 (1995) p. 227–238.

Poulle (1996a): Poulle, E., De l'effet des contresens (XIXe siècle): Gerbert horloger!, in: O. Guyotjeannin and E. Poulle, eds., *Autour de Gerbert D'Aurillac, le pape de l'an mil: album de documents commentés* (Paris: Ecole des chartes, 1996) p. 365–366.

Poulle (1996b): Poulle, E., Naissance de la légende scientifique (XIIe siècle): note sur l'autorité des traités de l'astrolabe in: O. Guyotjeannin and E. Poulle, eds., *Autour de Gerbert D'Aurillac, le pape de l'an mil: album de documents commentés* (Paris: Ecole des Chartes, 1996) p. 342–345.

Poulle (2000): Poulle, E., 'astrolabium', 'astrolapsus', 'horologium': enquête sur un vocabulaire, in: L. Callebat and O. Desbordes, eds., *Science antique, science médiévale (autour d'Avranches 235)* (Hildesheim: Olms-Weidmann, 2000) p. 437–448.

Proclus (1873): *Procli Diadochi in primum Euclidis elementorum librum commentarii*, ed. by. G. Friedlein (Leipzig, 1873, repr. Hildesheim: Olms, 1992).

Proctor (2005): Proctor, David, The construction and use of the astrolabe, in: K. van Cleempel, ed., *Astrolabes at Greenwich. A cataloge of the astrolabes in the National Maritime Museum, Greenwich* (Oxford: Oxford University Press, 2005) p. 15–22.

Psautier of Bianca of Castilla (1910): *Psautier de saint Louis et de Blanche de Castille, 50 planches reproduisant les miniatures, initiales, etc. du manuscript 1186 de la Bibliothèque de l'Arsenal*, ed. by H. Martin, in: *Les Joyaux de l'Arsenal*, vol. 1 (Paris: [1910]).

Ps.-Beda (1904a): Ps.-Beda, Libellus de astrolabio, in: J.-P. Migne, ed., *Patrologiae cursus completus. Series latina* 90 (Paris, 1862, repr. Paris: Migne, 1904) c. 955–960.

Ps.-Beda (1904b): Ps.-Beda, Libellus de mensura horologii, in: J.-P. Migne, ed., *Patrologiae cursus completus. Series latina* 90 (Paris, 1862, repr. Paris: Migne, 1904) c. 951–954.

Ps.-Herman of Reichenau (1853a): Ps.-Hermannus Contractus, De utilitatibus astrolabii libri duo, l. 1 (= J), in: J.-P. Migne, ed., *Patrologiae cursus completus. Series latina* 143 (Paris: Migne, 1853) c. 390–404.

Ps.-Herman of Reichenau (1853b): Ps.-Hermannus Contractus, De utilitatibus astrolabii libri duo, l. 2 (= 'de horologio viatorum' (hv, k)), in: J.-P. Migne, ed., *Patrologiae cursus completus. Series latina* 143 (Paris: Migne, 1853) c. 405–412.

Ps.-Messahalla (1929): Ps.-Messahalla, De compositione astrolabii, ed. by R. T. Gunter, in: R. A. Gunther, *Early science in Oxford, vol. 5: Chaucer and Messahalla on the astrolabe* (Oxford, 1929, repr. London: Dawson of Pall Mall, 1968) p. 195–231.

Ptolemy (1907): Ptolemy, Planisphaerium, in: *Claudii Ptolemaei opera quae exstant omnia, II, Opera astronomica minora*, edited by J. L. Heiberg (Leipzig: Teubner, 1907) p. 227–259

Puigvert I Planagumà (2000): Puigvert I Planagumà, Gemma, Textes communs au manuscrit ACA Ripoll 225 et au manuscrit Avranches 235, in: L. Callebat and O. Desbordes, eds., *Science antique, science médiévale (autour d'Avranches 235)* (Hildesheim: Olms-Weidmann, 2000) p. 171–187.

Rau/Schaldach (1994): Rau, Herbert and Karlheinz Schaldach, Vertikalsonnenuhren des 6.–14. Jahrhunderts, in: A. von Gotstedter, ed., *Ad radices: Festband zum fünfzigjährigen Bestehen des Instituts für Geschichte der Naturwissenschaften der J. W. Goethe-Universität Frankfurt am Main* (Stuttgart: Steiner, 1994) p. 273–290.

Richer (1839): Richer, Historiarum libri IIII, ed. by G.-H. Pertz, *Monumenta Germaniae Historica Scriptores* 51 (Hannover: Hahn, 1839) p. 561–657.

Richer (1967 and 1964): Richer, *Histoire de France (888–995)*, 2 vols, ed. and transl. by R. Latouche (Paris: Soc. d'Éd. "Les Belles Lettres", 1964 and 1967).

Rigutti (2001): Rigutti, Mario, *Atlante del cielo*. Second edition (Florence: Giunti, 2001).

Rossi (1983): Rossi, Paolo, *Clavis universalis. Arti della memoria e logica combinatoria da Lullo a Leibniz* (Bologna: Il Mulino, 1983).

Roman de Thébes (1966): *Le Roman de Thébes*, published by Guy Raynaud de Lage, 2 vols. (Paris: Champion, 1966).

Rotman (1998): Rotman, Brian, The technology of mathematical persuasion, in: T. Lenoir, ed., *Inscribing science. Scientific texts and the materiality of communication* (Stanford: Stanford University Press, 1998) p. 55–69 and 378–379.

Rotman (2000): Rotman, B., *Mathemaics as sign. Writing, imagining, counting* (Stanford: Stanford University Press, 2000).

Sā'id al-Andalusī (1991): Sā'id al-Andalusī, *Science in the medieval world: 'Book of the categories of nations'*, transl. and ed. by S. I. Salem and A. Kumar (Austin: University of Texas Press, 1991).

Saliba (1991): Saliba, George, A sixteenth century drawing of an astrolabe made by Khafīf Ghulām 'Alī b. 'Īsā (c. 850 A. D.), *Nuncius* 6/2 (1991) p. 109–119.

Sambursky (1973): P. Sambursky, John Philoponus, *Dictionary of scientific biography* 7 (1973) p. 134–139.

Samsó (1979): Samsó, Julio, The early development of astrology in al-Andalus, *Journal for the history of Arabic science* 3 (1979) p.228–243, repr. in: J. Samso, *Islamic astronomy and medieval Spain* (Aldershot: Variorum, 1994).

Samsó (1991): J. Samsó, Andalusian astronomy: its main characteristics and influence in the Latin West, in: J. Samso, *Islamic astronomy and medieval Spain* (Aldershot: Variorum, 1994) p. 1–23.

Samsó (1992): Samsó, J., *Las ciencias de los antiguos en al-Andalus* (Madrid: Ed. MAPFRE, 1992).

Samsó (2000): Samsó, J., Maslama al-Majriti and the star table in the treatise 'De mensura astrolabii', in: M. Folkerts and R. Lorch, eds., *Sic itur ad astra. Studien zur Geschichte der Mathematik und Naturwissenschaften. Festschrift für den Arabisten Paul Kunitzsch zum 70. Geburtstag* (Wiesbaden: Harrassowitz, 2000) p. 506–522.

Sarnowsky (1983): Sarnowsky, Jürgen, Zur Messung von Zeit und Bewegung: einige spätscholastische Kommentare zum Ende des vierten Buchs der aristotelischen Physik, in: A. Zimmermann, ed., *Mensura – Mass, Zahl, Zahlensymbolik im Mittelalter*, vol. 1 (Berlin: de Gruyter 1983) p. 153–161.

Saxl/Meier (1953): Saxl, Fritz and Hans Meier, *Catalogue of astrological and mythological illuminated manuscripts of the Latin Middle Ages. III: Manuscripts in english libraries*, ed. by H. Bober, 2 vols. (London: Warburg Institute, Univerity of London, 1953).

Schmale (1999): Schmale, Franz-Josef, Bern, *Lexikon des Mittelalters* 1 (1999) c. 1970–1971.

Schöning (1991): Schöning, Udo, *Thebesroman – Eneasroman – Trojaroman: Studien zur Rezeption der Antike in der französischen Literatur des 12. Jahrhunderts* (Tübingen: Niemeyer, 1991).

Schuler (2000): Schuler, Stefan, Pourquoi lire Vitruve au moyen âge? Un point de rencontre entre savoir antique et savoir médiéval, in: L. Callebat and O. Desbordes, eds., *Science antique, science médiévale (autour d'Avranches 235)* (Hildesheim: Olms-Weidmann, 2000) p. 319–341.

Schulteß (1893): Schulteß, Karl, *Die Sagen über Silvester II. (Gerbert)* (Hamburg: Lüderitz, 1893).

Schum (1887): Schum, W., *Beschreibendes Verzeichnis der amplonianischen Handschriftensammlung zu Erfurt*, (Berlin: Weidmann, 1887).

Schwartz (1981): Schwartz, Robert, Imagery – There's more to it than meets the eye, in: N. Block, ed., *Imagery* (Cambridge MA: The MIT Press, 1981) p. 109–130.

Sebokht (1899): Sebokht, Sévère, Le traité sur l'astrolabe plan de Sévère Sebokht écrit au VIIe siècle, ed. by G. M. F. Nau, in: *Journal asiatique* 13 (1899) p. 56–101, 238–303.

Selin (2000): Selin, Helaine, ed., *Mathematics across cultures. The history of non-western mathematics* (Dordrecht: Kluwer, 2000).

Sergeyeva/ Karpova (1978): Sergeyeva, N.D. and L. M. Karpova, al-Farghani's proof of the basic theorem of stereographic projection, in: Jordanus de Nemore, *Jordanus de Nemore and the mathematics of astrolabes: 'de plana spera'*, ed. by R. B. Thomson (Toronto: Pontifical institute of medieval studies, 1978) p. 210–217.

Sesiano (1999): Sesiano, Jacques, John of Sevilla, *Lexikon des Mittelalters* 5 (1999) c. 605–606.

Sezgin (1978): Sezgin, Fuat, *Geschichte des arabischen Schrifttums. Bd. VI: Astronomie bis ca. 430 H* (Leiden: Brill,1978).

Sezgin/Neubauer (2003): Sezgin, F. and Eckhard Neubauer, *Wissenschaft und Technik im Islam II: Katalog der Instrumentensammlung des Institutes für Geschichte der Arabisch-Islamischen Wissenschaften* (Frankfurt a. M.: Inst. für Gesch. der ar.-isl. Wiss., 2003) .

Shelby (1976): Shelby, Lon R., The 'secrets' of medieval masons, in: B. S. Hall and D. C. West, eds., *On premodern technology and science. A volume of studies in honor of Lynn White, jr.* (Malibu: Undena Publ., 1976) p. 201–219.

Sinisgalli/Vastoia (1992): Sinisgalli, Rocco and Salvatore Vastoia, *Il planisfero di Tolomeo* (Firenze: Cadmo, 1992).

Skubiszewski (1990): Skubiszewski, Piotr, L'intellectuel et l'artiste face a l'oeuvre a l'époque romane, in: J. Hamesse and C. Muraille-Samaran, eds., *Le travail au Moyen Âge: une approche interdisciplinaire* (Louvain-la-Neuve: Collège Erasme,1990) p. 263–321.

P. H. Smith (2004): Smith, Pamela H., *The body of the artisan: art and experience in the scientific revolution* (Chicago: Chicago University Press, 2004).

de Solla Price (1955): de Solla Price, Derek J., A computerized checklist of astrolabes, 2 parts, *Archives internationales d'histoire des sciences* 34 (1955) p. 243–263 and p. 263–281.

de Solla Price (1973): de Solla Price, D. J. et al., *A computerized checklist od astrolabes* (New Haven: 1973).

Sot (1996): Sot, Michel, Le cursus scolaire de Gerbert d'après Richer, in: O. Guyotjeannin and E. Poulle, eds., *Autour de Gerbert D'Aurillac, le pape de l'an mil: album de documents commentés* (Paris: Ecole des chartes, 1996) p.243–248.

Stautz (1994): Stautz, Burkhard, Die früheste bekannte Formgebung der Astrolabien, in: A.von Gotstedter, ed., *Ad radices: Festband zum fünfzigjährigen Bestehen des Instituts für Geschichte der Naturwissenschaften der J. W. Goethe-Universität Frankfurt am Main* (Stuttgart: Steiner,1994) p. 315–328.

Stautz (1997): Stautz, B., *Untersuchungen von mathematisch-astronomischen Darstellungen auf mittelalterlichen Astrolabien islamischer und europäischer Herkunft* (Bassum: Verl. für Gesch. der Naturwiss. und der Technik, 1997).

Stautz (1999): Stautz, B., *Die Astrolabiensammlung des Deutschen Museums und des Bayerischen Nationalmuseums* (Munich: Oldenbourg, 1999).

Sternagel (1966): Sternagel, Peter, *Die artes mechanicae im Mittelalter. Begriffs- und Bedeutungsgeschichte bis zum Ende des 13. Jahrhunderts* (Kallmünz: Laßleben, 1966).

Sternberg (1996): Sternberg , Robert J., What is mathematical thinking?, in: R. J. Sternberg and T. Ben-Zeev, eds., *The nature of mathematical thinking* (Mahwah: Erlbaum, 1996) p. 303–318.

Sternberg/Ben-Zeev (1996): Sternberg, R. J. and Talia Ben-Zeev, eds., *The nature of mathematical thinking* (Mahwah: Erlbaum, 1996).

Stevens (1993): Stevens, Wesley M., A double perspective on the Middle Ages, in: R. G. Mazzolini, ed., *Non-verbal communication in science prior to 1900* (Firenze: Olschki, 1993) p. 1–28.

Stevens (1995): Stevens, W. M., Paleographical studies of letter forms on the mater and tympana of astrolabe AI. 86–31, *Physis* 32 (1995) p. 253–301.

Stiefel (1985): Stiefel, Tina, 'Impious men': Twelfth-century attempts to apply dialectic to the world of nature, in: P. O. Long, ed., *Science and technology in medieval society* (New York: New York Acad. of Sciences, 1985) p. 187–203.

Stock (1983): Stock, Brian, *The implications of literacy. Written language and models of interpretation in the Eleventh and Twelfth centuries* (Princeton: Princeton University Press, 1983).

Struve (1999): Struve, Tilman, Hermann von Reichenau, *Lexikon des Mittelalters* 4 (1999) c. 2167–2169.

Sturlese (1993): Sturlese, Loris, *Die deutsche Philosophie im Mittelalter. Von Bonifatius bis zu Albert dem Großen (748–1280)* (Munich: Beck, 1993).

von Stuckrad (2003): von Stuckrad, Kocku, *Geschichte der Astrologie von den Anfängen bis zur Gegenwart* (Munich: Beck, 2003).

Tachau (1998): Tachau, Katherin H., God's compass and 'vana curiositas': scientific study in the Old French 'Bible Moralisée', *The art bulletin* 80 (1998) p. 7–33.

Tannery (1922): Tannery, Paul, ed., Une correspondance d'Ecolâtres du XIe siècle, in: P. Tannery, *Memoire scientifiques V: Sciences exactes au moyen âge (1887–1921)* (Toulouse: Privat, 1922) p. 103–111 and 229–263.

Telesko (2000): Telesko, Werner, 'Quid est ergo 'tempus'?' Überlegungen zu den Verbindungslinien zwischen Zeitbegriff, Heilsgeschichte und Typologie in der christlichen Kunst des Hochmittelalters, *Mediävistik* 13 (2000) p. 87–116.

Tezmen-Siegel (1985): Tezmen-Siegel, Jutta, *Die Darstellungen der septem artes liberales in der bildenden Kunst als Rezeption der Lehrplansgeschichte* (Munich: tuduv-Verlagsgesellschaft, 1985).

Thompson (1929): Thompson, James Westfall, The introduction of Arabic science into Lorraine in the tenth century, *Isis* 12 (1929) p.184–193.

Thorndike (1923–1958): Thorndike, Lynn, *A history of magic and experimental science.* 8 vols. (New York: McMillan, 1923–1958).

Thorndike (1949): Thorndike, L.,*The sphere of Sacrobosco and its commentators* (Chicago: Chicago University Press, 1949).

Thorndike (1965): Thorndike, L., *Michael Scot* (London: Nelson, 1965).

Tihon (1995): Tihon, Anna, Traités byzantins sir l'astrolabe, *Physis* 32 (1995) p. 323–354.

Toneatto (1995): Toneatto, Lucio, *Codices artis mensoriae. I manoscritti degli antichi opuscoli latini d'agrimensura (V–XIX sec.).* 3 vols. (Spoleto: Centro Italiano di Studi sull'Alto Medioevo, 1995).

Toomer (1975): Toomer, G. J., Ptolemy, *Dictionary of scientific biography* 11 (1975) p. 186–206.

Tosi (1985): Tosi, Michele, ed., *Gerberto: scienza, storia e mito* (Bobbio: Editrice degli A.S.B., 1985).

Toulze (1996): Toulze, Françoise, *Astronomie, mythe et vérité (Vitruve, De architectura, IX et Pline l'Ancien, Naturalis Historia, II)*, in: B. Bakhouche, A. Moreau and J.-C. Turpin, eds., *Les astres. Les correspondances entre le ciel, la terre et l'homme. Les 'survivances' de l'astrologie antique* (Montpellier: Séminaire d'Ètudes des Mentalités Antiques, 1996) p. 29–59.

Turnbull (2000a): Turnbull, David, *Masons, tricksters and cartographers. Comparative studies in the sociology of scientific and indigenous knowledge* (Amsterdam: Harwood Academic, 2000).

Turnbull (2000b): Turnbull, D., Rationality and the disunity of the sciences, in: H. Selin, ed., *Mathematics across cultures. The history of non-western mathematics* (Dordrecht: Kluwer, 2000) p. 37–54 .

A. J. Turner (1984): Turner, Anthony John, *The Time Museum: Catalogue of the collection*, 1, 1 (Rockford: The Time Museum, 1985).

A. J. Turner (1989): Turner, A. J., Sun-dials: history and classification, *History of science* 27 (1989) p. 303–318, repr. in: A. J. Turner, *Of time and measurement* (Brookfield: Variorum, 1993)).

A. J. Turner (2000): Turner, A. J., The anaphoric clock in the light of recent research, in: M. Folkers and R. Lorch, eds., *Sic itur ad astra. Studien zur Geschichte der Mathematik und Naturwissenschaften. Festschrift für den Arabisten Paul Kunitzsch zum 70. Geburtstag* (Wiesbaden: Harrassowitz, 2000) p. 536–547.

A. J. Turner (2005): Turner, A. J., From brass to text: the European astrolabe in literatue and print, in: K. Van Cleempoel, ed., *Astrolabes at Greenwich. A catalogue of the astrolabes in the Netional Maritime Museum, Greenwich* (Oxford: Oxford University Press, 2005) p. 31–40.

G. L'E. Turner (1995): Turner, Gerard L'E., The craftmanship of the 'carolingian' astrolabe, IC 3042, *Physis* 32 (1995) p. 421–432.

G. L'E. Turner (2000): Turner, G. L'E., A critique of the use of first point of Aries in dating astrolabes, in: M. Folkers and R. Lorch, eds., *Sic itur ad astra. Studien zur Geschichte der Mathematik und Naturwissenschaften. Festschrift für den Arabisten Paul Kunitzsch zum 70. Geburtstag* (Wiesbaden: Harrassowitz, 2000) p. 548–554.

Ullman (1964): Ullman, Berthold Louis, Geometry in the medieval quadrivium, in: *Studi di bibliografia e di storia in onore di T. De Marinis*, vol. 4 (Verona: Biblioeca Apostolica Vaticana, 1964) p. 263–285.

Van Cleempoel (2005a): Van Cleempoel, Koenraad, ed., *Astrolabes at Greenwich: a catalogue of the astrolabes in the National Maritime Museum, Greenwich* (Oxford: Oxford University Press, 2005).

Van Cleempoel (2005b): Van Cleempoel, K., Representations of astrolabes in western art, in: K. Van Cleempoel, ed., *Astrolabes at Greenwich. A catalogue of the astrolabes in the National Maritime Museum, Greenwich* (Oxford: Oxford University Press, 2005) p. 99–111.

Van de Vyver (1931): Van de Vyver, André, Les premières traductions latines (Xe–XIe s.) de traités arabes sur l'astrolabe, in: *1er congrès international de geógraphie historique. Tome II: Mémoires* (Bruxelles: Secrétariat général du 1er Congrès Internationale, 1931) p. 266–290.

Van de Vyver (1936): van de Vyver, A., Les plus anciennes traductions latines médiévales (Xe–XIe siècles) de traités d'astronomie et d'astrologie, *Osiris* 1 (1936) p. 658–691.

Verdier (1969): Verdier, Philippe, L'iconographie des arts libéraux dans l'art du Moyen Âge jusqu'à la fin du quinzième siècle, in: *Arts libéraux et philosophie au Moyen Âge* (Montreál: Institut d'Ètudes Médiévales, 1969) p. 305–354.

Vernet/Samsó (1996): Vernet, Juan and Julio Samsó, The development of arabic science in Andalusia, in: R. Rashed et al.,eds., *Encyclopedia of the history of arabic science*, vol. 1 (London: Routledge, 1996) p. 243–283.

Vitruvius (1931 and 1934): Vitruvius, *On architecture*, ed. and transl. by F. Granger, 2 vols. (Cambridge MA: Harvard University Press, 1931 and 1934).

Vogt/Schramm (1970): Vogt, Joseph and Matthias Schramm, Synesios vor dem Planisphaerium, in: K. Gaiser, ed., *Das Altertum und jedes neue Gute* (Stuttgart: Kohlhammer, 1970) p. 265–311.

Wählin (1931a): Wählin, Theodor, Astrolabe clocks and some thoughts regarding the age and development of the astrolabe, in: R. T. Gunther, *Astrolabes of the world* (London, 1931, repr. London: Holland, 1976) p. 540–559.

Wählin (1931b): Wählin, T.,The availability of the astrolabe for the construction of sundial, in: R. T. Gunther, *Astrolabes of the world* (London, 1931, repr. London: Holland, 1976) p. 550–560.

Wantzloeben (1911): Wantzloeben, Sigfrid, *Das Monochord als Instrument und System entwicklungsgeschichtlich dargestellt* (Halle a. d. S.: Karres, 1911).

Ward (1981): Ward, Francis A. B., *A catalogue of European scientific instruments in the department of medieval and later antiquities of the British Museum* (London: British Museum Publications, (1981).

Waugh (1973): Waugh, Albert E., *Sundials: their theory and construction* (New York: Dover, 1973).

Welborn (1931): Welborn, Mary Catherine, Lotharingia as a center of Arabic and scientific influence in the eleventh century, *Isis* 16 (1931) p. 188–199.

White (1978a): White, Lynn, jr., Introduction. The study of medieval technology, 1924–1974: Personal reflections, in: L. White jr., *Medieval religion and technology. Collected essays* (Berkeley: University of California Press, 1978) p. XI–XXIV.

White (1978b): White, L., jr., Medical astrologers and late medieval technology, in: L. White jr., *Medieval religion and technology. Collected essays* (Berkeley: University of California Press, 1978) p. 297–315.

Wiesenbach (1991): Wiesenbach, Joachim, Wilhelm von Hirsau. Astrolab und Astronomie im 11. Jahrhundert, in: K. Schreiner, ed., *Hirsau: St. Peter und Paul 1091–1991. Teil 2: Geschichte, Lebens- und Verfassungsformen eines Reformklosters* (Stuttgart: Theiss, 1991) p. 109–156.

Wiesenbach (1994): Wiesenbach, J., Der Mönch mit dem Sehrohr, *Schweizerische Zeitschrift für Geschichte* 44 (1994) p. 367–388.

Wigoder (1971): Wigoder, Geoffrey, Abraham bar Hiyya, *Encyclopaedia Judaica* 2 (1971) c. 130–133.

William of Conches (1980): Wilhelm von Conches, *Philosophia*, ed., transl. and comm. by G. Maurach (Pretoria: University of South Africa, 1980).

Wintroub (2000): Wintroub, Michael, Astolabes, in: A. Hessenbruch, ed., *Reader's guide to the history of science* (London: Fitzroy Dearborn, 2000) p. 44–46.

Witthöft (1983): Witthöft, Harald, Maßgebrauch und Meßpraxis in Handel und Gewerbe des Mittelalters, in: A. Zimmermann, ed., *Mensura. Maß, Zahl, Zahlensymbolik im Mittelalter*, vol. 1 (Berlin: de Gruyter, 1983) p. 234–260.

Woepcke (1858): Woepcke, Franz, Über ein in der Königlichen Bibliothek zu Berlin befindliches Astrolabium, *Abhandlungen der Königlichen Akademie der Wissenschaften zu Berlin* (1858) p. 1–31, repr in : Fuat Sezgin, ed., *Franz Woepcke – Études sue les mathématiques arabo-islamiques*, vol. 2 (Frankfurt am Main: Institut für Geschichte der Arabisch-Islamischen Wissenschaften, 1986) p. 131–165.

Wood (2000): Wood, Leigh N., Communicating mathematics across culture and time, in: H. Selin, ed., *Mathematics across cultures. The history of non-western mathematics* (Dordrecht: Kluwer, 2000) p. 1–12 .

Zimmermann (1986): Zimmermann, Albert, Maß und Zahl im philosophischen Denken des Mittelalters, in: H. Witthöft et al., eds., *Die historische Metrologie in den Wissenschaften* (St. Katharinen: Scripta Mercaturae, 1986) p. 7–18.

Yates (1966): Yates, Frances A., *The art of memory* (London, 1966, repr. London: Routledge and Paul, 1972).

Yates (1980): Yates, F. A., Architecture and the art of memory, *Architectural assosiation quarterly* 12 (1980) p. 4–9.

Zinner (1923): Zinner, Ernst, Das mittelalterliche Lehrgerät für Sternkunde zu Regensburg und seine Beziehungen zu Wilhelm von Hirsau, *Zeitschrift für Instrumentenkunde* 43 (1923) p. 278–282.

Zinner (1964): Zinner, E., *Alte Sonnenuhren an europäischen Gebäuden* (Wiesbaben: Steiner, 1964).

Zinner (1967): Zinner, E., *Deutsche und niederländische astronomische Instrumente des 11.–18. Jahrhunderts* (Munich: Beck, 1967).

Zuccato (2005): Zuccato, Marco, Gerbert of Aurillac and a tenth-century Jewish channel for the transmission of the Arabic science to the West, *Speculum* 80 (2005) p. 742–763.

NAME INDEX

Abbo, abbot of Fleury (940/945–1004) 69, 88, 117, 177–178, 191–192, 200
Abraham (patriarch) 26, 79, 170, 174, 210–213, 220
Abraham ibn Ezra (d. 1167) 216
Abraham bar Hiyya (Savasorda) (d. ca. 1136) 216–217
Abū l-Qāsim Aḥmad ibn aṣ-Ṣaffār (d. 1035) 67
Abū Ma'shar (Abumasar) (786–866) 216
Abū Yusuf Ḥasdāy ben Isḥāq ben Shaprūṭ (fl. 940) 68, 183
Adelard of Bath (fl. 1116–1142) 216, 219–220
Alexis, emperor of Byzantium (r. 1081–1118) 65
Ambrosius Theodosius Macrobius see: Macrobius, Ambrosius Theodosius
Ancius Manlius Severus Boethius see: Boethius, Ancius Manlius Severus
Anna Comnena (1083–ca. 1153/54) 65, 81
Apollonius of Perga (ca. 250–175 B.C.) 48
Ascelin of Augsburg (early eleventh century) 70, 82, 88, 128, 138, 148, 154, 238
Augustinus of Hippo (354–430) 64

Beda Venerabilis (673/674–735) 165, 181
Bern, abbot of Reichenau (ca. 978–1048) 69
Boethius (d. 524), Ancius Manlius Severus (d. 524) 19, 64, 104, 172–173, 189
Buridan, Jean see: Jean Buridan
Byrtferth of Ramsey (ca. 960–1012) 117, 180 (n. 516), 242

Capella, Martianus see: Martianus Capella
Cassiodor (d. after 580) 19
Cicero, Marcus Tullius (106–43 B.C.) 104, 148, 191
Calcidius (fl. ca. 300–350) 104, 113–114, 137
Constantinus Africanus (d. 1087) 66
Constantine of Fleury (fl. 988–996) 69

Cusanus, Nicolaus see: Nicolaus Cusanus

Damiani, Petrus see: Petrus Damiani

Euclid (fl. ca. 300 B.C.) 32, 112, 115, 159, 188–190, 216, 219, 222, 243

Flavius Josephus, see: Josephus, Flavius
Fulbert, bishop of Chartres (ca. 960–1028) 69–70, 81, 94, 112

Gerbert of Aurillac, pope Sylvester II (ca. 950–1003) 68–69, 81, 83–84, 87–88, 93–94, 107, 117, 165, 168, 172–173, 182–183, 192–193, 221–222, 224, 229, 232, 238

Helperic of Auxerre (9th c.) 165, 212
Henry V, king of England (r. 1413–1422) 225
Herman of Carinthia (fl. 1138–1143) 129, 216, 219, 240
Herman of Reichenau (1013–1054) 70, 81–84, 87–89, 128, 138, 148, 154, 163, 214, 217, 221–222, 235, 238
Hermannus Contractus see: Herman of Reichenau
Hyginus (fl. 28 B.C. –10 A.D.) 212
Hypatia of Alexandria (ca. 370–415) 63

Ingelard (fl. ca. 1031–1060) 210–212, 230
Isidor of Sevilla (ca. 570–636) 170

Jean Buridan (d. ca. 1360) 176
Jean Fusoris (ca. 1365–1436) 225
Johannes Philoponos of Alexandria (ca. 520–550) 19, 64
John XXIII, pope (r. 1410–1415) 221
John, abbot of Gorze (ca. 900–974) 68, 83
John of Holywood see: John of Sacrobosco
John of Sacrobosco (fl. ca. 1250) 37
John of Sevilla (fl. 1133–1153) 216, 240–241
Josephus, Flavius (ca. 37-97) 170

al-Khwārizmī, Muḥammad ibn Mūsā see: Muḥammad ibn Mūsā al-Khwārizmī

SUBJECT INDEX

SUDHOFFS ARCHIV • BEIHEFTE
Herausgegeben von
Peter Dilg, Menso Folkerts, Gundolf Keil, Fritz Krafft, Rolf Winau

44. Michael Segre / Eberhard Knobloch
    Der ungebändigte Galilei
    Beiträge zu einem Symposion
    2001. 128 S., kt.
    ISBN 978-3-515-07208-3

45. Ralf Vollmuth
    Traumatologie und Feldchirurgie
    an der Wende vom Mittelalter
    zur Neuzeit
    Exemplarisch dargestellt anhand der
    „Großen Chirurgie" des Walther Hermann
    Ryff
    2001. 352 S., 51 Abb., geb.
    ISBN 978-3-515-07742-2

46. Heng-an Chen
    Die Sexualitätstheorie und „Theoretische
    Biologie" von Max Hartmann in der
    ersten Hälfte des zwanzigsten Jahrhun-
    derts
    2003. 308 S., kt.
    ISBN 978-3-515-07896-2

47. Andreas Mettenleiter
    Adam Christian Thebesius (1686–1732)
    und die Entdeckung der Vasa Cordis
    Minima
    Biographie, Textedition, medizinhistori-
    sche Würdigung und Rezeptionsgeschichte
    2001. 580 S., kt.
    ISBN 978-3-515-07917-4

48. Kerstin Springsfeld
    Alkuins Einfluß auf die Komputistik
    zur Zeit Karls des Großen
    2002. 418 S., kt.
    ISBN 978-3-515-08052-1

49. Alois Kernbauer
    Die „klinische Chemie" im Jahre 1850
    Johann Florian Hellers Bericht über seine
    Studienreise in die deutschen Länder, in die
    Schweiz, nach Frankreich und Belgien im
    Jahre 1850
    2002. X, 192 S., kt.
    ISBN 978-3-515-08122-1

50. Gerhard Klier
    Die drei Geister des Menschen
    Die sogenannte Spirituslehre in der
    Physiologie der Frühen Neuzeit
    2002. 212 S., kt.
    ISBN 978-3-515-08196-2

51. Raphaela Veit
    Das Buch der Fieber des Isaac Israeli
    und seine Bedeutung im lateinischen
    Westen
    Ein Beitrag zur Rezeption arabischer
    Wissenschaft im Abendland
    2003. 335 S., kt.
    ISBN 978-3-515-08324-9

52. Andreas Frewer
    Bibliotheca Sudhoffiana
    Medizin und Wissenschaftsgeschichte in
    der Gelehrtenbibliothek von Karl Sudhoff
    2003. 406 S., geb.
    ISBN 978-3-515-07883-2

53. Doris Schwarzmann-Schafhauser
    Orthopädie im Wandel
    Die Herausbildung von Disziplin und
    Berufsstand in Bund und Kaiserreich
    (1815–1914)
    2005. 396 S., geb.
    ISBN 978-3-515-08500-7

54. Alfons Labisch u. Norbert Paul (Hg.)
    Historizität
    Erfahrung und Handeln – Geschichte
    und Medizin
    2004. 303 S., kt.
    ISBN 978-3-515-08507-6

55. Hans-Uwe Lammel
    Klio und Hippokrates
    Eine Liaison littéraire des 18. Jahrhunderts
    und die Folgen für die Wissenschaftskultur
    bis 1850 in Deutschland
    2005. 505 S., kt.
    ISBN 978-3-515-08511-3

56. Florian Mildenberger
    Umwelt als Vision
    Leben und Werk Jakob von Uexkülls
    (1864–1944)
    2007. 320 S. m 4 Abb., kt.
    ISBN 978-3-515-09111-4

57. Arianna Borrelli
    Aspects of the Astrolabe
    'architectonica ratio' in tenth- and elev-
    enth-century Europe
    2008. 270 S. m zahlr. Abb., kt.
    ISBN 978-3-515-09129-9

FRANZ STEINER VERLAG STUTTGART

ISSN 0341-0773